# Hydrogen

## Physics and technology

I0033550

Online at: https://doi.org/10.1088/978-0-7503-5172-0

# Hydrogen

Physics and technology

**Sindhunil Barman Roy**

*UGC-DAE Consortium for Scientific Research, Indore, India*
*and*
*Ramakrishna Mission Vivekananda Educational and Research Institute, Belur, India*

**IOP** Publishing, Bristol, UK

© IOP Publishing Ltd 2024

All rights reserved. No part of this publication may be reproduced, stored in a retrieval system or transmitted in any form or by any means, electronic, mechanical, photocopying, recording or otherwise, without the prior permission of the publisher, or as expressly permitted by law or under terms agreed with the appropriate rights organization. Multiple copying is permitted in accordance with the terms of licences issued by the Copyright Licensing Agency, the Copyright Clearance Centre and other reproduction rights organizations.

Certain images in this publication have been obtained by the author(s) from the Wikipedia/ Wikimedia website, where they were made available under a Creative Commons licence or stated to be in the public domain. Please see individual figure captions in this publication for details. To the extent that the law allows, IOP Publishing disclaim any liability that any person may suffer as a result of accessing, using or forwarding the image(s). Any reuse rights should be checked and permission should be sought if necessary from Wikipedia/Wikimedia and/or the copyright owner (as appropriate) before using or forwarding the image(s).

Permission to make use of IOP Publishing content other than as set out above may be sought at permissions@ioppublishing.org.

Sindhunil Barman Roy has asserted his right to be identified as the author of this work in accordance with sections 77 and 78 of the Copyright, Designs and Patents Act 1988.

ISBN    978-0-7503-5172-0 (ebook)
ISBN    978-0-7503-5170-6 (print)
ISBN    978-0-7503-5173-7 (myPrint)
ISBN    978-0-7503-5171-3 (mobi)

DOI    10.1088/978-0-7503-5172-0

Version: 20240601

IOP ebooks

British Library Cataloguing-in-Publication Data: A catalogue record for this book is available from the British Library.

Published by IOP Publishing, wholly owned by The Institute of Physics, London

IOP Publishing, No.2 The Distillery, Glassfields, Avon Street, Bristol, BS2 0GR, UK

US Office: IOP Publishing, Inc., 190 North Independence Mall West, Suite 601, Philadelphia, PA 19106, USA

*To Gopa and many other unknown scientists, whose contributions to society remain unknown and unsung.*

# Contents

Preface                                                                      xiv

Acknowledgements                                                              xv

Author biography                                                             xvi

**0    Introduction**                                                       **0-1**
       References                                                            0-6

**Part I    Physics of hydrogen**

**1    Quantum physics of hydrogen atom**                                    **1-1**

1.1    Hydrogen atomic spectrum                                              1-1

1.2    Quantum physics of hydrogen atom                                      1-2

       1.2.1   Radii of hydrogen atom orbits                                 1-6

       1.2.2   Ground and excited states of the hydrogen atom               1-6

       1.2.3   Spectra of the hydrogen atom                                  1-7

       1.2.4   Ionization potential, binding energy, and excitation potential   1-8

       1.2.5   Limitations of Bohr's model                                   1-10

1.3    Wave-particle duality of the electron and the Heisenberg uncertainty   1-10
       principle

1.4    Schrödinger's wave equation of the hydrogen atom                      1-11

       1.4.1   Fine structures of the hydrogen spectra                       1-14

       Reference                                                             1-16

**2    Covalent bonding and the hydrogen molecule**                          **2-1**

2.1    Born–Oppenheimer approximation                                        2-2

2.2    Heitler–London theory of hydrogen molecule                           2-2

2.3    Beyond the Heitler–London theory of the hydrogen molecule            2-6

2.4    Physical origin of the covalent chemical bond in the hydrogen molecule   2-8

       2.4.1   Electrostatic potential energy approach                       2-9

       2.4.2   Kinetic energy approach                                       2-12

       2.4.3   Energy analysis of the $H_2^+$ and $H_2$ molecules within the kinetic   2-13
               energy approach

2.5    Ortho-hydrogen and para-hydrogen                                      2-21

2.6    Summary                                                              2-23

       References                                                           2-24

**3    Strong Coulomb repulsion in the hydrogen molecule and the**      **3-1**
**Hubbard model**

3.1   An approximate representation of electron interactions in a hydrogen      3-2
molecule

3.2   Ionized hydrogen molecule $H_2^+$                                          3-2

3.3   Hydrogen molecule $H_2$                                                    3-3

      References                                                                 3-7


**4    Thermodynamic properties of fluid hydrogen**                             **4-1**

4.1   Nuclear spin: ortho-hydrogen and para-hydrogen                            4-3

4.2   The quantum law of corresponding states                                   4-8

4.3   Some experimental results of the properties of liquid hydrogen           4-10

      4.3.1   Pressure–volume–temperature isotherms and thermodynamic          4-10
              properties

      4.3.2   Thermal and electrical conductivity                              4-11

4.4   The equations of state                                                   4-15

      4.4.1   The regression process                                           4-17

      4.4.2   Ideal-gas contributions to the reduced Helmholtz                 4-17
              free energy

      4.4.3   Residual contribution to the reduced Helmholtz                   4-18
              free energy

      4.4.4   Fixed-point properties and vapor pressures                       4-19

      4.4.5   Comparison of calculated data to experimental data               4-20

      References                                                               4-21


**5    Exotic properties of dense hydrogen**                                   **5-1**

5.1   Hydrogen under pressure                                                   5-1

5.2   Phase I                                                                   5-4

      5.2.1   Rotational disorder in phase I                                    5-5

      5.2.2   Vbrational localization transitions                              5-5

5.3   Symmetry breaking and phase II of hydrogen                               5-6

      5.3.1   Direct investigation of phase II crystal structure               5-7

5.4   Phase III: symmetry breaking at higher pressure                          5-8

5.5   Phase IV and phase V of solid hydrogen                                    5-9

5.6   Phase VI: metallic hydrogen                                              5-10

      References                                                               5-12

**6    Hydrogen in various solid matrix**                                    **6-1**

6.1    Physically bound hydrogen                                              6-1

    6.1.1    Nanoporous carbon materials                             6-2

    6.1.2    Metal–organic frameworks                                6-5

    6.1.3    Covalent organic framework                              6-6

    6.1.4    Porous aromatic framework                               6-7

    6.1.5    Nanoporous organic polymers                             6-8

6.2    Chemically bound hydrogen                                             6-9

    6.2.1    Hydrogen–metal systems                                 6-10

    6.2.2    Physical properties of hydrogen–metal systems          6-20

6.3    Different classes of metal hydrides                                   6-23

    6.3.1    Metal hydrides from elemental metals                   6-23

    6.3.2    Metal hydrides from alloys                             6-25

    6.3.3    Transition metal hydride complexes                     6-27

    6.3.4    Amorphous metal hydrides                               6-28

    References                                                     6-28


**7    Solid proton conductor**                                              **7-1**

7.1    Electrolytes and fuel cells                                           7-1

7.2    Solid proton conductor                                               7-3

7.3    Materials structure and proton conductivity                          7-4

7.4    Different classes of solid proton conductors                         7-6

    7.4.1    Disordered type hydrous systems                        7-6

    7.4.2    Anhydrous hydrogen-containing systems                  7-9

7.5    Proton-conducting oxides                                             7-11

7.6    Hybrid organic networks                                             7-13

    7.6.1    Metal–organic framework based ionic conductor         7-14

    7.6.2    Covalent-organic frameworks based ionic conductors    7-15

    References                                                     7-17


**8    Superconductivity in hydrogen-based systems**                        **8-1**

8.1    Bardeen–Cooper–Schrieffer theory of superconductivity                8-2

    8.1.1    Superconducting transition temperature                 8-2

    8.1.2    Superconducting energy gap                             8-4

8.2    Strong coupling superconductivity                                    8-5

8.3    Various superconductors                                              8-7

8.4 Possible superconducting state in metallic hydrogen    8-8

8.5 Superconductivity in hydrides    8-9

8.6 Structure and superconductivity of hydrides from first principles    8-12

8.7 Developments on the experimental front    8-14

     8.7.1 Discovery of superconductivity in hydrogen sulfide    8-15

     8.7.2 Lanthanide hydrides and beyond    8-16

     References    8-18

**9    Hydrogen fusion    9-1**

9.1 Properties of the nucleus    9-1

     9.1.1 Nuclear radius    9-1

     9.1.2 Nuclear spin    9-2

     9.1.3 Stability of nucleus    9-3

9.2 Nuclear forces    9-3

9.3 Binding energy    9-4

9.4 Nuclear fusion    9-4

     9.4.1 Fusion in the Sun    9-5

     9.4.2 Fusion in the laboratory    9-6

     9.4.3 Lawson criterion    9-6

     References    9-7

**Part II    Hydrogen technology**

**10    Applications of hydrogen    10-1**

10.1 Hydrogen for power systems and energy storage    10-1

10.2 Hydrogen as a transportation fuel    10-4

     10.2.1 Road    10-4

     10.2.2 Railway    10-5

     10.2.3 Shipping    10-5

     10.2.4 Aviation    10-6

10.3 Production of hydrocarbon fuels    10-7

10.4 Refining of crude oil and petroleum products    10-8

10.5 Production of ammonia    10-8

10.6 Metallurgical industries    10-9

     References    10-9

## 11   Methods of hydrogen production      11-1

11.1 Environmental cleanliness and hydrogen color coding    11-2

11.2 Hydrogen production from fossil fuels    11-5

     11.2.1   Hydrocarbon reforming methods    11-5

     11.2.2   Hydrocarbon pyrolysis    11-8

11.3 Hydrogen production from renewable sources    11-10

     11.3.1   Biomass process    11-10

     11.3.2   Biological methods    11-11

     11.3.3   Electrolysis    11-16

     11.3.4   Thermolysis    11-17

     11.3.5   Photo-electrolysis    11-18

11.4 Comparison of various $H_2$ production processes    11-20

11.5 High-temperature conversion in nuclear power plants    11-22

     References    11-23

## 12   Methods of hydrogen storage      12-1

12.1 Large-scale storage    12-2

     12.1.1   Liquid hydrogen storage    12-2

     12.1.2   Underground storage    12-3

12.2 Small-scale storage    12-5

     12.2.1   Compressed hydrogen gas storage    12-5

     12.2.2   Cryogenic and cryo-compressed hydrogen    12-7

12.3 Solid-state storage    12-10

     12.3.1   Physisorption    12-10

     12.3.2   Chemisorption    12-18

     12.3.3   Liquid organic and circular carriers    12-23

     References    12-24

## 13   Hydrogen safety and integrity      13-1

13.1 Properties of hydrogen    13-1

13.2 Hydrogen hazards    13-3

     13.2.1   Physiological hazards    13-3

     13.2.2   Physical hazards    13-3

     13.2.3   Chemical hazards    13-3

     13.2.4   Explosion phenomena    13-3

13.3 Hydrogen integrity phenomena    13-4

13.3.1  Hydrogen damage                                           13-4

13.3.2  Low-temperature embrittlement                             13-6

13.3.3  Thermal contraction                                       13-6

13.4  Safety comparisons of hydrogen, methane, and gasoline       13-8

References                                                        13-8


**14    Hydrogen transport and distribution                       14-1**

14.1  Hydrogen transport via ammonia                              14-1

14.1.1  Synthesis of ammonia                                      14-2

14.1.2  Ammonia decomposition                                     14-2

14.1.3  Advantages and disadvantages of ammonia                  14-3

14.2  Liquid organic hydrogen carrier                             14-3

14.2.1  Characteristic properties of LOHC systems                 14-4

14.3  Transport of gaseous and liquid hydrogen                    14-7

14.3.1  Road and rail transportation of hydrogen                  14-8

14.3.2  Ocean transportation of hydrogen                          14-10

14.3.3  Pipeline transportation of hydrogen                       14-11

References                                                        14-12


**15    Hydrogen energy conversion technologies                   15-1**

15.1  Flame combustion                                            15-2

15.1.1  Hydrogen ICE                                              15-4

15.1.2  Turbines and jet engines                                  15-5

15.2  Steam generation by hydrogen/oxygen combustion             15-5

15.3  Catalytic hydrogen combustion                               15-7

15.3.1  Fixed-bed reactor                                         15-9

15.3.2  Monolithic reactor                                        15-10

15.3.3  Microchannel reactor                                      15-10

15.4  Electrochemical conversion                                  15-12

15.5  Energy conversions involving metal hydrides                15-16

References                                                        15-19


**16    Hydrogen nuclear fusion technology                        16-1**

16.1  Magnetic confinement fusion                                 16-3

16.2  Inertial confinement fusion                                 16-5

References                                                        16-8

**17   Hydrogen in semiconductor technology**                                    **17-1**

17.1  A brief history of hydrogen in semiconductors                              17-2

17.2  Monoatomic hydrogen                                                        17-3

    17.2.1  Electronic structure and transition levels                     17-3

    17.2.2  Alignment of hydrogen levels                                    17-4

    17.2.3  Effects on doping and passivation                               17-5

17.3  Hydrogen molecules and molecular complexes                                 17-6

    17.3.1  Complexes with impurities and point defects                     17-7

    17.3.2  Defect and impurity engineering                                 17-8

17.4  Hydrogen on semiconductor surfaces                                         17-8

17.5  Summary                                                                    17-9

    References                                                               17-10


**18   Road towards hydrogen economy**                                           **18-1**

    References                                                               18-5


**Appendix A: Schrödinger wave equation for the hydrogen atom**                  **A-1**

**Appendix B: Liquefaction of hydrogen**                                         **B-1**

# Preface

Hydrogen is the first element of the periodic table and forms about 90% of matter in the visible Universe. It is found in a combined form with other elements in compounds such as water or mineral substances, hydrocarbons, and biological molecules. It is also the primary component of the Sun and other stars. The very name hydrogen refers to the early interest in the chemistry of hydrogen. Water is formed when hydrogen is combusted, hence the name 'hydrogen'—the water generator.

Hydrogen is very much in the focus of academics and scientists with interests in quantum mechanics in the small-scale to cosmology in the very large-scale. In technology, hydrogen is presently one of the most promising options for storing energy from renewables and transporting energy with hydrogen-based fuels over long distances.

Beyond its journey from the simplest element to a very promising energy material, hydrogen in its condensed material form is predicted to have very interesting physical properties, starting from exotic metallic behavior to superfluidity and high-temperature superconductivity. Various phases of condensed hydrogen are currently under intense experimental scrutiny and newer features are being discovered.

In modern times, hydrogen is profusely used in industries such as oil refining, ammonia production, methanol production, and steel production. In the transport sector, there is currently much promise for the implementation of hydrogen fuel cell cars, which will depend on the competitiveness of fuel cell costs and the easy availability of refueling stations. The shipping and aviation sector also represents an opportunity for hydrogen-based fuels. In the power sector, hydrogen is one of the promising options for storing renewable energy.

The science and technology of hydrogen has so far been traditionally taught in chemical and industrial engineering departments of academic institutions, and the existing books on the subject are more or less tuned in that direction. However, with the worldwide effort of tapping into hydrogen's potential to play a key role in tackling critical energy challenges, the subject has become interdisciplinary. New R&D centers are being established worldwide dedicated to energy studies. It is quite clear now that quantum effects are going to play a big role in this exciting energy technology. So, there is a need for a book that will be interdisciplinary in nature, and of interest to the advanced undergraduate, postgraduate, and doctoral students of physics, chemistry, materials science, and engineering and energy studies. To this end, this book aims to cover the journey of hydrogen from the simplest element to a condensed matter with exotic physical properties and with enormous technological promises as a clean energy source and energy carrier.

# Acknowledgements

I acknowledge financial support from the Department of Atomic Energy, Government of India in the form of a Raja Ramanna Fellowship during the period of writing of this book. I thank the UGC-DAE Consortium for Scientific Research (UGC-DAE CSR), Indore, and the Ramakrishna Vivekananda Educational and Research Institute (RKMVERI), Belur for the kind hospitality and necessary support for the completion of this book. I also thank various faculty members and students of UGC-DAE CSR, Indore, and RKMVERI, Belur for many stimulating discussions on condensed matter physics in general. I thank the editorial team of the Institute of Physics (IOP), in particular Caroline Mitchell, Isabelle Defillion and Betty Barber, for enthusiastically providing the necessary support at various stages of this book. Last, but not least, I thank my wife Gopa Barman Roy and nephew Soumyanil Barman Roy for their relentless support and encouragement.

# Author biography

## Sindhunil Barman Roy

**Sindhunil Barman Roy** received his BSc (Hons.) degree in physics (in 1977) from North Eastern Hill University, Shillong, and his MSc degree in physics (in 1980) from Jadavpur University, Kolkata. He obtained his PhD degree from the Indian Institute of Technology, Kanpur in 1985. Subsequently, he was a postdoctoral research associate at Imperial College, London and the University of Florida, Gainesville, between 1986 and 1991. He was a staff scientist at the Raja Ramanna Centre for Advanced Technology, Indore from 1992 to 2018, a professor at the Homi Bhabha National Institute, Mumbai between 2007 and 2018, and an emeritus professor and Raja Ramanna Fellow at the UGC-DAE Consortium for Scientific Research, between 2018 and 2023. He is presently a distinguished professor at the Ramakrishna Vivekananda Educational and Research Institute, Belur. His research interests span both basic and applied aspects of magnetism and superconductivity, and he has published more than 220 research papers in international peer reviewed journals and has published two technical books. He served on the editorial board of the Superconductor Science and Technology journal between 2008 and 2014. He is a recipient of the Homi Bhabha Science and Technology Award and the Raja Ramanna Fellowship of the Department of Atomic Energy, India, and is a fellow of the Institute of Physics, UK.

**IOP** Publishing

# Hydrogen
Physics and technology
**Sindhunil Barman Roy**

# Chapter 0

## Introduction

'H Stands for Hydrogen … and Humility.' (J Rigden)

All matter in the Universe was formed in one explosive event 13.7 billion years ago—the Big Bang [1, 2]. During the first moments after the Big Bang, the Universe was extremely hot and dense. With the subsequent cooling of the Universe, the conditions became just conducive enough to give rise to the building blocks of matter—the quarks and electrons. The quarks combined to produce protons and neutrons a few millionths of a second later. These protons and neutrons then combined to produce nuclei within minutes. It then took 380 000 years for the emergence of neutral hydrogen atoms by trapping the electrons in orbits around nuclei. A hydrogen atom consists of a single proton and a single electron, which is the simplest atom in nature.

In the primordial mix of the early Universe, protons outnumbered alpha particles by about 11 to 1, and the deuteron was a mere fraction in the mix [3]. As a result, the formation of atoms resulted in an atomic mix of about 92% hydrogen, 8% helium, and a fraction of a percent deuterium. Today, approximately 15 billion years after the Big Bang, the temperature of the Universe has dropped to three degrees above absolute zero. There are galactic systems spread across the far reaches of the visible Universe. Each of these galaxies consists of stars and dust clouds. All these stars and dust clouds consist of about 90% hydrogen atoms and 9% helium atoms. The Sun shines and stars twinkle because of this composition, which was set in the early Universe approximately 14 billion years ago.

It is important to note here that stars and galaxies do not encompass the complete picture of the Universe. As per astronomical and physical calculations, the visible Universe comprises only of a tiny amount (4%) of the Universe [1]. In fact, about 26% is made of a yet unknown type of matter, called 'dark matter'. This dark matter

© IOP Publishing Ltd 2024

does not emit any light or electromagnetic radiation of any kind, unlike stars and galaxies. As a result, it is possible to detect dark matter only through its gravitational effects. There is an even more mysterious form of energy called 'dark energy', which accounts for about 70% of the mass-energy content of the Universe [1]. This idea emerged from the observation that all galaxies appear to be moving away from each other at an accelerating pace, which in turn indicates that some invisible extra energy is at the root of this phenomenon. We will not continue with this subject any further here and rather focus on hydrogen, which is the main subject of interest in this book.

The Sun is a typical star and is fueled by the fusion of hydrogen. Approximately 600 million tons of hydrogen are fused into helium every second in the core of the Sun. This fusion of hydrogen atoms releases an enormous amount of energy that slowly propagates from the core to the surface of the Sun. The temperature at the surface of the Sun reaches about 5800 K, and in turn provides life-giving warmth to the Earth, 92 million miles away. To quote John Rigden [3] here: 'The world as we know it is a consequence of the balance between the number of hydrogen nuclei and the number of helium nuclei, established in the early moments after the Big Bang. Perhaps it is preferable to say that the world is a consequence of the basic laws that produced this particular blend of hydrogen and helium. Did the laws of nature exist before the origin of the Universe? Did the laws of nature take their present form at the instant of the Big Bang? One-millionth of a second after the Big Bang? No one can say.' It may not be out of context here to say that the creation of the Universe remains a profound question in modern science and philosophy [4].

Focusing on the subject of hydrogen, Rigden continues: 'Looking back, however, we can say the following: if the weak force had been just a little weaker, the free neutron would decay a little more slowly and, as a result, the Universe would have started as predominantly helium rather than hydrogen. A world without hydrogen is a world without water, a world without carbohydrates, a world without proteins—a world without life. So, take your pick. We can say that the world is the way it is because the laws of nature are the way they are, or we can say that the world is the way it is because hydrogen is the way it is. Whichever you select, one or the other is a matter of preference. Either way, the little hydrogen atom commands the stage on which the long and enchanting drama of our Universe, the story of galaxies, stars, planets, and life, unfolds.' The readers are encouraged to take a look at this very interesting book by Rigden, 'Hydrogen: The Essential Element', which provides a brief history of twentieth-century physics with the use of the hydrogen atom as a guide. The primary objective of the study of basic physics is to learn how the world around us works. However, the knowledge gathered through the pursuit of basic research often leads to very practical societal applications. Hydrogen indeed provides a striking example of how basic science ultimately leads to very profound technological developments.

While hydrogen is the most abundant element in the visible Universe, it accounts for only about 0.14% of Earth's crust by weight. It can originate from volcanos mixed with other gases. Hydrogen can also be traced in underground inclusions of potash salt beds. Hydrogen can be set free through various pyrolitic processes, such as the formation of fossil fuels in the geological past. Under ambient conditions, hydrogen gas is a loose collection of hydrogen molecules. Each hydrogen molecule consists of a pair of hydrogen atoms, a diatomic molecule, $H_2$. The earliest known important chemical property of hydrogen is that it burns with oxygen to form water, $H_2O$. The name hydrogen originates from Greek words meaning 'water maker'. Hydrogen exists in enormous quantities as part of the water in oceans, ice packs, rivers, lakes, and the atmosphere. Hydrogen is also present as part of numerous carbon compounds, in all animal and vegetable tissue, and in petroleum. In a way, hydrogen is involved in the processes of life; for instance, in photosynthesis and energy conversion in living cells.

Since the nineteenth century, coal and oil have been our main and preferred sources of energy. The two hundred years of global consumption of fossil fuels has, however, caused a significant depletion of their supply and created various environmental consequences. The world needs clean energy right now for two main reasons: (i) to meet an ever-increasing worldwide demand for energy, and (ii) to reduce carbon emissions, which are warming the global atmosphere. There have been considerable discussions lately of solar and wind power, and also nuclear power as alternative energy sources. So far, hydrogen has only played a small supporting role in the energy sector, but this scenario is changing quite rapidly.

Hydrogen is a potentially carbon-free fuel and only produces water vapor when consumed in a fuel cell. It is also an energy carrier, which can store, transport, and deliver energy produced from other sources. The potential applications of hydrogen today include energy storage and transportation, heat generation, liquid fuels (e.g., biofuels and synfuels), and the industrial sector (e.g., metals refining). Hydrogen-based technology is becoming quite common in our daily lives too. For instance, hydrogen in metals is used in batteries, sensors, ferromagnets, switchable mirrors, and heat pumps. Hydrogen is also widely used in the petrochemical industry.

Hydrogen can be produced from primary energy sources through various technologies of thermochemical, electrochemical, and biological origins. Furthermore, hydrogen can be converted to and produced from other secondary energy sources, such as electricity and heat. This opens up the possibility of mutual conversion among these secondary energy sources. Hydrogen utilization encompasses a broad range of oxidation technologies, which include turbine combustion, internal combustion engines, fuel cells, and fuel mixing [5]. It is also possible to use hydrogen as an effective energy storage system, such as batteries. Conventional batteries have the drawbacks of self-discharge and degradation of capacity after the storage period and cycle. In contrast, hydrogen can store energy for a longer period and also maintain its high energy density.

Several renewable energy sources (e.g., solar, wind, etc) can be used in hydrogen production. However, the variable and intermittent natures of these sources are the major problems. This calls for innovation, research, and development of cost-effective hydrogen production methods and infrastructure development, taking into account variability in the renewable supply, i.e., solar, wind, and others, and varying energy demand [5]. Currently, more than 100 hydrogen production technologies are available, with over 80% of these technologies being focused on the steam conversion of fossil fuels and 70% of them are based on natural gas steam reforming [5]. However,

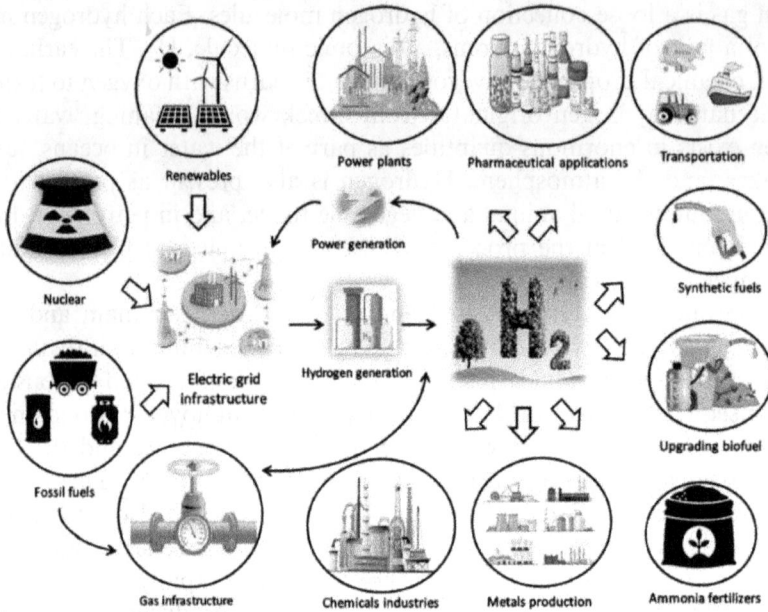

**Figure 0.1.** Schematic presentations of various hydrogen production routes and various possible applications of hydrogen. Reproduced from [5]. CC BY 4.0.

a broader range of hydrogen generation processes (e.g., methane pyrolysis and seawater electrolysis using alternative energy sources) needs to be considered to minimize environmental pollution. Figure 0.1 shows a schematic diagram presenting various hydrogen production routes, along with various possible applications of hydrogen. The massive future of environmentally-friendly hydrogen deployment is expected to establish an economically competitive hydrogen economy. Goldman Sachs predicts that the hydrogen generation market may exceed 11 trillion USD by 2050, driven by global demand for zero-carbon emissions [6].

Starting with this general introduction, this book is presented in two parts. Part I of this book will have nine chapters covering the basic scientific aspects of hydrogen. Part II of this book will have nine chapters and will give an exposition of various components of hydrogen technology. In addition, there will be two appendices, one on the detailed quantum mechanical aspects of the hydrogen atom and the other on the technical details of the hydrogen liquefaction process. Hydrogen is a quantum material and hydrogen technology, both present and future, is a quantum technology. This is highlighted throughout the book. However, efforts have been made to present the texts in a form that is accessible to advanced undergraduate students, fresh graduate students, and experimental scientists, keeping the number of equations to a bare minimum but at the same without compromising on the rigor of quantum physics.

Chapter 1 starts with an introduction to the hydrogen atomic spectrum, which eventually leads to the development of the quantum mechanics of hydrogen atoms.

Chapter 2 introduces the concept of covalent bonding and a detailed understanding of its mechanism, which is instrumental in the formation of the hydrogen molecule. The existence of two different types of hydrogen molecules, ortho-hydrogen and para-hydrogen, is described. Chapter 3 elaborates on the important Coulomb interaction between electrons in a hydrogen molecule. Chapter 4 will extend the studies of the previous chapters to fluid hydrogen, namely the macroscopic state of gaseous and liquid hydrogen. The thermodynamic properties of fluid hydrogen will be discussed in detail. While the interaction between hydrogen atoms is a purely quantum mechanical effect, which leads to the formation of one of the strongest bonds in chemistry, namely the covalent H–H bond; however, the intermolecular bonding is very weak. Extreme conditions are required to bring the hydrogen molecules together and bind them into a solid state. Chapter 5 will cover a variety of exciting and interesting phenomena that have been revealed in dense hydrogen in low-temperature and high-pressure environments. Hydrogen can combine with solid materials either physically or chemically. In chapter 6, we will focus on the scientific aspects related to hydrogen in various solid matrices. This information will be relevant for various classes of materials important for hydrogen energy technology to be discussed in part II of this book. Chapter 7 will be devoted to the solid proton conductor, which is a solid material (crystalline or amorphous) that allows the passage of electrical current through the bulk of the material exclusively by the movement of protons, $H^+$ ion. A proton-conducting solid electrolyte with high conductivity would be advantageous for the construction of a hydrogen fuel cell. In 1968, Neil Ashcroft proposed theoretically that hydrogen would be a high-temperature superconductor when metalized under pressure [7]. While the existence of superconductivity in metallic hydrogen is yet to be established, hydride compounds have turned out to be a fertile source of high-temperature superconductivity, and currently it is a topic of much interest. Chapter 8 provides an exposition of superconductivity in hydrogen-based systems. Interestingly, the superconductivity in this compound can be understood within the framework of the classical Bardeen–Cooper–Schrieffer (BCS) theory of superconductivity. A short introduction to superconductivity along with a presentation on BCS theory is provided in this chapter to make it self-contained. Nuclear fusion is a process in which one or more light nuclei fuse to generate a relatively heavier nucleus, which leads to some mass deficiency that is released as energy. This is the main source of energy in the Sun that it radiates to the Universe, including the Earth. Chapter 9 discusses the nuclear fusion of hydrogen atoms.

After providing an idea of the various applications of hydrogen in chapter 10 at the beginning of part II of the book, the other chapters in this part deal with various facets of hydrogen technology needed for such applications. Different methods of hydrogen production are presented in chapter 11, while chapter 12 deals with the various ways in which hydrogen can be stored before its utilization. Safety and integrity are matters of profound interest in the usage of hydrogen in different sectors. These matters are discussed in chapter 13. The transport of hydrogen in various forms and its distribution are essential parts of hydrogen technology, and chapter 14 deals with these aspects. Hydrogen is a prominent energy carrier. In chapter 15, we will discuss how hydrogen can be converted into other forms of energy in more ways and more efficiently than other energy fuels. Futuristic hydrogen nuclear fusion technology, mimicking the Sun on the Earth, is at a crossroads today. Chapter 16 provides a short introduction to this very interesting and important technology, and outlines its current status. Hydrogen plays an important role in semiconductor technology, and chapter 17 presents the major ways of using hydrogen in the semiconductor industry. Finally, in chapter 18 the readers are provided with a formal introduction to the term 'hydrogen economy', what it means, and its present status.

## References

[1] The early universe, CERN Accelerating Science; https://home.cern/science/physics/early-universe

[2] Weinberg S 1977 *The First Three Minutes: A Modern View of the Origin of the Universe* (New York: Basic Books)

[3] Rigden J S 2002 *Hydrogen: The Essential Element* (Cambridge, MA: Harvard University Press)

[4] Nasadiya Sukta, https://en.wikipedia.org/wiki/Nasadiya_Sukta

[5] Osman A I *et al* 2021 *Environ. Chem. Lett.* **20** 153

[6] Green hydrogen: the next transformational driver of the utility industries, *Equity Research* September 2020. https://www.goldmansachs.com/intelligence/pages/gs-research/green-hydrogen/report.pdf

[7] Ashcroft N W 1968 *Phys. Rev. Lett.* **21** 1748

# Part I

Physics of hydrogen

Hydrogen
Physics and technology
**Sindhunil Barman Roy**

# Chapter 1

## Quantum physics of hydrogen atom

### 1.1 Hydrogen atomic spectrum

Isaac Newton in 1666 realized that white light was composed of the colors of the rainbow. He transmitted sunlight sequentially through a narrow slit and a prism, to project a colored spectrum onto a wall. The same effect is noticed in the sky in the form of a rainbow. Rene Descartes and others had previously suggested that the white light became colored when refracted, and the color depended on the angle of refraction. Newton, however, could get back the white light with the help of a second prism, thus strengthening the idea that the white light was composed of separate colors. He further took a monochromatic component from the spectrum generated by one prism and then transmitted that through another prism, and demonstrated that no further colors were generated. Newton concluded that white light was made up of all the colors of the rainbow, and different color components was refracted through slightly different angles on transmitting through a prism. This separates the components of white light into the observed spectrum.

William Wollaston discovered in 1802 that there were many small gaps in the solar spectrum. The rainbow of color had many thin dark lines. Beginning in 1814 Joseph von Fraunhofer investigated the solar spectrum more systematically, and he discovered nearly six hundred dark lines in the spectrum. Modern techniques can now detect many thousands of lines in sunlight. These lines are typical spectral absorption lines and are attributed to the absorption of light in the outer layers of the Sun. In general, an absorption spectrum is obtained when light passes through a gas. This spectrum appears as black lines only at specific wavelengths in the background of the continuous spectrum of white light. These missing wavelengths of the incoming radiation are absorbed by the gas.

Bunsen, Kirchhoff, and others found that when gases are heated to incandescence, they emitted light with a series of sharp wavelengths. When analyzed by a simple prism or sophisticated spectrometer the spectrum appears as colorful lines on a black background. These lines are characteristic of the atomic composition of the gas. Each chemical element has its characteristic emission spectrum. The positions of the emission lines of an element are the same as the positions of its absorption lines. This indicates that atoms of a specific element absorb radiation only at specific wavelengths and the radiation emitted by atoms of each element has the same wavelengths as the radiation they absorb.

Anders Ångström first observed the spectrum of hydrogen atoms in 1853. Ångström measured the four visible spectral lines of hydrogen with wavelengths 656.21, 486.07, 434.01, and 410.12 nm. Based on these numbers Balmer in 1885 proposed a simple formula for predicting the wavelength of any of the lines in atomic hydrogen:

$$\lambda = b\left(\frac{n_2^2}{n_2^2 - 4}\right) \tag{1.1}$$

Here $b = 364.56$. The first four wavelengths of equation (1.1) with $n_2^2 = 3, 4, 5, 6$ matched well with the experimental obtained spectral lines of hydrogen. The integer $n_2^2$ extends theoretically to infinity and the series (known as Balmer series) represents a monotonically increasing energy and frequency of the absorption lines. Balmer's phenomenological formula was further generalized by Rydberg to determine the wavelengths of any of the lines in the hydrogen emission spectrum. Rydberg's phenomenological equation in terms of the inverse wavelength or the wavenumber $\bar{\nu}$ is expressed as:

$$\bar{\nu} = R_H\left(\frac{1}{n_1^2} - \frac{1}{n_2^2}\right) \tag{1.2}$$

Here $R_H = 1.097\,37 \times 10^{-7}\,\mathrm{m}^{-1}$ (or $2.18 \times 10^{-18}$ J) is the Rydberg constant and $n_1$ and $n_2$ are integers with $n_2 > n_1$.

The various combinations of $n_1$ and $n_2$ can be substituted into equation (1.2) to calculate the wavelength of any lines in the hydrogen emission spectrum. For the Balmer series $n_1 = 2$, and $n_2$ can be any whole number between three and infinity. The spectral lines for $n_1 = 1$ are in the UV region, and these lines form the Lyman series. Thus, the spectral lines are grouped into series according to values of $n_1$. Further series are Paschen series with $n_1 = 3$, Brackett series with $n_1 = 4$, Pfund series with $n_1 = 5$, and Humphreys series with $n_1 = 6$.

## 1.2 Quantum physics of hydrogen atom

The Rydberg phenomenological formula for hydrogen provided the exact positions of the spectral lines as they are observed experimentally in a laboratory. However, the explanation of the origin of the spectral lines had to wait for a complete understanding of the atomic structure. It was long known that matter is made of atoms, and according to nineteenth-century science, atoms were the smallest indivisible quantities of matter. This scientific view was however, negated by the subsequent discoveries of subatomic particles, such as electrons, protons, and neutrons.

The electron was discovered by J J Thompson in 1897. Subsequently, in 1904 Thompson proposed the first model of atomic structure, in which an atom was comprised of negatively charged electrons embedded in an unknown positively charged matter like plums in a pudding. In 1909, Rutherford along with Ernest Marsden and Hans Geiger used $\alpha$-particles in their famous scattering experiment (which is known as the Geiger–Marsden experiment) that disproved Thomson's 'plum-pudding' model. In the Geiger–Marsden experiment $\alpha$-particles were incident on a thin gold foil and were scattered by gold atoms inside the foil. The outgoing $\alpha$-particles were detected by a 360° scintillation screen surrounding the gold target. A small flash of light was observed on the screen when a scattered particle struck the screen. Rutherford and coworkers determined what fraction of the incident $\alpha$-particles was scattered and what fraction were not deflected at all by counting the scintillations observed at the screen at various angles with respect to the direction of the incident beam. No back-scattered $\alpha$-particles were expected if the plum-pudding model were correct. However, the results of the Rutherford experiment revealed that while a finite fraction of $\alpha$-particles emerged from the foil unscattered as if the foil was not in their way a significant fraction of $\alpha$-particles were back-scattered toward the source. Such a result is expected only when most of the mass and the entire positive charge of the gold atom were confined within a tiny space inside the atom. Based on the experimental observations, in 1911 Rutherford proposed a nuclear model of the atom with a positively charged nucleus of negligible size but containing almost the entire mass of the atom. The negatively charged electrons were located within the atom but at a relatively large distance from the nucleus. Around 1920 Rutherford coined the name proton for the nucleus of hydrogen. He also postulated the presence of a neutron, the hypothetical electrically neutral particle, to mediate the binding of positive protons in the nucleus of heavier elements. The neutron was experimentally discovered in 1932 by James Chadwick. Rutherford is credited with the discovery of the atomic structure, especially the idea of the atomic nucleus, but his model could not explain the Rydberg formula for the hydrogen emission lines.

In 1913, Niels Bohr proposed a phenomenological theory of hydrogen atoms to explain the discrete wavelengths of the hydrogen spectrum. Bohr's model is a combination of the classical mechanics of planetary motion with the quantum concept of electromagnetic radiation. We recall here that Newton's universal law of gravitation has a similar formulation to Coulomb's law of electrostatics representing the attraction between two opposite charges in the sense that the gravitational force and the electrostatic force are both decreasing as $1/r^2$, where $r$ is the separation distance between the bodies or charges. Classically, if the electron moves around the nucleus in a planetary fashion, then it will then undergo centripetal acceleration and, as per Maxwell's laws of electromagnetism, an accelerating charge particle will radiate energy. This loss of energy will cause it to spiral down into the nucleus. Thus, a planetary configuration of hydrogen atoms as such will not be stable. Bohr intuitively proposed that the puzzle of hydrogen spectra may be solved if the following assumptions are made [1]:

1. In a hydrogen atom the electron revolves around a dense nucleus consisting of a proton in circular orbits.

2. Electron orbit around the nucleus takes only particular values of radius. In such special orbits, an electron does not radiate energy in the form of an electromagnetic wave as per Maxwell's laws. Such orbits are known as stationary orbits.

3. An electron is allowed to make transitions from one orbit with energy $E_n$ to another orbit with energy $E_m$. The electron moves to a higher-energy orbit when an atom absorbs electromagnetic radiation. On the other hand, when an electron transits to a lower-energy orbit, the atom emits electromagnetic radiation. These inter-orbit electron transitions with the simultaneous absorption or emission of electromagnetic radiation take place instantaneously. The allowed electron transitions need to satisfy the Einstein–Planck equation: $h\nu = |E_n - E_m|$. Here $h\nu$ is the energy of either an emitted or absorbed electromagnetic radiation with frequency $\nu$ and $h = 6.626\,070\,15 \times 10^{-34}$ m$^2$ kg s$^{-1}$ is the Planck constant. Any change in the energy of an electron is quantized in the hydrogen atom.

4. The angular momentum $l_n$ of the electron in the $n$th orbit can take only discrete values $l_n = n\hbar$, where $n = 1, 2, 3, ....$ Here $\hbar = h/2\pi$. The angular momentum of electron is quantized. If we denote the radius of the $n$th orbit and the speed of electron in this orbit by $r_n$ and $v_n$, respectively, then the first quantization condition can be expressed as $m_e v_n r_n = n\hbar$. Here $m_e$ is the mass of an electron.

The last assumption is known as Bohr's quantization rule, and the first three assumptions are called Bohr's postulates [1].

These three postulates along with the quantization rule of the early quantum theory of the hydrogen atom allow one to derive important properties of the hydrogen atom, such as its energy levels, its ionization energy, and the sizes of electron orbits, along with the Rydberg formula and the value of the Rydberg constant. Bohr's theory can also be applied to hydrogen-like ions with just one electron, such as He$^+$, Li $^{++}$, Be $^{+++}$, etc.

We shall now use Bohr's postulates to calculate the allowed energies of the hydrogen atom for different allowed orbits of the electron. Let us assume an electron with a negative charge $e$ moves with a constant speed $v$ along a circular orbit of radius $r$ with the center at the hydrogen nucleus containing a proton with positive charge $e$. The force felt by the electron is the electrostatic Coulomb attraction due to the nucleus, which is expressed as:

$$F = \frac{e^2}{4\pi\epsilon_0 r^2} \tag{1.3}$$

Here $\epsilon_0$ is the permittivity of the free space.

The electron feels a centripetal acceleration of magnitude $v^2/r$ towards the nucleus. If the mass of the electron is $m$, then from Newton's law we can express:

$$\frac{mv^2}{r} = \frac{e^2}{4\pi\epsilon_0 r^2} \tag{1.4}$$

Using equation (1.4), the radius of the electron orbit $r$ can be expressed as:

$$r = \frac{e^2}{4\pi\epsilon_0 mv^2} \tag{1.5}$$

Now using Bohr's quantization rule, the angular momentum of an electron can be written as:

$$mvr = n\frac{h}{2\pi} \tag{1.6}$$

Here $n$ is a positive integer.

Eliminating $r$ from equations (1.5) and (1.6), we can get an expression for the velocity $v$ of the electron as:

$$v = \frac{e^2}{2\epsilon_0 hn} \tag{1.7}$$

Substituting $v$ from equation (1.7) back to equation (1.6), we can get an expression for $r$ as:

$$r = \frac{\epsilon_0 h^2 n^2}{\pi m e^2} \tag{1.8}$$

From equation (1.8) we can see that there is an allowed orbit for each value of $n$ and the radii of the allowed orbits are proportional to $n^2$. The kinetic energy $K$ of the electron in the $n$th orbit of the hydrogen atom can be expressed using equation (1.7) as:

$$K = \frac{1}{2}mv^2 = \frac{me^4}{8\epsilon_0^2 h^2 n^2} \tag{1.9}$$

Assuming the potential energy of the electron–nucleus pair of the hydrogen atom to be zero when the electron and nucleus are widely separated, the potential energy $V$ of the hydrogen atom can be expressed as:

$$V = -\frac{e^2}{4\pi\epsilon_0} = -\frac{me^4}{4\epsilon_0^2 h^2 n^2} \tag{1.10}$$

With the assumption that the kinetic energy of the hydrogen nucleus is negligible because of its relatively large mass, the total energy $E$ of the hydrogen atom is expressed as:

$$E = K + V = -\frac{me^4}{8\epsilon_0^2 h^2 n^2} \tag{1.11}$$

Using equations (1.7)–(1.11) it is possible to obtain various parameters of the hydrogen atom when the electron is in the $n$th orbit. The hydrogen atom is said to be in the $n$th energy state in this situation.

### 1.2.1 Radii of hydrogen atom orbits

From equation (1.8) the radius of the smallest circular orbit allowed to the electron in a hydrogen atom can be expressed as:

$$r_1 = \frac{\epsilon_0 h^2}{\pi m e^2} \tag{1.12}$$

By substituting the values of $e$, $m$, $h$, and $\epsilon_0$ in equation (1.12), we obtain the radius $r_1 = 53 \times 10^{-12}$ m or 0.053 nm. This length is generally represented by the term $a_0$ and is called the *Bohr radius*. This is a suitable unit to measure lengths in atomic physics. In general, the radius of the $n$th orbit of an electron in a hydrogen atom is represented as:

$$r_n = n^2 a_0 \tag{1.13}$$

### 1.2.2 Ground and excited states of the hydrogen atom

The total energy of the hydrogen atom when the electron revolves in the smallest allowed orbit $r_1 = a_0$ can be calculated using equation (1.11), as:

$$E_1 = -\frac{m e^4}{8 \epsilon_0^2 h^2} \tag{1.14}$$

This is the lowest energy state of the hydrogen atom and is also known as the ground state. By putting the values of various constants in equation (1.14), one can obtain the value of the ground state energy as $E_1 = -13.6$ eV. One can also see from equation (1.11) that the energy of the hydrogen atom in the $n$th energy state is proportional to $1/n^2$. Thus, we can express $E_n$ as:

$$E_n = \frac{E_1}{n^2} = -\frac{13.6 \text{ eV}}{n^2} \tag{1.15}$$

Figure 1.1 shows schematically the allowed orbits, along with the corresponding energies of the hydrogen atom. The lowest energy state or the ground state corresponds to the lowest circular orbit. We note here that the energy expressed in equation (1.11) is negative, and hence a larger magnitude means lower energy. The zero energy here corresponds to the energy state when the hydrogen nucleus and the electron are widely separated. The energy of the first excited state of a hydrogen atom is $-3.4$ eV.

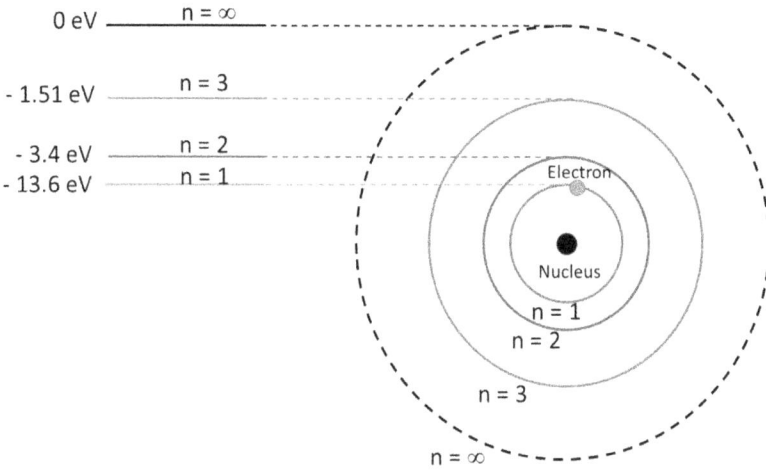

**Figure 1.1.** Schematic representation of the allowed orbits along with the corresponding energies of a hydrogen atom in Bohr's atomic model.

### 1.2.3 Spectra of the hydrogen atom

We shall now try to understand the experimentally obtained hydrogen spectra with the help of Bohr's model of atomic hydrogen. If an electron jumps from the $n_2$th orbit to $n_1$th orbit (where $n_2 > n_1$), the energy of the hydrogen atom changes from $E_{n_2}$ to $E_{n_1}$. The atom will radiate energy in the process. The corresponding inverse wavelength of the emitted radiation can be expressed as:

$$\frac{1}{\lambda} = \frac{E_{n_2} - E_{n_1}}{hc}$$

$$= \frac{me^4}{8\epsilon^2 h^3 c}\left[\frac{1}{n_2^2} - \frac{1}{n_1^2}\right] \tag{1.16}$$

$$= R\left[\frac{1}{n_1^2} - \frac{1}{n_1^2}\right]$$

Here $R = \frac{me^4}{8\epsilon^2 h^3 c}$ is known as **Rydberg constant**. With the insertion of the values of different constants, the Rydberg constant $R$ turns out to be $1.0973 \times 10^7$ m$^{-1}$. Equation (1.16) is in excellent agreement with Rydberg's phenomenological formula (1.2) obtained from the experimental results. The energy of the hydrogen atom can be expressed in terms of the Rydberg constant as:

$$E = -\frac{Rhc}{n^2} \tag{1.17}$$

Often the energy of the atom is expressed in *rydberg* units, where the energy of 1 *rydberg* means $-13.6$ eV.

If a hydrogen atom undergoes a transition from the energy state $n = 2$ to the energy state $n = 1$, then the wavelength of the resultant emitted radiation is given by:

$$\frac{1}{\lambda} = \left[1 - \frac{1}{4}\right]$$

$$= 121.6 \text{ nm}$$

(1.18)

On the other hand, if a hydrogen atom undergoes a transition from the energy state $n = \infty$ to the energy state $n = 1$, then the wavelength of the resultant emitted radiation is given by:

$$\frac{1}{\lambda} = [1 - 0]$$

$$= 91.2 \text{ nm}$$

(1.19)

All the transitions ending at the $n = 1$ state, i.e., ground state, of the hydrogen atom correspond to the wavelengths grouped between 121.6 nm and 91.2 nm. These spectral lines belong to the Lyman series. Similarly, it can be shown that the transitions from the higher-energy states to $n = 2$ energy state generate radiation with wavelengths lying between 656.3 nm and 365 nm, which belong to the visible region and form the Balmer series. The transitions to the $n = 3$ energy state from the higher-energy state give rise to the Paschen series, where the wavelengths lie between 1875nm and 822 nm. Thus, the grouping of wavelengths emitted generated from hydrogen gas can be explained with the help of Bohr's model of the hydrogen atom. The energy level diagram and spectral series of the hydrogen atom are presented in figure 1.2.

### 1.2.4 Ionization potential, binding energy, and excitation potential

If an energy excess of 13.6 eV is applied to a hydrogen atom in its ground state, then the total energy of the atom will be positive. In this situation, the electron is free to move anywhere and not bound to the nucleus anymore. The atom is now ionized with its electron being detached from the nucleus and moving independently with some kinetic energy.

Ionization energy is the minimum energy needed to ionize an atom, and the ionization energy of a hydrogen atom in its ground state is 13.6 eV. The ionization potential is the potential difference through which an electron needs to be accelerated to acquire ionization energy. Thus, the ionization potential of a hydrogen atom in its ground state is 13.6 V.

The energy released when the constituents of an atom are brought from infinity to form the atom is called the binding energy of the atom. It may also be defined in another way as the energy required to separate the constituents of an atom over large distances. Assuming that an electron and a proton are initially at rest and then brought together to form an atom, an energy of 13.6 eV will be released. Thus, the binding energy of a hydrogen atom is 13.6 eV, which is the same as its ionization energy.

## (a)

$E$(eV)

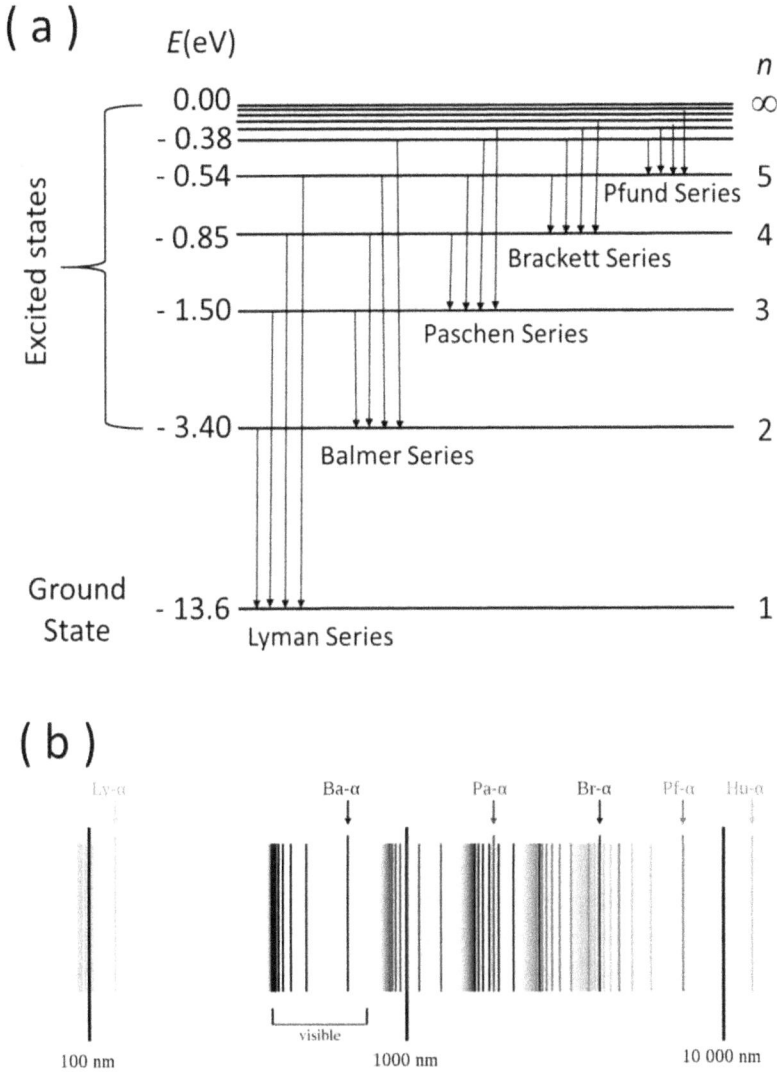

**Figure 1.2.** (a) Schematic representation of energy level diagram of hydrogen atom. (b) Spectral series of hydrogen on a logarithmic scale. This Hydrogen spectrum image has been obtained by the author(s) from the Wikimedia website where it was made available under a CC BY-SA 3.0 licence. It is included within this book on that basis. It is attributed to OrangeDog.

Excitation energy is the energy needed to take the atom from its ground state to an excited state. One can estimate from equation (1.17) that energy of 10.2 eV is required to take a hydrogen atom from the ground state to its first excited state. Thus, the excitation energy of a hydrogen atom in its first excited state is 10.2 eV. The excitation potential is the potential through which an electron needs to be accelerated to get the excitation energy. The excitation potential of a hydrogen atom in its first excited state is therefore 10.2 V.

### 1.2.5 Limitations of Bohr's model

While Bohr's model of a hydrogen atom was an improvement over Rutherford's nuclear model, it failed to account for finer details of hydrogen spectra, such as multiple closely spaced spectral lines revealed by modern and more sophisticated spectroscopic techniques. At least seven components with slightly different wavelengths have been revealed in what was originally known as the 656.3 nm line [1]. Moreover, Bohr's model was unable to explain the splitting of the spectral lines in the presence of an electric field (i.e., Stark effect) and magnetic field (i.e., Zeeman effect). It also cannot explain the formation of hydrogen molecules out of hydrogen atoms.

The Bohr model of hydrogen is semiclassical. It combines the classical concept of electron orbits with the new concept of quantization. In Bohr's postulates, it is rather arbitrarily assumed that the orbits are quantized, and in those orbits the electron disobeys the classical law of electromagnetism and does not radiate energy. Accelerating electrons do not spiral into the nucleus by radiating energy as expected within the realm of classical electromagnetism.

## 1.3 Wave-particle duality of the electron and the Heisenberg uncertainty principle

The limitations of Bohr's model subsequently led to the formulation of a more appropriate and general quantum theory of atoms. Two important concepts are basic to the formulation of this new quantum theory:

1. Wave-particle duality of matter.
2. Heisenberg uncertainty principle.

In 1924 Louis de Broglie introduced the concept that matter exhibited dual behavior—both particle and wavelike properties. In the same sense that the quantum of radiation photon has momentum as well as wavelength, electrons, and for that matter any material particle, should also have momentum and wavelength. In addition, de Broglie proposed the following relation between wavelength ($\lambda$) and momentum ($p$) of a material particle.:

$$\lambda = \frac{h}{mv} = \frac{h}{p} \tag{1.20}$$

Here $h$ is the Planck constant, $m$ is the mass of the particle, $p$ its velocity, and $p$ its momentum.

In 1927, C J Davisson and L H Germer experimented to demonstrate the wave nature of the electron. This experiment, which is now popularly known as the Davisson–Germer experiment, showed that electron beams can undergo diffraction when passed through atomic crystals. This demonstrated that electrons behaving as waves could exhibit interference and diffraction. It may be noted here that according to de Broglie's theory, every object has a wavelike character associated with it. However, for macroscopic objects, due to the large value of mass $m$ and inherently small value of Planck constant $h$, their wave properties are negligibly small and hence cannot be detected. On the other hand, for electrons and other subatomic particles with small masses, the associated wavelengths can be determined experimentally with an appreciable degree of accuracy.

In 1927 Wener Heisenberg introduced the *uncertainty principle*, which is a consequence of the wave-particle duality of matter and radiation. In the case of an electron, the *uncertainty principle* states that it is impossible to simultaneously determine the exact position and exact momentum (or velocity) of an electron in motion. This can be expressed mathematically as:

$$\Delta x \times \Delta p_x \geqslant \frac{\hbar}{2}$$

$$\Delta x \times m\Delta v_x \geqslant \frac{\hbar}{2} \qquad (1.21)$$

$$\Delta x \times \Delta v_x \geqslant \frac{\hbar}{2m}$$

Here $\Delta x$ denotes uncertainty in position, $\Delta p_x$ denotes uncertainty in momentum, and $\Delta v_x$ denotes uncertainty in velocity. This indicates that if the position of the electron is known with a high degree of accuracy (i.e., $\Delta x$ is small), then the velocity of the electron will be uncertain (i.e., $\Delta v_x$ is large), and vice versa.

Heisenberg's uncertainty principle rules out the existence of definite trajectories of electrons. According to Heisenberg's uncertainty principle for an electron (with mass $9.11 \times 10^{-31}$ kg) $\Delta v . \Delta x \geqslant \frac{\hbar}{2\pi m} = 10^{-4} m^{-1} s^{-1}$. This means that if one can determine the exact location of an electron with an uncertainty of $10^{-8}$m, then the uncertainty in the velocity of the electron would be so large that the semiclassical picture of the electron moving in a fixed Bohr orbit will not hold good anymore. A Bohr orbit is a clearly defined path with the position and velocity of an electron being exactly known at the same time. Heisenberg's uncertainty principle thus indicates that the definite statement of position and momentum of electrons needs to be replaced with the concept of *probability* that the electron has a certain value of position and momentum at an instant of time. A proper model of the Hydrogen atom needs to take into account the wave-particle duality of electrons and be consistent with Heisenberg's uncertainty principle.

## 1.4 Schrödinger's wave equation of the hydrogen atom

De Broglie's equation (equation (1.20)) associates a wavelength $\lambda$ with an electron. The particle property of an electron can be understood through this wave property of the electron, which is represented in terms of a wave function $\psi(\mathbf{r}, t)$. This wave function varies continuously in space and at any instant it may be extended over a large part of space $(\mathbf{r})$. This does not necessarily indicate that the electron is spread over that large part of space. If one puts an instrument to detect the electron at any point, either a whole electron will be detected or none. The question is then, where will the electron be found? The answer lies within the wave function $\psi(\mathbf{r}, t)$. There is a probability to find the electron wherever $|\psi(\mathbf{r}, t)|^2 \neq 0$.

It may be noted that the wave function is a mathematical function whose value depends on the coordinates of the electron in the atom, and when an electron is in a particular energy state the wave function corresponding to that energy state contains all information about the electron. The wave function of a hydrogen atom or hydrogen-like ions with one electron is called *atomic orbital*. We shall see shortly that the wave functions are characterized by a set of *quantum numbers*. It should be made clear here that *orbit* and *orbital* are not synonymous. We have seen above that the Bohr semiclassical orbit is a circular path with the nucleus at the center, in which the electron is supposed to move. However, the Heisenberg uncertainty principle prohibits a precise description of such a path, and the existence of an orbit cannot be demonstrated experimentally. In contrast, an *atomic orbital* is a quantum mechanical concept, which corresponds to one-electron wave function $\psi$ in an atom. While $\psi$ as such has no straightforward physical meaning, the quantity $|\psi|^2$ termed as *probability density* provides very important information. We shall see below that wave function $\psi$ provides not only information about the position of the electron but also many other properties, including of course the energy states.

The wave function $\psi(\mathbf{r}, t)$ of the electron and the allowed energies $E$ of a hydrogen atom can be determined by solving the wave equation introduced by Erwin Schrödinger during 1925–26, which is a linear partial differential equation. A detailed formalism and the solution of Schrödinger's equation is provided in appendix A, and here will present only the essential details in the present context of the hydrogen atom. The Schrödinger equation can be written as [1]:

$$-\frac{h^2}{8\pi^2 m}\left[\frac{\partial^2\psi}{\partial x^2} + \frac{\partial^2\psi}{\partial y^2} + \frac{\partial^2\psi}{\partial z^2}\right] - \frac{e^2\psi}{4\pi\epsilon_0\mathbf{r}} = E\psi \tag{1.22}$$

Here $(x, y, z)$ corresponds to a point at a distance $\mathbf{r}$ from the nucleus at the origin. The first term on the left-hand side represents the kinetic energy of the electron, and the second term represents the potential energy of the electron in the field of the nucleus consisting of a single proton. The term $E$ on the right-hand side represents the energy of the atom. The discrete energy levels, spectral lines etc arise from the motion of the electron around the proton. However, the effects of the motion of the proton are ignored because the proton mass $m_p$ is much larger than the electron mass $m_e$.

There can be an infinite number of solutions of equation (1.22). These wave functions $\psi(\mathbf{r})$, which satisfy equation (1.22), are characterized in terms of three parameters $n$, $l$, and $m_l$, which are known as *quantum numbers*. The parameter $n$ is known as the principal quantum number, $l$ is the orbital angular quantum number, and $m_l$ is the magnetic quantum number. It can be shown (see appendix A) that associated with each solution $\psi_{nlm_l}$ is a unique value of energy $E$ of the hydrogen atom. The energy $E_n$ corresponding to the wave function $\psi_{nlm_l}$ depends only on the quantum number $n$ and can be expressed as [1]:

$$E_n = -\frac{me^4}{8\epsilon_0^2 h^2 n^2} \tag{1.23}$$

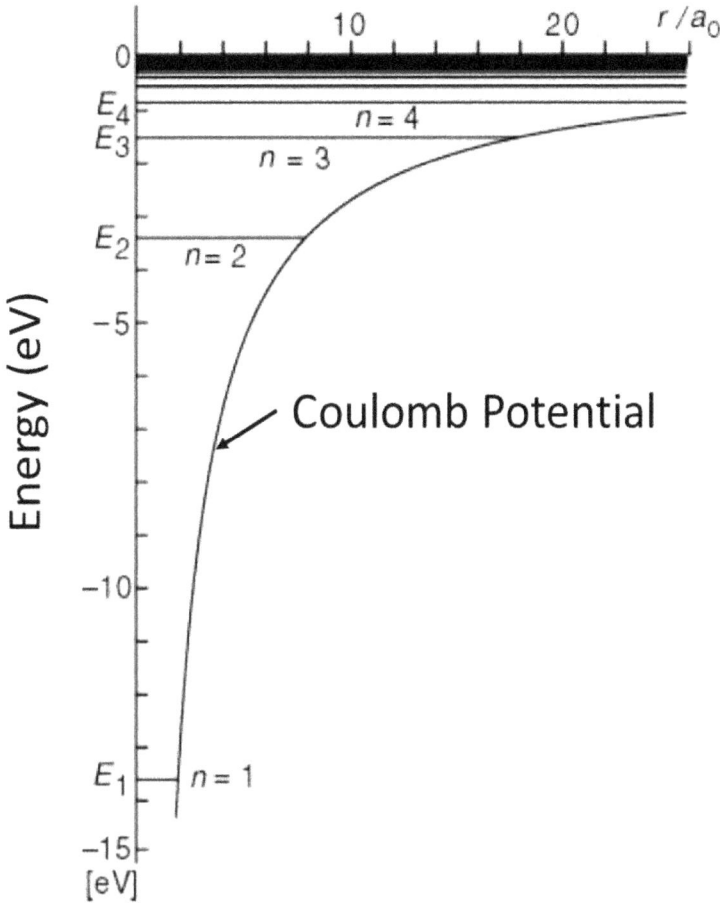

**Figure 1.3.** The energy eigenvalues of hydrogen atom obtained as a solution of Schrödinger's wave equation.

We can see that these energies are identical to the allowed energies of hydrogen atoms calculated earlier within Bohr's model (see equation (1.11)). This explains the success of Bohr's model in obtaining quantitatively the main spectral lines of hydrogen. For each principal quantum number $n$ there are $n$ values of orbital quantum number $l$, and for each $l$ there are $2l+1$ values of magnetic quantum number $m_l$ namely $m_l = -l, -l + 1, -l + 2, ...l - 1, l$.

Figure 1.3 presents the energy levels of the hydrogen atom, in which the heights of the horizontal levels represent the energy eigenvalues obtained by solving Schrödinger's wave equation (1.22) and using equation (1.23) for energy expression. The lowest possible energy or $n = 1$ ground state of a hydrogen atom is $-Rhc = -13.6$ eV. The corresponding radial wave function of the electron in this ground state is [1]:

$$\psi_{000}(r, \theta, \phi) = R_{10}(r) Y_0^0(\theta, \phi) = \frac{2}{\sqrt{4\pi}} \left(\frac{1}{a_0}\right)^{3/2} e^{-r/a_0} \tag{1.24}$$

This wave function given in equation (1.24) is spread over large distances in space, and the presence of electrons is felt wherever $\psi \neq 0$. If the electron is experimentally detected at some point, then it will, however, be detected at a single point only. The probability of finding an electron in a small element of volume $dV$ is $|\psi(\mathbf{r})|^2$. It is possible to calculate the probability $P(r)dr$ of finding the electron at a distance $r$ and $dr$ from the nucleus. Here the function $P(r)$ is termed as a radial probability density, which in the ground state given by equation (1.24) can be expressed as:

$$P(r) = \frac{4}{a_0^3} r^{-2} e^{-2r/a_0} \tag{1.25}$$

Figure 1.4(a) shows a plot of radial probability density $P(r)$ presented as a function of $r$. It may be noted that $P(r)$ is maximum at $r = a_0$. This indicates that the electron is more likely to be found near $r = a_0$ than at any other distance from the nucleus. This is in clear contrast with Bohr's model where it is stated that in the ground state of a hydrogen atom, the electron moves in a circular orbit of radius $a_0 = 0.053$ nm. In the more advanced framework of quantum mechanics, the very idea of the electron orbit is invalid. It may, however, be at least satisfying that the probability of finding the electron is maximum where according to Bohr's model the electron should exist. But this comparison cannot be extended too far. For example, figure 1.4(b) shows the radial probability density $P(r)$ in the first excited state $n = 2$, $l = 0$ and $m_l = 0$, which shows two maxima one near $r = a_0$ and the other near $r = 5.4a_0$. In Bohr's model of the hydrogen atom, the electron in the first excited state should be in the orbit at a distance $r = 4a_0$.

A very important property of electrons is that every electron has an intrinsic angular momentum called *spin*, whose component along any given direction is $h/\pi$ or $-h/\pi$. This spin angular momentum is very different from the orbital angular momentum that arises from the motion of the electron. The complete wave function of an electron is also dependent on the state of electron spin. A spin quantum number $m_S$ characterizes the spin part of the electron wave function $\psi(\mathbf{r})$, which takes the value $m_S = \pm 1/2$. An electron wave function is thus completely defined by four quantum numbers $n$, $l$, $m_l$, and $m_S$, and each wave function corresponds to a quantum state. In a quantum state with $n = 1$, one has $l = 0$ and $m_l = 0$, but $m_S$ can take values $+1/2$ or $-1/2$. Hence, two quantum states correspond to $n = 1$. It can easily be shown that for $n = 2$ there are eight quantum states, and for $n = 3$ there are 18 quantum states, and so on. In general, there are $2n^2$ quantum states corresponding to a particular $n$, and the quantum states belonging to a particular principal quantum number $n$ are called together a *major shell* [1].

### 1.4.1 Fine structures of the hydrogen spectra

Very importantly, electrons obey the Pauli exclusion principle, which states that there cannot be more than two electrons in a particular quantum state, hence the

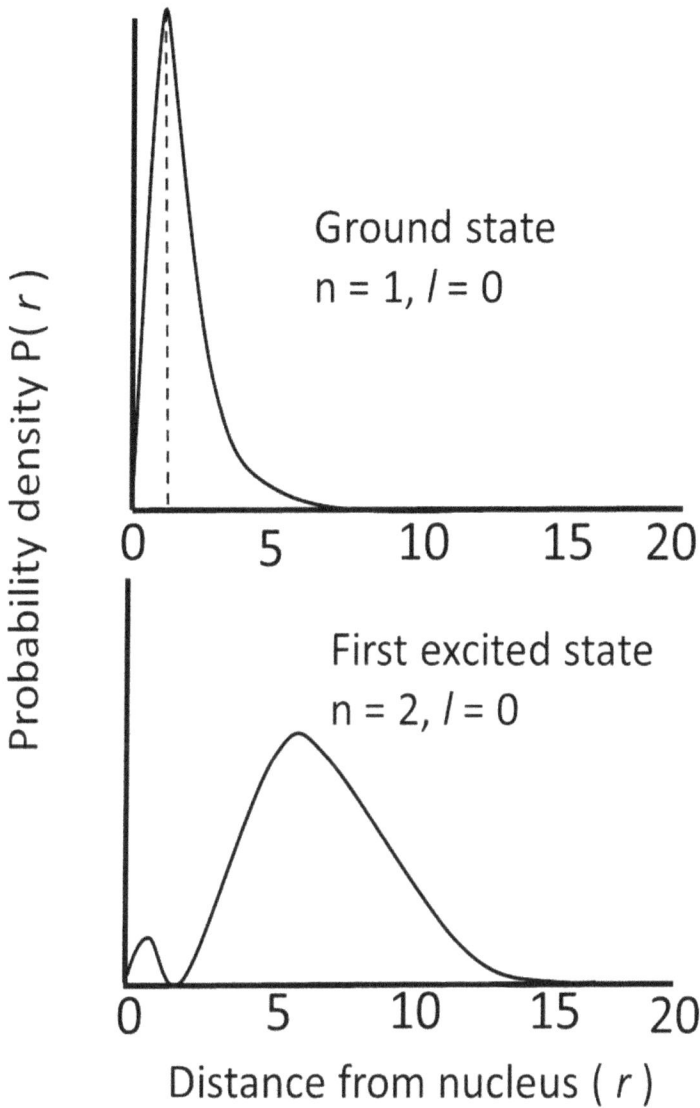

**Figure 1.4.** Radial probability density $P(r)$ as a function of distance $r$ from the nucleus for hydrogen atom: (a) ground state $n = 1$, (b) first excited state $n = 2$.

electron shell corresponding to $n = 1$ cannot contain more than two electrons. The hydrogen atom has only one electron. The spin of this electron plays an important role in the fine structures of hydrogen spectra, the existence of which cannot be explained within the realm of the semiclassical Bohr's model. The interaction between the electron spin magnetic moment and the orbital angular momentum of the electron affects the energy levels of the atomic electron. This effect can be visualized as a magnetic field caused by the orbital motion of the electron interacting with the spin magnetic moment of the electron. This effective magnetic field can be expressed in terms of the electron orbital angular momentum. This spin-orbit interaction energy contribution depends upon the relative orientation of its orbital and spin angular momentum, and causes the splitting of the energy states. The mechanism of this splitting of energy states will be discussed in appendix A. The atomic spectral lines can also be split by the application of an external magnetic field. It is then called the Zeeman effect.

## Reference

[1] Verma H C 2020 *Concepts of Physics* **vol 2** (New Delhi: Bharati Bhawan)

# Chapter 2

# Covalent bonding and the hydrogen molecule

The idea of the covalent bond plays a very important role in chemistry and materials science. The method of quantum chemistry has made it possible to understand and resolve the covalent bond with ever-increasing accuracy. This is mainly due to the availability of precise quantitative molecular electronic structure computations. Interestingly, there is still a lot of debate, discussion, controversy, and misinformation regarding the physical basis of the covalent band. A discussion of the covalent bonding models and the debate that took place over the last 100 years will be both interesting and educational in the context of the molecules of main focus in this book, namely the simple diatomic hydrogen molecule $H_2$ and the hydrogen-ion molecule $H^{+2}$.

The concept of shared electron pairs corresponding to chemical bonds was first presented by G N Lewis in a seminal work published more than a century ago in 1916 [1]. In order to construct a chemical universe, Lewis defined the quantum unit known as the 'electron pair bond' [2]. Lewis's idea originated from an even earlier study by R Abegg [3], which claimed that a difference between maximal positive and negative valences of an element was frequently found to be eight. Lewis used this rule to develop the 'octet rule' in his theory of cubic atoms, which states that electrons in nonpolar atoms or molecules are located at the eight corners of a cube. A stable configuration of an atom is obtained when eight electrons encircle the atom.

The electrons of an atom as well as the electrons shared by other atoms can give rise to this octet. The process of bond formation continues until an octet of electrons is achieved. The theories of Abegg and Lewis were further expanded upon by I Langmuir, who established the principle that electrons of the atoms tend to surround their nuclei in successive layers of 2, 8, 18, and 32 electrons [4]. Furthermore, he introduced the term 'covalent bond' to describe a pair of shared electrons. Lewis' theory of the shared electron pair bond undoubtedly made a significant contribution to the development of the octet rule in its present form. However, it may be noted that in contrast to this nearly universal perception found in introductory chemistry textbooks, the octet rule was not invented by Lewis alone [5]. The readers are referred to an interesting essay by Shaik [2] covering these early activities leading to the concept of the chemical bond.

## 2.1 Born–Oppenheimer approximation

It is possible to solve Schrödinger's equation exactly for a hydrogen atom. However, this is not possible for a hydrogen molecule because it is made up of two nuclei and two electrons. This necessitates for the molecular structure theories to start with certain simplifications. The first step in this direction is the Born–Oppenheimer approximation, which makes the assumption that the nuclei can be considered to be stationary because they are much heavier than electrons, and hence they move more slowly. The electrons then move in the field of the apparently static nuclei. Using the Born–Oppenheimer approximation, Schrödinger's equation is solved for the wave function of the electrons alone by treating the nuclei as fixed at arbitrary places. The approximation is quite good for the ground state of the hydrogen molecule [6]. Calculations indicate that the nuclei in $H_2$ move through just about 1 pm while the electron has traveled 1000 pm. Thus, there is very little error in assuming that the nuclei are stationary.

Using the Born–Oppenheimer approximation, one can solve Schrödinger's equation for the electrons at that nuclear separation by first choosing an internuclear separation in a diatomic molecule. The calculation is then repeated by considering a new separation, and so on. This allows one to create a molecular potential energy curve (see figure 2.1) and investigate how the energy of the molecule varies with bond length. Since the kinetic energy of the assumed to be stationary nuclei is zero, this is known as a potential energy curve [6]. The equilibrium bond length $R_e$ (which represents the internuclear separation at the minimum of energy curve) and the bond dissociation energy $D_0$ (which is directly correlated with the depth $D_e$, of the minimum below the energy of the infinitely widely separated and stationary atoms) can both be identified in the potential energy curve. These important parameters can be obtained once the potential energy curve is computed or determined experimentally by using spectroscopic techniques.

## 2.2 Heitler–London theory of hydrogen molecule

In a seminal work published in 1927, Walter H Heitler and Fritz London studied the hydrogen molecule problem theoretically within the framework of quantum mechanics and looked into the origin of its binding energy [7]. They began with the fundamental notion that a hydrogen molecule consisted of two hydrogen atoms. Heitler and London first approximated that the spatial wave function $\Psi$ of the hydrogen molecule was represented with products of the known eigenfunctions of the hydrogen atom, $\psi_1$ and $\phi_2$, where $\psi_1$ was the eigenfunction of electron 1 at nucleus $a$ and $\phi_2$ was the eigenfunction of electron 2 at nucleus $b$. This estimation was based on the assumption that the hydrogen atoms were initially separated at an infinite internuclear distance. Let $R$ denote the internuclear distance, $r_{12}$ denote the distance between the two electrons, $r_{a1}$ ($r_{b1}$) denote the distance between the nucleus $a$

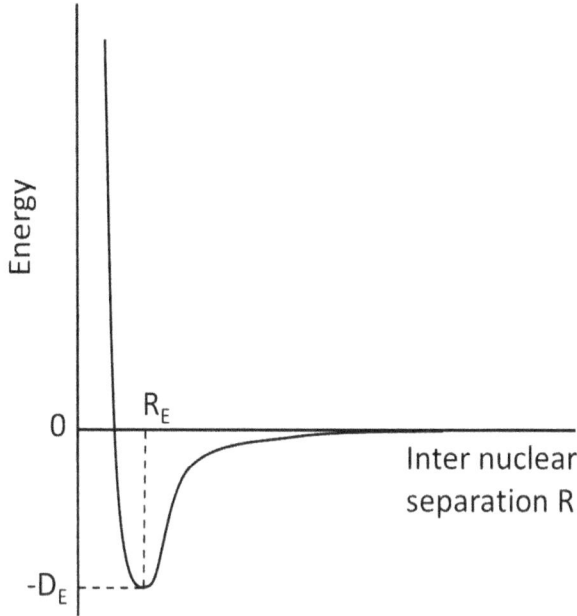

**Figure 2.1.** Schematic representation of the molecular potential energy curve.

($b$) and the electron 1, $r_{a2}$ ($r_{b2}$) denote the distance between the nucleus $a(b)$ and the electron 2, and $e$ denote the electronic charge. Following the Born–Oppenheimer approximation, the electronic part of Schrödinger's equation can be written as:

$$\hat{H}\Psi(r_1, r_2, R) = E\Psi(r_1, r_2, R) \tag{2.1}$$

Here, with the assumption of the value of $\hbar$ and electron mass $m_e$ to be 1, $\hat{H}$ is expressed as (see appendix A):

$$\hat{H} = -\frac{1}{2}\nabla_1^2 - \frac{1}{2}\nabla_1^2 - \frac{e^2}{r_{a1}} - \frac{e^2}{r_{b1}} - \frac{e^2}{r_{a2}} - \frac{e^2}{r_{b2}} + \frac{e^2}{r_{12}} + \frac{e^2}{R} \tag{2.2}$$

The internuclear separation $R$ is treated as a parameter, allowing one to solve Schrödinger's equation for each value of $R$. The interaction between the atoms may be ignored to start with. However, as the atoms approach toward one another, the interaction becomes important to take into account. This interaction between two electrons and two nuclei was treated as the perturbation of the system. Heitler and London considered that the perturbation covered the possibility for the exchange of electrons in addition to the standard electrical Coulomb interaction between electrons. Since the electrons are identical particles, it is also necessary to take account of the situation that electron 1 may be close to nucleus $b$ and electron 2 may be close to nucleus $a$. Thus, $\phi_1\psi_2$ is also an acceptable approximation for the molecular wave function $\Psi$ as $\psi_1\phi_2$, and the correct representation would be a linear combination of $\psi_1\phi_2$ and $\phi_1\psi_2$ [8]:

$$\Psi_\alpha = \frac{1}{2 + 2S}(\psi_1\phi_2 + \psi_2\phi_1) \tag{2.3}$$

$$\Psi_\beta = \frac{1}{2 - 2S}(\psi_1\phi_2 - \psi_2\phi_1) \tag{2.4}$$

where $S$ is the overlap integral and given by $S = \int \psi_1\phi_1\psi_2\phi_2\,d\tau$. Then putting the wave functions expressed in equations (2.3) and (2.4), Heitler and London obtained two different energy levels, $E_\alpha$ and $E_\beta$, for the hydrogen molecule as solutions of Scrödinger's equation [8]:

$$E_\alpha = E_{11} - \frac{E_{11}S - E_{12}}{1 + S} = \frac{E_{11} + E_{12}}{1 + S} \tag{2.5}$$

$$E_\beta = E_{11} + \frac{E_{11}S - E_{12}}{1 - S} = \frac{E_{11} - E_{12}}{1 - S} \tag{2.6}$$

Here $E_{11}$ and $E_{12}$ are integrals, which take the following form [8]:

$$E_{11} = \int \left[ \left( \frac{e^2}{r_{12}} + \frac{e^2}{R} \right)\frac{\psi_1^2\phi_2^2 + \psi_2^2\phi_1^2}{2} - \left( \frac{e^2}{r_{a1}} + \frac{e^2}{r_{b2}} \right)\frac{\psi_2^2\phi_1^2}{2} - \left( \frac{e^2}{r_{a2}} + \frac{e^2}{r_{b1}} \right)\frac{\psi_1^2\phi_2^2}{2} \right] d\tau \tag{2.7}$$

$$E_{12} = \int \left( \frac{2e^2}{r_{12}} + \frac{2e^2}{R} - \frac{e^2}{r_{a1}} - \frac{e^2}{r_{a2}} - \frac{e^2}{r_{b1}} - \frac{e^2}{r_{b2}} \right)\frac{\psi_1\phi_2\psi_2\phi_1}{2}\,d\tau \tag{2.8}$$

After derivation of the mathematical expression for the energy of the hydrogen molecule $H_2$, Heitler and London looked for the physical meaning of $E_\alpha$ and $E_\beta$, and their components $E_{11}$ and $E_{12}$. It became apparent that $E_{11}$ was related to the Coulomb interaction in the existing charge distribution. This integral can be solved analytically as a function of the internuclear distance $R$. However, $E_{12}$ does not have a simple classical interpretation. It is quite difficult to calculate all the integrals involved in $E_{12}$, in particular the one known as the *exchange integral* $\int \psi_1\phi_2\psi_2\phi_1/r_{12}\,d\tau$. Heitler and London considered only the upper limit of the *exchange integral* and got approximate graphs of $E_\alpha$ and $E_\beta$, which are reproduced in figure 2.2. At any internuclear distance, the atoms' repulsion was represented by $E_\beta$ in this energy curve. Conversely, $E_\alpha$ exhibited repulsion at closer ranges and attraction at greater ones. This results in a minimum value of $E_\alpha$ when the internuclear distance $R$ is 0.8 Å. From this energy diagram, the corresponding dissociation energy or the binding energy $E_\alpha$ is around 2.4 eV.

The Heitler–London theory could explain the attraction between the two non-polar hydrogen atoms without considering perturbation by polarization. The theory introduced a characteristic quantum-mechanical effect associated with the exchange of electrons between two hydrogen atoms participating in the formation of $H_2$ molecules. The exchange effect is represented by the integral $E_{12}$ and it affects the energy level $E_\beta$ as the van der Waals repulsion, i.e., elastic reflection of two hydrogen

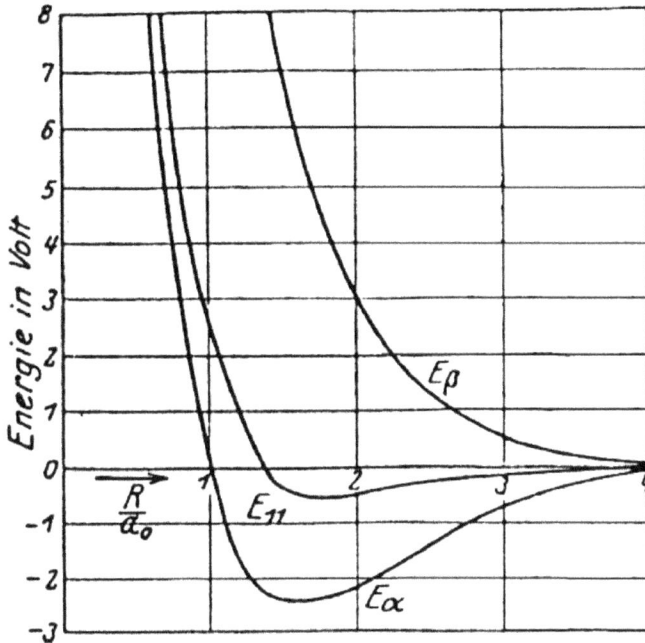

**Figure 2.2.** Energy diagram obtained by Heitler and London for the hydrogen molecule. $E_\alpha$ represents nonpolar attraction, $E_\beta$ elastic reflection and $E_{11}$ Coulomb interaction. Reprinted from [7], Copyright (1927), with permission from Springer.

atoms [8]. It contributes to $E_\alpha$ as the strength of the molecular binding, i.e., the chemical bond.

So far, the wave function $\Psi$ of the hydrogen molecule we have talked about is the spatial wave function and the Pauli exclusion principle dictates that the total wave function, which is a combination of the spatial wave function and spin wave function $\chi$, should be antisymmetric in $H_2$ molecule. Thus, for the energy state $E_\alpha$, electrons need to have opposite spin orientations, i.e., $\chi$ is antisymmetric, because $\psi_\alpha$ is symmetric. In the same argument, $E_\beta$ is an energy state in which the electrons are in the same orientation, i.e., parallel, because $\psi_\beta$ is antisymmetric. Thus, the electron spin state is a useful indicator of the formation of the molecule. The antiparallel spin state gives rise to attraction and in turn bonding, whereas the parallel spin state leads to the unstable excited state or the antibonding state. In other words, the chemical bond results from the pairing of electrons of different spin orientations, and valence is predicated from this pairing [8].

With the above rather mathematical description of the quantum-mechanical exchange interaction, it may also be worth discussing the physical origin of exchange energy. The term exchange may be conceptualized in the following manner. When the two hydrogen atoms are in near proximity, it is likely that the electron in each hydrogen atom moves around the corresponding nucleus. However, electrons are indistinguishable, and hence one needs to consider the possibility that the two electrons exchange places so that the electron in atom 1 moves around the proton in atom 2 and the electron in atom 2 moves around the proton in atom 1. This exchange of electrons takes place at a very high frequency, about $10^{18}$ times per second in the hydrogen molecule [9]. Now, the Pauli exclusion principle states that two electrons can be in the same energy state only if they have opposite spins. If in the pair of hydrogen atoms in $H_2$ molecule the two electrons have opposite spin, the hydrogen atoms can come so close together that their two electrons can have the same velocity and occupy very nearly the same small region of space, i.e., have the same energy [9]. On the other hand, the two electrons in a $H_2$ molecule will tend to stay far apart if their spins are parallel. This process introduces an additional term, that of the *exchange energy*, into the expression for the total energy of the two atoms in $H_2$ molecule. It can be said that if the spins are antiparallel, the sum of the electrostatic and exchange forces is attractive and a stable hydrogen molecule forms. The two hydrogen atoms repel one another if the electron spins are parallel. The ordinary Coulomb electrostatic energy is therefore modified by the electron spin orientations, which means that the exchange force is effectively electrostatic [9].

The calculated value of binding energy 2.4 eV of $H_2$ molecule with Heitler–London theory was considerably lower than the empirical value of binding energy 4.2 eV available at that time. Subsequently, there were several attempts to make the Heitler–London approach acceptable quantitatively as well as qualitatively without using empirical information [8]. Those calculations of the exchange integral involved newer physical and chemical insights to narrow the gap between theory and experiment, and also the development of different kinds of numerical computational schemes. In one such attempt, Rosen [10] suggested that the electronic cloud of a hydrogen atom would be polarized towards the other binding atom. This consideration of the polarization effect led to an improvement in the value of the binding energy to 4.02 eV. The other weakness of the Heitler–London approach was that the possibility of ionic configurations in $H_2$ molecule was ignored. The addition of ionic terms to the wave functions originally proposed by Heitler and London led to a further improvement in the binding energy of 0.0031 eV [11].

## 2.3 Beyond the Heitler–London theory of the hydrogen molecule

In a companion paper [12] to the original Heitler–London theory, London in 1928 obtained the contour diagrams of the electron densities in hydrogen molecules associated with the antisymmetric and symmetric solutions to the Heitler–London equations. London employed the concept of the density $\rho(r)$ introduced earlier by Schrödinger [13] and calculated the electron densities by integrating $\Psi^*\Psi(r_1, r_2)$ over the coordinates of one of the electrons. Figures 2.3 and 2.4 present electron densities calculated by London. In the spatially antisymmetric state, the electron densities are pushed outward as if the hydrogen atoms would separate if possible (see figure 2.3). Figure 2.4 presents the electron density for the symmetric solution of two hydrogen atoms, which are in a state of homopolar binding. In this case, the two densities seem to draw closer and tend to merge into one. In the words of Richard Feynman [14]: 'In a $H_2$ molecule for example the Heitler–London symmetrical solution can easily

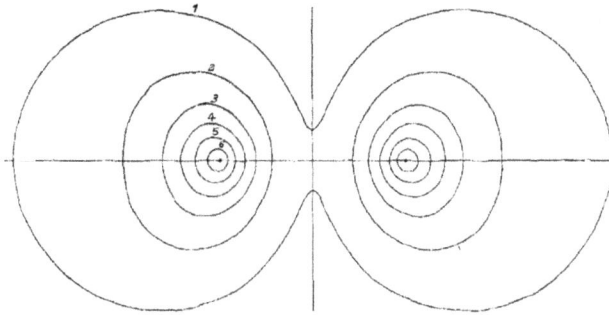

**Figure 2.3.** Electron density distributions for the hydrogen molecule obtained by London from the Heitler–London antisymmetric wave function representing the lowest excited energy state. Reprinted from [12], Copyright (1928), with permission from Springer.

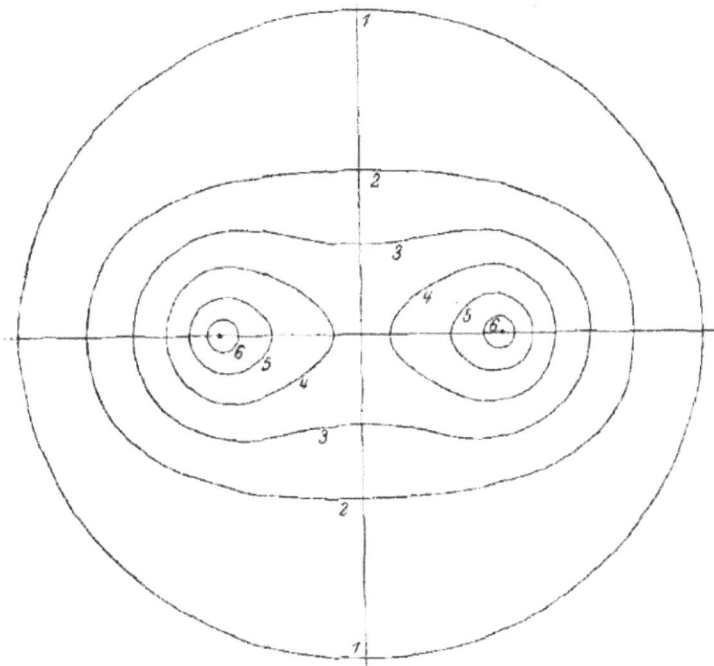

**Figure 2.4.** Electron density distributions for the hydrogen molecule obtained by London from the Heitler–London symmetric wave function representing the ground energy state. Reprinted from [12], Copyright (1928), with permission from Springer.

permit charge concentration between the nuclei and hence it is the only solution which is symmetrical that leads to strong attraction, and the formation of a molecule, as is well known.'

The work of Heitler–London was further extended by Linus Pauling [15, 16], who used the pair bonding ideas of Lewis together with the Heitler–London theory to introduce two other key concepts in *valence bond* (VB) theory, namely *resonance* and *hybrid-orbital*, and turned it into a hugely successful method for understanding an enormous range of chemical phenomena. In *valence bond* theory, the wave function does not take account of any ionic configuration in which two electrons can reside in the same atom, and it is assumed that there will always be one electron centered at atom 1 and the other at atom 2. The process of exchanging the positions of electrons between the two hydrogen atoms was likened to quantum-mechanical resonance. In this approach, the Coulomb repulsion between electrons is reduced by keeping the two electrons well separated, and a covalent bond is formed between hydrogen atoms by taking advantage of the attraction of electrons to each of the nuclei. But this happens at the expense of the kinetic energy of electrons, which would have been lower in an ionic configuration. In this case, more importance is implicitly given to the Coulomb repulsion between electrons than the energy gain due to electron delocalization.

Around the same time, Hund and Mulliken introduced the method of *molecular orbitals*, where the electrons are placed in wave functions extending over the whole molecule. The Hund–Mulliken method was also a very successful method of understanding most of the observed phenomena of quantum chemistry. Within the molecular orbital approach, the bonding molecular orbital for hydrogen molecule is occupied by two electrons of opposite spin, and the probability of finding both the electrons at the same site is 50%. The antisymmetric wave function is slightly higher in energy than the symmetric wave function. This is because an electron in the antisymmetric state is excluded from a small region of space where the antisymmetric wave function changes its sign. An electron in the symmetric state is, however, allowed in this region. The exclusion from a region shortens the range of positions allowed for the electrons in the antisymmetric state and, through the uncertainty principle, there is an increase in the kinetic energy for such electrons. Thus, in the molecular orbital theory, the kinetic energy of the electrons is optimized, while the Coulomb repulsion between the electrons remains considerably large. The chemists traditionally preferred molecular orbital theory more, especially when dealing with multicenter bonds, and more generally for problems of electronic excitation, reactivity, and transition metal chemistry [6]. This is mostly because of the computational ease using sophisticated numerical techniques and the advent of modern computer facilities. However, since the 1990s there has been a renewed interest in the interest valence bond theory and it has been demonstrated to be the natural approach to several important chemical problems [17, 18].

## 2.4 Physical origin of the covalent chemical bond in the hydrogen molecule

Even with readily available methods for quantitative molecule electronic structure calculations of ever-increasing accuracy and complexity, a complete understanding of the physics of the covalent chemical bond is still a matter of great debate [18]. The physical origin of covalent bonds can be explained in two very different ways: one relies on electrostatic energy, while the other considers the quantum dynamical effect. These two distinct qualitative models of covalent bonding will be discussed here. We will see that a dynamical viewpoint of the covalent bonding mechanism is made possible by quantum mechanics. This viewpoint offers a clear understanding of atomic reactivity, the synthesis of molecules, the fundamental role of kinetic energy, and the significant but secondary role of electrostatics. We will closely follow this relatively recent review work by Nordholm and Bacskay [19], which is highly instructional and interesting.

### 2.4.1 Electrostatic potential energy approach

In the standard textbooks of chemistry, students are taught to visualize and think about covalent bonding between two hydrogen atoms as a static phenomenon. The electrons are shared between two atoms and are located in-between the bonded atoms, and they act as *electronic glue*. The only role of quantum theory is to replace the concept of shared electrons in Lewis's theory with electron clouds, which are localized in the binding regions of the hydrogen molecule. A solution to Schrödinger's equation is generated to work out the ground state for a molecule and its energy. The wave function for the electrons is then squared to obtain a probability density for the distribution of electrons around the atoms. Thus, it is possible to visualize the electron density and see the molecule as a set of joined-up fuzzy balls. The ground state is said to be stationary. This means that the physical properties such as electron densities do not change with time. Within this framework of stationary objects, the origin of covalent bonding is correlated to the electron density difference between a molecule and its constituent atoms, and thus to the electrostatic interactions and the changes that occur as a molecule forms

In 1933 John Slater in a landmark paper [20] attributed the force acting on the nuclei to the gradients of the potential energy surface. Later on, Feynman [14] showed this force to be the electrostatic force acting on the nuclei. Feynman's electrostatic theorem [14] demonstrated that bonding originated from the accumulation of electron density between the nuclei, which exerted an attractive force sufficient enough to overcome the force of repulsion between the nuclei. The Ehrenfest force acting on the electron density and the Feynman force acting on the nuclei are the necessary forces involved in the physics of chemical bonding, and the virial theorem provides a unified statement of these forces by relating them to the total energy of the molecule and its kinetic and potential components [21]. According to the virial theorem, for any system of charges at equilibrium, molecule, or atoms, the ratio of potential to kinetic energy is exactly $-2$. The same ratio is expected to be valid for the potential and kinetic components of the binding energy.

Slater derived the molecular virial theorem for a diatomic molecule beginning with infinitely separated atoms. He differentiated Schrödinger's equation concerning an electronic coordinate $x_i$ and followed that with multiplication by $x_j \psi^*$. The result can be rearranged to yield [21]:

$$\sum_i -\left(\frac{\hbar^2}{8\,m}\right)\sum_j \left[x_j\left(\psi^*\frac{\partial^3\psi}{\partial x_i^2 \partial x_j} - \frac{\partial^2\psi^*}{\partial x_i^2}\frac{\partial\psi}{\partial x_j}\right)\right] + \left[\sum_j x_j\left(\frac{\partial V}{\partial x_j}\right)\right]\psi^*\psi = 0 \quad (2.9)$$

The above step is followed by integration over the coordinate space. The first term is integrated by parts and this is performed by using the identity [20]:

$$\sum_j x_j \left( \psi^* \frac{\partial^3 \psi}{\partial x_i^2 \partial x_j} - \frac{\partial^2 \psi^*}{\partial x_i^2} \frac{\partial \psi}{\partial x_j} \right) = -2\psi^* \frac{\partial^2 \psi}{\partial x_i^2} + \frac{\partial}{\partial x_i}\left[ \psi^{*2} \frac{\partial}{\partial x_i}\left( \frac{\sum_j (x_j \partial \psi / x_j)}{\psi^*} \right) \right] \qquad (2.10)$$

This identity can be proved by performing the differentiations indicated on the right-hand side and with a little manipulation [20]. When the right-hand side is integrated over the coordinates, the derivative integrates to zero. This is because if we assume the system to be a closed one, then we take the limits of integration at infinity, where $\psi = 0$. We are then left with the expression:

$$\sum_i -\left( \frac{\hbar^2}{8m} \right) \int \psi^* \frac{\partial^2 \psi}{\partial x_i^2} d\tau = -\frac{1}{2} \int \sum_j x_j \left( \frac{\partial V}{\partial x_j} \right) \psi^* \psi d\tau = V \qquad (2.11)$$

The left-hand side of equation (2.11) represents the mean kinetic energy $T$. Identifying $F_j = -\frac{\partial V}{\partial x_j}$, the term on the right-hand side defines the virial of all the forces acting on the electrons, the virial $V$ [21]. Slater [20] applied this result to study the formation of a diatomic hydrogen molecule from the separated hydrogen atoms. He used empirical potential energy functions to calculate the forces $F_j$ acting on the nuclei for geometries removed from the equilibrium atomic separation $R_{Eq}$. Let the total electronic energy, the sum of the electronic kinetic and potential energy $(T + V)$ be represented by $E$. Then, the force exerted by the nucleus 2 on the other is $-dE/dR$ and the force $F$, which leads to an equilibrium state is $dE/dR$. We now drop the average signs for convenience and rewrite equation (2.11) as:

$$T = \frac{1}{2}V - R\frac{dE}{dR} \qquad (2.12)$$

Solving equation (2.12) simultaneously with the equation $T + V = E$ one can write [20]:

$$T = -E - R\frac{dE}{dR} = -E + RF(R) \qquad (2.13)$$

and

$$V = 2E + R\frac{dE}{dR} = 2E - RF(R) \qquad (2.14)$$

The term $F(R) = dE/dR$ represents the Feynman force [14] on a nucleus at separation $R$.

We rewrite equations (2.13) and (2.14) in terms of the changes in the energies relative to their values at infinite separation to give the relation between $\Delta E(R)$, the kinetic energy $\Delta T(R)$, and the potential energy $\Delta V(R)$:

$$\Delta T = -\Delta E + RF(R) \qquad (2.15)$$

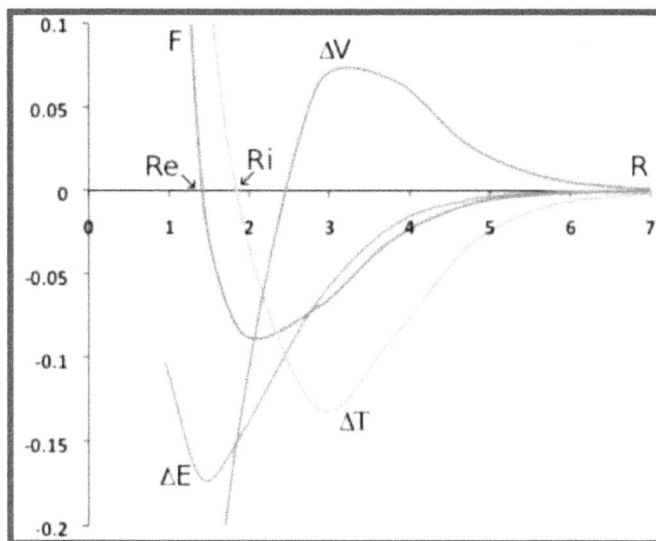

**Figure 2.5.** Schematic representation for the variation in total energy $\Delta E$, kinetic energy component $\Delta T$, potential energy component $\Delta V$, and Feynman force $F$ with internuclear separation R for the ground state of the hydrogen molecule $H_2$ in atomic units. Reprinted with permission from [21]. Copyright (2011) American Chemical Society.

and

$$\Delta V = 2\Delta E - RF(R) \qquad (2.16)$$

Figure 2.5 presents the variation in $\Delta E$, $\Delta T$ and $\Delta V$ with internuclear separation $R$ for the ground state of the hydrogen molecule in atomic units. Slater in 1933 studied this behavior of the change in kinetic energy $\Delta T$ and potential energy $\Delta V$ using empirical molecular potential energy functions in conjunction with the virial theorem [20]. When the Feynman force is attractive, the quantities $RF(R)$ in equations (2.15) and (2.16) are negative and have a stabilizing effect on the formation of the molecule. It is only in the presence of an attractive force that $\Delta T$ and $\Delta V$ can change sign [21]: $\Delta T < 0$ when $|RF(R)| > -\Delta E$ and $\Delta T = 0$ when $|RF(R)| = \Delta E$; $\Delta V > 0$ when $-RF(R) > 2|\Delta E|$; and $\Delta V = 0$ when $2|\Delta E| = -RF(R)$. Thus, potential energy $\Delta V$ becomes negative before kinetic energy $\Delta T$ becomes positive. When the Feynman force is repulsive, $RF(R) > 0$ and $\Delta T > 0$ and $\Delta V < 0$ [21].

The behavior in the range $\Delta T < 0$ and $\Delta V > 0$ is correlated with the changes in the electron density that accompany the initial approach of the atoms from a large distance apart. Electron density is removed from the immediate vicinities of both nuclei where kinetic energy $T(r)$ is maximally positive and potential energy $V(r)$ is maximally negative, and is accumulated as a diffused distribution in the binding region. This gives rise to the relaxation of the gradients in Feynman force $F$ and causes a further reduction in $T$. The initial decrease in kinetic energy is a result of the creation of attractive Feynman forces on the nuclei resulting from the accumulation of density in the internuclear region. This is the essential first step in bond formation, and this continues up to the equilibrium separation where $\Delta E = \Delta T$. It is important to note that the decrease in kinetic energy is found only when attractive Feynman forces act on the nuclei, and thus it does not occur at the equilibrium separation $R_e$.

Summarizing the discussion above, the energy lowering that leads to the bond formation between the atoms in a diatomic molecule is the result of the decrease in potential energy due to the attractive interaction between the nuclei and the electronic charge that is accumulated in the bond region between the atoms. The diatomic molecule is free and exists in a stationary energy state, which is determined by the appropriate Schrodinger's equation. The charge distribution in such a state is independent of time. Thus, on the basis of quantum mechanics, it is argued that the interaction between the nuclei and the electronic charge can be interpreted from the electrostatic viewpoint, although this charge distribution is dependent on the electronic motion. This essentially static picture of interacting charge distributions is very appealing in its simplicity, and it appears to be a straightforward extension of Lewis's theory. This electrostatic view appears to be consistent with the virial theorem, and the attractive component of the binding energy is due to the electro-static potential energy, whereas the kinetic component is repulsive.

### 2.4.2 Kinetic energy approach

An important phenomenon in quantum mechanics is the existence of zero-point energy. Localized particles are never completely at rest and always have some kinetic energy. In an atom (molecule) the kinetic energy of an electron will depend on the electron's confinement in the potential energy well created by the nucleus (nuclei) and other electrons. In a simplistic picture, it is essentially a particle-in-a-box problem. In the static potential energy approach discussed above, this important aspect of the electron's kinetic energy contribution towards the covalent bonding is not addressed adequately. Hellmann in 1933 proposed that covalent bonding was a quantum-mechanical effect due to the lowering of the ground-state kinetic energy associated with the delocalization of the motions of valence electrons between atoms in a molecule [22]. This kinetic view was largely ignored in the scientific community, mostly because it went against the already accepted simpler electrostatic explanation of covalent bonding. Furthermore, Hellmann [22] had based his argument on the statistical Thomas–Fermi model, which was subsequently found to be unsuitable for describing covalent bonding [19]. In addition, Hellmann's theory could not resolve the apparent conflict with the virial theorem.

The kinetic theory was revived by Ruedenberg and coworkers in 1960 through their analysis on the basis of the quantum-mechanical variation principle [23, 24]. These investigations revealed that covalent bonding was indeed a quantum-mechanical effect as was suggested originally by Hellmann. The critical component of bonding is interatomic electron delocalization, which is the quantum-mechanical term for electron sharing [19]. The process of the stabilization of electron delocal-ization in the ground state of a hydrogen molecule is associated with the combination and constructive interference of the atomic orbitals of the hydrogen atoms in a hydrogen molecule to form molecular orbitals. This results in the bonding of hydrogen atoms by a net decrease in the kinetic energy of the hydrogen molecule. An additional effect comes into play, namely intra-atomic orbital contraction, when the internuclear distance becomes smaller than about twice the equilibrium separation. This orbital contraction causes further stabilization by a decrease in the potential energy while the kinetic energy is increased. These energy shifts ensure that the virial theorem is satisfied [19].

It is interesting to mention here that almost three decades after Feynman supported the electrostatic view of covalent bonding in hydrogen molecules, in 1939 Feynman in his famous *Lectures on Physics* [25] explained covalent bond formation in $H_2^+$ ion molecule and by extension in hydrogen and other molecules as the consequence of a flip-flop motion of electrons between bonded atoms. This flip-flop motion caused a corresponding drop in the kinetic energy of electrons as a molecule forms. Based on Heisenberg's uncertainty principle, $\Delta x \Delta p \geqslant \hbar$, Feynman argued that the lower energy of the electron in the bonding state of the hydrogen molecule is a consequence of delocalization or increase in $\Delta x$. This resulted in a drop in kinetic energy, which is proportional to $\Delta p^2$ without a significant increase in its potential energy. The opposite situation would take place for the repulsive antibonding state. Treating the $H_2^+$ ion molecule as a simple two-state system of bonding and antibonding molecular orbitals, Feynman used time-dependent quantum theory to show that if the bonding electron in $H_2^+$ is localized, then it must then oscillate between the two nuclei. The frequency of that oscillation is directly related to the energy difference between the delocalized molecular orbitals. These molecular orbitals are stationary states and their energy difference is approximate twice the bond energy [19]. Feynman termed this dynamical picture of interatomic electron oscillation the *flip-flop mechanism* of covalent bonding, which could also be visualized in terms of wave function delocalization combined with energy splitting between molecular orbitals formed from localized atomic orbitals [25].

We will now analyze the covalent bonding in $H_2^+$ and $H_2$ molecules within the kinetic energy picture. In this analysis, we will use the notations for energy and wave functions used in the article by Nordholm and Bacskay [19].

### 2.4.3 Energy analysis of the $H_2^+$ and $H_2$ molecules within the kinetic energy approach

The simplest molecular wave functions for the $H_2^+$ molecule involving just one electron or $H_2$ molecule with archetypal Lewis pair are constructed from the exact atomic orbitals of a hydrogen atom:

$$\Psi(H_2^+; \zeta, R) = [2(1 + S_{ab})]^{-1/2}(\phi_a(\zeta) + \phi_b(\zeta)) \qquad (2.17)$$

and

$$\Psi(H_2; \zeta, R) = [2(1 + S_{ab}^2)]^{-1/2}(\phi_a(1)\phi_b(2) + \phi_b(1)\phi_a(2)) \qquad (2.18)$$

Here $\zeta$ is the orbital exponent and $R$ is the internuclear separation. The overlap integral $S_{ab}$ is dependent on $\zeta$ and $R$, and is expressed as:

$$S_{ab} = \langle \phi_a | \phi_b \rangle = \int \phi_a(\mathbf{r})\phi_b(\mathbf{r})d\mathbf{r} \tag{2.19}$$

The full wave functions for the ground states are obtained by multiplying the above spatial wave functions by the doublet and singlet spin eigenfunctions for $H_2^+$ and $H_2$, respectively [19]. The $H_2$ wave function in equation (2.18) is the archetypal valence bond wave function as originally proposed by Heitler and London [7] with coordinates of the two electrons being simply written as 1 and 2. Since a $H_2$ molecule smoothly dissociates into two H atoms, here it is preferred to have the molecular orbital wave function that allows a mixture of H atoms and $H^+/H^-$ ions as $R \to \infty$. In the case of $H_2^+$, the single electron wave function of equation (2.17) can be considered as a molecular orbital or valence bond type. The normalized atomic orbitals are just 1s-type atomic orbitals [19]:

$$\phi_a = \left( \frac{\zeta^3}{\pi} \right)^{1/2} \exp(-\zeta/r_a) \tag{2.20}$$

Here $r_a$ is the distance from nucleus $a$. The optimized orbital exponents $\zeta$ for $H_2^+$ and $H_2$ vary between the separated H atoms limit of 1.0 ($R = 1$), and the united $H^+$ and H atom limits of 2.0 and 1.688, respectively ($R = 0$), in atomic units [19]. The basic physics of covalent bonding is adequately understood with the help of these simplest of wave functions [19]. These minimal sets can, however, be improved upon for quantitative predictions with the inclusion of polarization functions, and further in the case of hydrogen molecule $H_2$ taking account for a greater degree of electron correlation.

### 2.4.3.1 Bonding energetics

Figure 2.6 shows the computed energy curves of $H_2^+$ and $H_2$ obtained by Nordholm and Bacskay [19] with $\zeta$ optimized at each distance (full lines), as well as fixed at the H atom value of 1.0 (dashed lines). The energy curves of $H_2^+$ and $H_2$ are qualitatively similar. The electrostatic potential energies for both molecules are repulsive at all distances with the exponent $\zeta$ fixed at 1. This is a clear indication that bonding takes place because of the decrease in kinetic energy. This is a consequence of the quantum-mechanical nature of the electron. A drop in kinetic energy is indicative of

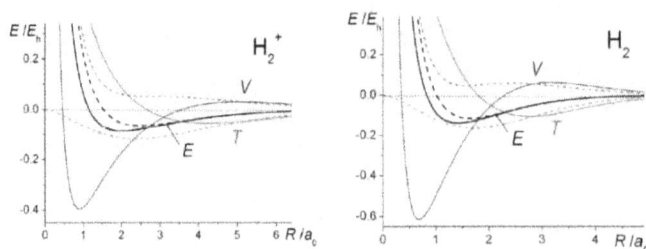

**Figure 2.6.** Total energy $E$, kinetic energy $T$, and potential energy $V$, all relative to the H and H + H atoms respectively, computed using minimal H 1s-atomic orbital basis with exponent $\zeta$ optimized (full lines) and fixed at the H atomic value of 1 (dashed lines). Reproduced from [19]. CC BY 4.0.

the electron having more room to move around. This is the conclusion that was reached by Hellmann [22] in 1933 and three decades later by Feynman [25]. As mentioned above, Feynman treated the $H_2^+$ and $H_2$ molecules as examples of two-state systems, where the back-and-forth flip of the electron(s) from one atom to the other, i.e., delocalization of electrons, produced the bonding in both systems. Greater stability of the molecule and shorter equilibrium bond lengths are achieved with the optimization of the orbital exponent $\zeta$ with increasingly larger values than 1.0 as $R$ decreases. This process of orbital contraction in the molecules is accompanied by an increase in the kinetic energy but a greater degree of drop in the potential energy. Thus, the virial theorem is satisfied precisely at the equilibrium configuration [19]. The overall effect of orbital contraction is to strengthen the bond, whereas the actual shifts in the total bond energies are minor and essentially intra-atomic in nature. The key to bonding is associated with a decrease in interatomic kinetic energy. This view conflicts with the electrostatic theory, which claims that the drop in potential energy as ensured by the virial theorem is due to the electrostatic interaction of the increased electronic charge in the interatomic region with the nuclei.

The orbital contraction and its effects on the equilibrium geometries and energies satisfying the virial theorem can be obtained to a very good approximation by a simple scaling procedure, that yields [19]:

$$\zeta = -\frac{V(1)}{2T(1)}, \ R_e(\zeta) = \frac{R_e(1)}{\zeta}, \ E[R_e(\zeta)] = \zeta^2 T(1) + \zeta V(1) \qquad (2.21)$$

Application of the above procedure yields $\zeta = 1.238$, an equilibrium distance $R_e(\zeta)$ of $2.01a_0$, and a total binding energy of $-0.086E_h$ for $H_2^+$. In the case of $H_2$, one obtains $\zeta = 1.166$, a $R_e(\zeta)$ of $1.41a_0$ and a total binding energy of $-0.139E_h$. Thus, for $H_2^+$ and $H_2$, orbital optimization is essentially a rescaling of the molecular wave functions and their constituent atomic orbitals. In other words, the delocalization of electrons, hence bond formation, is effectively between contracted atoms, as suggested by Ruedenberg [23] in 1960.

### 2.4.3.2 Molecular density and delocalization

To explore further the phenomenon of covalent bonding, we will now discuss the change in electron density that takes place as a molecule, $H_2^+$ or $H_2$, forms from the constituent atoms and/or nuclei. The molecular density $\rho(\zeta, R) = n_{el}\Psi^2(\zeta, R)$ at any internuclear separation $R$ and 1s-atomic orbitals with exponent $\zeta$ and where $n_{el}$ is the number of electrons, i.e., 1 or 2 for $H_2^+$ or $H_2$, respectively, can be decomposed into quasi-classical atomic $\rho_{qc}$ and interference $\rho_I$ contributions [19]:

$$\rho = \rho_{qc} + \rho_I \qquad (2.22)$$

and,

$$\rho_{qc}(\zeta) = \frac{n_{el}}{2}[\phi_a^2(\zeta) + \phi_b^2(\zeta)] \qquad (2.23)$$

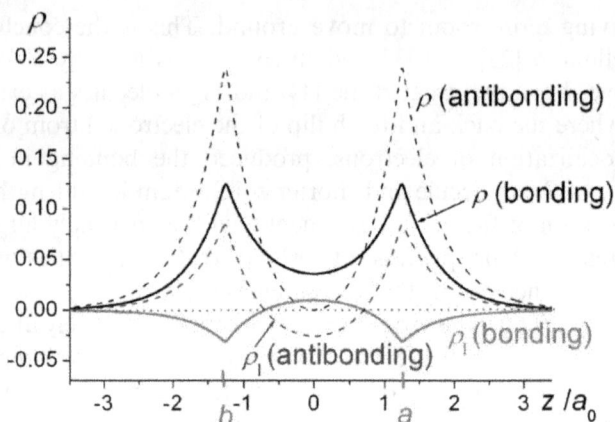

**Figure 2.7.** Total electron densities ($\rho$) of bonding (solid-black line) and antibonding (dashed-black line) states of $H_2^+$ at $R = 2.5a_0$ with $\zeta = 1$ as functions of internuclear coordinate $z$. The corresponding interference contributions ($\rho_I$) are shown in red and blue, respectively. Reproduced from [19]. CC BY 4.0.

Figure 2.7 presents one-dimensional plots of the densities of $H_2^+$ as functions of the internuclear coordinate $z$ and the resulting interference densities. The wave function of the bonding state (equation (2.17)) is an in-phase combination of the atomic orbitals $\phi_a$ and $\phi_b$. Their constructive interference results in a buildup of electron density in the bond region in-between the nuclei with a negative interference contribution close to the nuclei. In the antibonding state, a negative interference takes place in the bond region with an increased density around the nuclei. It may be noted that the very existence of an antibonding state is a quantum effect. Its origin is associated with the quantum-mechanical wave nature of the electrons. Its energy at $R = 2.5a_0$ is 0.209 $E_h$ above that of the separated atoms H + H$^+$. It is a repulsive state because of its large 0.316 $E_h$ kinetic energy, despite an attractive $-0.106$ $E_h$ potential energy contribution (both relative to H + H$^+$) [19]. The large kinetic energy is a consequence of the node in the wave function, i.e., a region of large gradients (in an absolute sense), consistent with the quantum nature of the electron [19].

The interference of the atomic orbitals at any internuclear distance $R$ is related to their overlap integral $S_{ab}$, which plays a crucial role in the energy expressions of both $H_2^+$ and $H_2$. In particular, the kinetic energies are [19]:

$$T(H_2^+) = \frac{T_{aa} + T_{bb} + 2T_{ab}}{2(1 + S_{ab})} = \frac{T_{aa} + T_{ab}}{1 + S_{ab}} \tag{2.24}$$

$$T(H_2) = \frac{2T_{aa} + 2T_{bb} + 4T_{ab}S_{ab}}{2(1 + S_{ab}^2)} = \frac{2(T_{aa} + T_{ab}S_{ab})}{1 + S_{ab}^2} \tag{2.25}$$

Here off-diagonal matrix element $T_{ab} = \langle \phi_a | \hat{T} | \phi_b \rangle$ and from symmetry the diagonal element $T_{aa} = T_{bb}$. As per calculations, the diagonal term $T_{aa}$ is significantly larger in magnitude than the off-diagonal term $T_{ab}$ at distances larger than $\sim 3a_0$

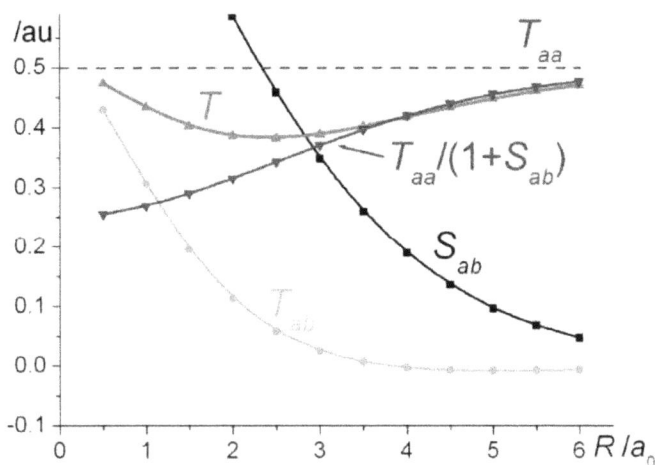

**Figure 2.8.** Bond length dependence of the overlap integral $S_{ab}$, kinetic energy matrix elements $T_{aa}$ and $T_{ab}$ for $H_2^+$. Reproduced from [19]. CC BY 4.0.

[19]. This indicates that the dominant contribution to the total kinetic energy in the case of $H_2^+$ comes from the quotient $T_{aa}/(1 + S_{ab})$. This is shown in figure 2.8, which presents the bond length dependence of the overlap integral $S_{ab}$, kinetic energy matrix elements $T_{aa}$ and $T_{ab}$ for $H_2^+$. The contribution of the off-diagonal kinetic coupling term $T_{ab}$ is essentially negligible in the region quite near to the equilibrium distance of 2.5 $a_0$ when the orbital exponent is fixed at the atomic value of 1.0 and the kinetic energy is effectively at its minimum. The drop in kinetic energy leading to binding is effectively due to the rapid increase in the overlap integral $S_{ab}$. The off-diagonal term $T_{ab}$ is no longer negligible in the region R $\leqslant \sim 3a_0$, and it leads to the repulsive kinetic energy contribution. The same conclusions apply in the case of $H_2$ molecule. However, the critical distance where the repulsive effect of $T_{ab}$ becomes non-negligible is smaller, $\sim 2.5a_0$. The effect of orbital contraction becomes notice-able at distances smaller than $\sim 4.5a_0$ in the case of $H_2^+$. This effect increases the magnitude of Taa (since $T_{aa} = \zeta^2/2$), and that in turn leads to an increase in total kinetic energy.

Figure 2.9 presents the interference density in $H_2^+$ molecule for optimized exponent $\zeta = 1.239$ and bond length $R_e = 2.0a_0$, with nuclei at $z = \pm 1a_0$, compared with interference densities for $\zeta = 1$ and $R_e = 2.5a_0$ and relative to H atoms with $\zeta = 1$ as functions of the internuclear coordinate $z$. Full geometry and optimization of the orbital exponent result in a larger difference between densities of molecules and contracted atoms or quasi-atoms in the bonding region than obtained with $\zeta = 1$, and also a correspondingly larger decrease in the density near the nuclei. In comparison to the uncontracted H atoms ($\zeta = 1$), the orbital contraction in the molecule causes a prominent increase in the density around the nuclei, and to a lesser extent also in the bond region [19]. The overall effect is a net contraction of the molecular wave function, and hence density. A further reduction in interatomic kinetic energy happens due to the additional interference density that originates

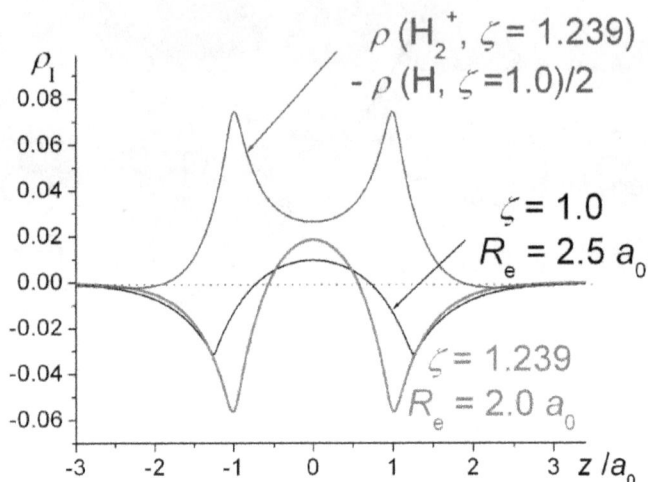

**Figure 2.9.** The interference density in $H_2^+$ molecule for optimized exponent $\zeta = 1.239$ and bond length $R_e = 2.0a_0$ (red line), with nuclei at $z = \pm 1a_0$, compared with interference densities for $\zeta = 1$ and $R_e = 2.5a_0$ (black line) and relative to H atoms with $\zeta = 1$ (blue line) as functions of the internuclear coordinate $z$. Reproduced from [19]. CC BY 4.0.

from orbital contraction. However, this takes place at the expense of the intra-atomic contribution. The process altogether lowers the total energy of the molecule.

In summary, the buildup of electron density in the internuclear region is a quantum-mechanical consequence of the constructive interference of electron waves that are specified in terms of atomic orbitals. This is a process that accompanies the formation of a covalent bond. The buildup of density in the bond region is not caused by, nor does it result in, a drop in potential energy. Constructive interference is a precondition of electron delocalization, which results in a decrease in kinetic energy. An increase in the orbital exponent $\zeta$ causes a tighter electron density around the nuclei, but this also increases the interference density in the bond region. The overall effect leads to a shorter and stronger bond. Electron delocalization leads to a drop in the quantum-mechanical kinetic energy of the electron and also an increased electrostatic attraction of the electrons to the nuclei (i.e., a lower potential energy). An increase in the orbital exponent gives rise to a tighter, mostly atomic, contribution to the density.

If one uses polarized atomic orbitals for the construction of molecular orbitals, then this will result in relatively large density shifts, as well as considerable spatial changes in the kinetic and potential energy densities. The resulting total energy changes will, however, be quite small. These changes are mainly those that free hydrogen atoms have when they are polarized, i.e., they are not directly associated with electron sharing [18].

### 2.4.3.3 The dynamics of electron delocalization in $H_2^+$ molecule

So far, we have implicitly visualized and thought about covalent bonding as a static phenomenon, where the shared electrons are located in-between the bonded atoms and act as an *electronic glue*. This is a straightforward interpretation of Lewis's structures, where the quantum mechanics are only entered to replace the shared electron theory of Lewis [1] with electron clouds that are localized in the binding regions of the molecule. In reality, quantum mechanics demands that electrons are constantly on the move, and this may not be obvious from the standard time-independent calculations that we have dealt with so far.

We will now discuss electron delocalization using the time-dependent quantum theory, which is at the heart of the dynamic description of covalent bonding [19]. This treatment concentrates on $H_2^+$, and following the procedure of Feynman [25] the $H_2^+$ molecule is considered as a two-state system described by a minimal set of normalized atomic orbitals $\psi_a$ and $\psi_b$. The eigenfunctions of the Hamiltonian are the bonding and antibonding molecular orbitals $\phi_g$ and $\phi_u$, which are expressed as [19]:

$$\psi_{g,u} = [2(1 \pm S_{ab})]^{-1/2}(\phi_a \pm \phi_b) \tag{2.26}$$

It is assumed that the nuclei $a$ and $b$ are sufficiently far apart for the overlap of the atomic orbitals to be neglected, i.e., $S_{ab} = 0$. The time evolution of any arbitrary time-dependent state $|\varphi(\mathbf{r}, t)\rangle$ of this system can be written in terms of the eigenfunctions $\psi_g$ and $\psi_u$ as:

$$|\varphi(\mathbf{r}, t)\rangle = \langle\psi_g(\mathbf{r})|\varphi(\mathbf{r}, 0)\rangle \exp(-iE_g t/\hbar) |\psi_g(\mathbf{r})\rangle + \langle\psi_u(\mathbf{r})|\varphi(\mathbf{r}, 0)\rangle \exp(-iE_u t/\hbar) |\psi_u(\mathbf{r})\rangle \tag{2.27}$$

Here $|\varphi(\mathbf{r}, 0)\rangle$ is the initial localized state of interest at t = 0, which is assumed to be the atomic orbital $|\phi\rangle$. Equation (2.27) can be written in a simpler form:

$$|\varphi(\mathbf{r}, t)\rangle = \frac{1}{\sqrt{2}}[\exp(-iE_g t/\hbar) |\psi_g(\mathbf{r})\rangle + \exp(-iE_u t/\hbar) |\psi_u(\mathbf{r})\rangle] \tag{2.28}$$

The decay and subsequent variation of the integrated probability of electron density $n_a(t)$ associated with nucleus $a$ in the Hilbert space spanned by $\{|\phi_i\rangle\}$ is described by the projection [19]:

$$\begin{aligned}
n_a(t) &= \langle\phi_a|\varphi(\mathbf{r_a}, t)\rangle\langle\varphi(\mathbf{r_a}, t)|\phi_a\rangle \\
&= \frac{1}{4}\{|\exp(-iE_g t/\hbar)|^2 + |\exp(-iE_u t/\hbar)|^2\} \\
&\quad + \frac{1}{4}\{\exp[i(E_u - E_g)t/\hbar] + \exp[-i(E_u - E_g)t/\hbar]\} \\
&= \frac{1}{2} + \frac{1}{4}\exp(i\Delta Et/\hbar) + \frac{1}{4}\exp(-i\Delta Et/\hbar)
\end{aligned} \tag{2.29}$$

or,

$$n_a(t) = \frac{1}{2}[1 + \cos(\Delta Et/\hbar)] \tag{2.30}$$

Here $\Delta E = E_u - E_g$. We can see that $n_a$ is a periodic function of time with a periodicity of $2\pi\hbar/\Delta E$. Thus, the transfer tunneling rate $\tau^{-1}$ of the electron from one atom to the other can be predicted as [19]:

$$\tau^{-1} = \frac{\Delta E}{\pi\hbar} \tag{2.31}$$

This is the standard formula for transfer rate in simple two-level systems [19]. Furthermore, it can be written [19]:

$$\Delta E(R) = E_u(R) - E_g(R) = 2B(R) \tag{2.32}$$

Here $B(R)$ is the positive binding energy of $H_2^+$ molecule and equation (2.32) is applicable for $R \geqslant 4.5\ a_0$. The electron transfer rate has a direct dependence on the binding energy at such distances:

$$\tau^{-1} = \frac{2B(R)}{\pi\hbar} \tag{2.33}$$

The transfer rate rapidly becomes larger than that given by equation (2.33) at shorter distances, where $\Delta E(R) > B(R)$.

Figure 2.10 presents computed transfer rates at various points of internuclear separations. At $R = 8a_0$, the height of the Coulomb barrier is $-0.5\ E_h$, which is the same as the energy of an H atom [19]. Therefore, at larger distances electron transfer occurs by tunneling. The transfer rate rapidly increases with the reduced distance.

Figure 2.11 presents the probability densities calculated from the electron numbers $n_a$ and $n_b$ using equations (2.29) and (2.30) at $R = 8a_0$, for a number of different times. If one starts with the electron fully localized on nucleus $a$ with density $|\phi_a(\mathbf{r})|^2$, then more and more of the electron density appears on nucleus $b$ at

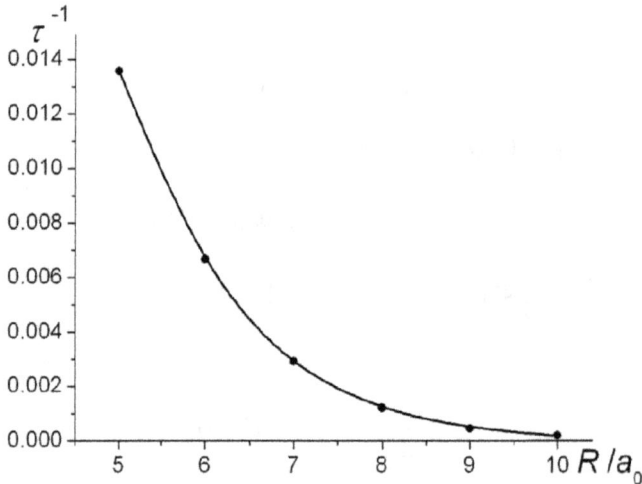

**Figure 2.10.** Electron transfer rates computed from equation (2.33) as a function of the internuclear separation. Reproduced from [19]. CC BY 4.0.

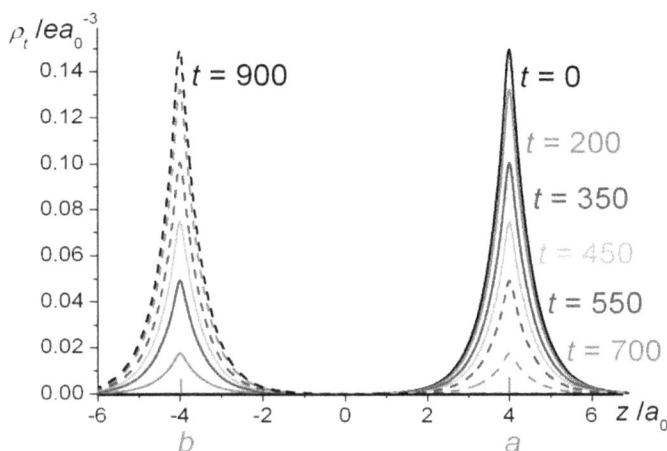

**Figure 2.11.** Time dependence of electron density $\rho_t(x, y, z)$ with $x = y = 0$ in $H_2^+$ molecule as a function of the internuclear coordinate $z$ at selected times in arbitrary units. Reproduced from [19]. CC BY 4.0.

subsequent times $t = 200$, $350$, $450$, $550$, and $700$ in arbitrary units (au) until at $t = 900$ au the transfer of density to $b$ is complete.

The analysis presented above drives the important point that the shared electron, which corresponds to the covalent bond in the $H_2^+$ molecule, is not localized. If it is assumed that at an initial time $t = 0$ the electron is associated with nucleus $a$, then one can see that it does not remain localized on that nucleus but moves to nucleus $b$ and back. In other words, the electron executes an oscillatory behavior between the atomic centers. This in turn indicates that the ground state, which is normally referred to as stationary, is not localized except in the limit of infinite separation $R$. At that separation, there are two states of equal energy that can be represented as left or right localized. The ground and first excited states at any finite $R$ are split by energy representing the rate of interatomic electron transfer of a localized electron. This means that delocalization or electron sharing as used in energy analysis of covalent bonding pioneered by Hellman [22] and Ruedenberg [23] is a dynamical process. The shared electron, in addition to moving in the proximity of one nucleus, will transfer to the other nucleus and back with a well-defined periodicity [19]. The electron probability density averaged over all possible phases of that motion is time-independent (or stationary), but as we know from its kinetic energy the electron is constantly moving.

## 2.5 Ortho-hydrogen and para-hydrogen

Heisenberg and Hund with the help of quantum theory had explained the existence of two different spin states for helium, namely ortho-helium and para-helium, and they postulated a similar possibility for the hydrogen molecule [26, 27].

Around the same time, Bonhoeffer and Harteck [28] and independently Eucken and Hiller [29] could experimentally observe changes in the properties of hydrogen samples kept at low temperatures for many hours. The hydrogen samples in these

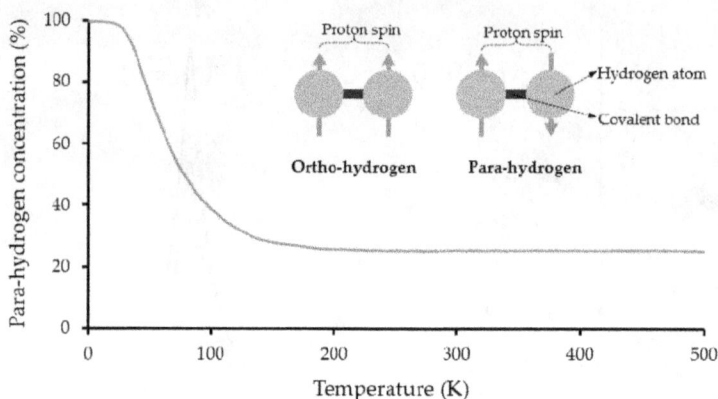

**Figure 2.12.** Schematic representation of the hydrogen molecule in ortho- and para-hydrogen forms, and temperature dependence of para-hydrogen concentration in equilibrium hydrogen. Reproduced from [32]. CC BY 4.0.

experiments were nearly isolated. It was concluded that a conversion of the hydrogen molecules from one spin state to another spin state possessing different property values was the only possible cause for the observed change. Following the nomenclature chosen by Heisenberg for the helium molecule, Bonhoeffer and Harteck [28] chose the names ortho-hydrogen and para-hydrogen for the two different types of hydrogen molecules.

Ortho-hydrogen and para-hydrogen differ in the magnetic interactions of the protons in the hydrogen molecule nuclei due to the spin state of the protons. Ortho-hydrogen represents the condition where the protons in the nuclei of both hydrogen atoms have the same spin direction; therefore, the resultant nuclear spin is one. On the other hand, para-hydrogen refers to the configuration in which both protons in the nuclei of hydrogen atoms have their spin anti-aligned; thus, the resultant spin is zero. Ortho-hydrogen and para-hydrogen have slightly different physical properties, including thermal, magnetic, and optical properties [30, 31], while having quite similar chemical properties. The inset of figure 2.12 shows a schematic representation of the hydrogen molecule in ortho- and para-hydrogen forms. The conversions between ortho- and para-hydrogen molecules do not occur naturally. They can be regarded as two distinct forms of the hydrogen molecule.

An assembly of hydrogen molecules at room temperature is generally a mixture of ortho-hydrogen and para-hydrogen in 3:1 ratio i.e. 75% ortho- and 25% para-hydrogen. Figure 2.12 shows the para-hydrogen concentrations in equilibrium hydrogen according to the temperature. The term equilibrium hydrogen is used to describe hydrogen with an equilibrated ortho-hydrogen–para-hydrogen ratio at the ambient temperature. The para-hydrogen form increases and the ortho-hydrogen form decreases with the lowering of temperature below room temperature, and the opposite trend is observed with the increase in temperature above room temperature. The physical properties, such as melting point and vapor pressure, of the two forms of hydrogen are slightly different. The melting point of para-hydrogen (for example) is 0.10 degree lower than that of a 3:1 mixture of ortho-hydrogen and para-hydrogen.

Ortho- and para-hydrogen forms can interconvert in the presence of a paramagnetic catalyst and based on temperature conditions. The usage of catalysts such as activated charcoal can also help in achieving equilibrium between these two isomeric forms. Para-hydrogen is energetically favored at low temperatures, and a cooling temperature below 25 K can result in the formation of 99% pure form. A large amount of energy is released when ortho-hydrogen gets converted to para-hydrogen. This energy needs to be removed when cooling the hydrogen, and this in turn means that the liquefaction of hydrogen is quite an energy intensive process. However, obtaining ortho-hydrogen in pure form is difficult.

## 2.6 Summary

To study the ground state for a hydrogen molecule (or for that matter any molecule) and its energy, one looks for a solution to Schrödinger's equation, and the result is a wave function for the electrons. A probability density for the distribution of electrons around the atoms is obtained by squaring the wave function. With the help of this electron density, one visualizes the molecule as a set of joined-up fuzzy balls. The ground state is said to be stationary. This means that the physical properties such as probability densities and electron densities are constants, i.e., do not change with time [18, 19]. Within such a framework, many chemists have been looking for the origin of covalent bonding in the electron density difference between a molecule and its constituent atoms. This gave rise to the bonding models based on electrostatics, which focused on electrostatic interactions and the changes that occur as a molecule forms Such models agree with the virial theorem whereby there is a decrease in the total potential energy as a molecule forms, despite the repulsive nature of the kinetic energy associated with the electron localization. Famous scientists, including Slater [20] and Feynman [14], supported this view that covalent bonding was due to favorable electrostatic interactions during molecule formation.

A different view was first proposed by Hellmann [22] and later on by Ruedenberg and collaborators [23, 24], which focused on the role of quantum-mechanical kinetic energy in the energy analysis of covalent bonding. The electrons are fuzzy clouds around the nuclei of the molecule, and in addition to the electrostatic potential energy they also have quantum-mechanical kinetic energy. The interatomic delocalization that is implicit in molecule formation invariably leads to a reduction in the kinetic energy of the electrons. This reduction in kinetic energy more than outweighs the antibonding effect of the potential energy. Within this framework, the kinetic energy of the shared electrons is the quantum effect that gives rise to covalent bonding. In addition to the interatomic changes, there are also intra-atomic energy shifts, which arise because of the orbital contractions that become prevalent in the equilibrium region. The changes in intra-atomic components of the kinetic and potential energies on contraction take place in the opposite sense to the interatomic ones, and the resultant total intra-atomic and interatomic energy changes are in accord with the virial theorem. The main objection towards the kinetic energy approach to understanding covalent bonding is thus circumvented with the recognition of the intra-atomic orbital contraction effects. The latter effects represent a minor contribution to bonding but are not the origin of bonding. Subsequently, it has been shown that the covalent bonding mechanism is not confined to systems of particles interacting via Coulombic potentials, and therefore the virial theorem (that holds for Coulombic particles) is not essential for bonding, nor is it the cause of bonding in Coulomb systems [19].

The fact that kinetic energy is the key to covalent bonding gives a clear indication that bonding is a dynamic phenomenon. Understanding covalent bonding from a dynamical point of view is complementary to that of an energy analysis because quantum mechanics offers a duality of representations in terms of either energy or dynamics. We have seen in equations (2.28)–(2.31), which describe the oscillation of the initially localized electron in $H_2^+$, that if we know the energy eigenfunctions and their energies, then the time-dependence can be predicted for all times. If we have a nondegenerate ground state which is delocalized, then the electron dynamics is delocalized. We have seen in section 2.4 that such phenomena give rise to a covalent bond in $H_2^+$ with a bond strength proportional to the energy spacing between the bonding and antibonding states. This bond strength is in turn proportional to the frequency of interatomic oscillation of the localized electron of minimal energy. In general, it is easy to get the impression that the valence electrons are static charges interacting with static nuclei screened by core electrons, but it needs to be realized that the electrons are never static entities. The electrons always move, and the key to understanding covalent bond is that the shared electrons move between the atomic centers taking part in the bond formation. The combination of the two concepts— energy and dynamics—allows the most complete understanding and description of the covalent bonding mechanism.

Finally, depending on the alignments of the spin directions in the protons in the nuclei of a hydrogen molecule, there are two types of molecular hydrogen, namely ortho-hydrogen and para-hydrogen. They have slightly different physical properties, including melting point and vapor pressure. Para-hydrogen is energetically favored at low temperatures and we will see later on that this property has interesting technological implications.

Certain sections of text in this chapter have been reproduced with permission from [19] CC BY 4.0. © 2020 by the authors.

# References

[1]  Lewis G N 1916 *J. Am. Chem. Soc.* **38** 762

[2]  Shaik S 2006 *J. Comput. Chem.* **28** 51

[3]  Abegg R and Anorg Z 1904 *Z. Anorg. Chem.* **39** 330

[4]  Langmuir I 1919 *J. Am. Chem. Soc.* **41** 868

[5]  Jensen W B 1984 *J. Chem. Edu.* **61** 191

[6]  Atkins P and de Paula J 2006 *Phys. Chem.* (New York: W. H. Freeman)

[7]  Heitler W and London F 1927 *Z. Phys.* **44** 455

[8]  Park B S 2009 *Hist. Stud. Nat. Sci.* **39** 32

[9] Cullity B D and Graham C D 2009 *Introduction to Magnetic Materials* (Hoboken: Willey)

[10] Rosen N 1931 *Phys. Rev.* **38** 2099

[11] Weinbaum S 1933 *J. Chem. Phys.* **1** 593

[12] London F 1928 *Z Phys.* **46** 455

[13] Schrödinger E 1926 *E. Ann. D. Phys.* **81** 109

[14] Feynman R P 1939 *Phys. Rev.* **56** 340

[15] Pauling L 1928 *Proc. Natl Acad. Sci. USA* **14** 359

[16] Pauling L 1960 *The Nature of the Chemical Bond* 3rd edn (Ithaca, NY: Cornell University Press)

[17] Gerratt J, Cooper D L, Karadakovc P B and Raimondid M 1997 *Chem. Soc. Rev.* **26** 87

[18] Bacskay G B and Nordholm S 2013 *J. Phys. Chem.* A **117** 7946

[19] Nordholm S and Bacskay G B 2020 *Molecules* **25** 2667

[20] Slater J 1933 *J. Chem. Phys.* **1** 687

[21] Bader R F W 2011 *J. Phys. Chem.* A **115** 12667

[22] Hellmann H 1933 *Z. Phys.* **85** 180

[23] Ruedenberg K 1962 *Rev. Mod. Phys.* **34** 326

[24] Feinberg M J, Ruedenberg K and Mehler E 1970 The origin of binding and antibinding in the hydrogen molecule ion *Advances in Quantum Chemistry* **vol 5** ed P O Löwdin (New York: Academic) p 27

[25] Feynman R P, Leighton R B and Sands Sands M 1965 *The Feynman Lectures on Physics, Quantum Mechanics* **vol III** (Reading, MA: Addison-Wesley) pp 10-1–10-12

[26] Heisenberg W 1927 *Z. Phys.* **41** 239

[27] Farkas A 1935 *Orthohydrogen, Parahydrogen and Heavy Hydrogen* (Cambridge: Cambridge University Press)

[28] Bonhoeffer K F and Harteck P 1929 *Z. Phys. Chem. Abt.* **B4** 113

[29] Eucken A and Hiller K 1929 *Z. Phys. Chem. Abt.* **B4** 142

[30] Giauque W F and Johnston H L 1928 *J. Am. Chem. Soc.* **50** 3221

[31] Brickwedde F G, Scott R B and Taylor H S 1935 *J. Chem. Phys.* **3** 653

[32] Aziz M 2021 *Energies* **14** 5917

# Chapter 3

## Strong Coulomb repulsion in the hydrogen molecule and the Hubbard model

In the previous chapter we have seen how the delocalization of electrons in a hydrogen molecule plays a leading role in the formation of a covalent bond in the hydrogen molecule. The electrons propagate within the molecule freely, and in this process the optimization of kinetic energy compensates for the effect of strong Coulomb repulsion between the electrons. The same picture involving the propagating of electron waves in solids gave birth to the electron band theory of solids, which is extremely successful in explaining the physical properties of semiconductors and nonmagnetic metals. However, the role of Coulomb repulsion energy between electrons cannot always be neglected, and this competition between Coulomb repulsion and reduction in the cost of kinetic energy plays a very important role in the electronic properties of many emerging classes of transition metal-based solid materials including the Mott insulators [1, 2].

The Hubbard model has been quite useful for the understanding of correlated electron motion in Mott insulators. The hopping kinetic energy $t$ and the intra-atomic Coulomb repulsion energy $U$ are the only qualitative characteristics of a correlated electron system. The qualitative trends brought about by the conflict between the localized or particle-like and delocalized or wave-like characteristics of an electron are represented by these two quantities. These features are captured well in the Hubbard model. In the same spirit, we will use the Hubbard model to understand the significant role of Coulomb repulsion energy between the electrons in a hydrogen molecule. This aspect was largely ignored in the treatment of covalent bond theories discussed earlier in chapter 2. Here we shall follow closely the treatments of the subject given in the book by Ashcroft and Mermin [3] and the article by Alvarez-Fernández and Blanco [4].

## 3.1 An approximate representation of electron interactions in a hydrogen molecule

The approximate representation of electron interactions using the Hubbard model is as follows. Each hydrogen ion is represented by a single localized orbital level, replacing the vast and complex set of discrete bound and continuum electron levels. The energy states are specified by the four possible electronic configurations of each ion: the level could either be empty, contain one electron with either of two spins, or two electrons of opposite spins according to the Pauli principle [4]. Therefore, the the Hubbard model in matrix representation of the Hamiltonian has two types of contributions: (i) diagonal terms in these states, and (ii) other off-diagonal terms that have nonvanishing matrix elements between just those pairs of states that differ only by a single electron, which has been removed without change in spin from a given ion to one of its neighbors. The electron–electron interactions that are described by the Coulomb potential are captured by this last term.

The full two-electron Hamiltonian for hydrogen molecule can be written as [4]:

$$H = h_1 + h_2 + V_{12} \tag{3.1}$$

Here $h_1$ and $h_2$ are one-electron atomic hydrogen Hamiltonians and $V_{12}$ stands for the Coulomb repulsion potential between the two electrons when they are found to be on the same hydrogen atom.

## 3.2 Ionized hydrogen molecule $H_2^+$

Let us now consider a hydrogen molecule in which two hydrogen atoms are located at $\mathbf{R}$ $\mathbf{R}'$ and described in the spatial representation by their orbital electronic wavefunctions $|\phi(\mathbf{R})\rangle$ and $|\phi(\mathbf{R}')\rangle$, respectively. If there is no electron on the hydrogen atom, i.e., an empty level $|\phi(\mathbf{R}, 0)\rangle$, then the energy is zero. In case there is one electron of either spin direction up ($\uparrow$) or down ($\downarrow$) in the hydrogen atomic level $|\phi(\mathbf{R}, \uparrow)\rangle_{up}$ or $|\phi(\mathbf{R}, \downarrow)\rangle_{down}$, its energy is designated by $E_0$. If there are two electrons of opposite spins in the hydrogen atom $|\phi(\mathbf{R}, \uparrow\downarrow)\rangle_{singlet}$, then the energy is $2E_0 + U$. The positive energy term $U$ stands for the intra-atomic Coulomb repulsion between the two localized electrons on the hydrogen atom. The Hubbard model for a hydrogen molecule comprises two such orbitals $|\phi(\mathbf{R})\rangle$ and $|\phi(\mathbf{R}')\rangle$ (see figure 3.1), which represent electrons localized at hydrogen atoms located at $\mathbf{R}$ and $\mathbf{R}'$, respectively. It is assumed for the sake of simplicity that these two states are orthogonal $\langle\phi(\mathbf{R})|\phi(\mathbf{R}')\rangle = 0$.

We will now take up the case of an ionized $H_2^+$ molecule with two protons and one electron in the Hamiltonian expressed in equation (3.1). If the one-electron Hamiltonians, $h_1$ and $h_2$, are diagonal in $|\phi(\mathbf{R})\rangle$ and $|\phi(\mathbf{R}')\rangle$, then the stationary states would correspond to a hydrogen atom and a proton. From figure 3.1 we can see that there is a probability that the electron can be transferred by tunneling from one atom to the other. This leads to an ionized hydrogen molecule. The amplitude for tunneling is represented by the off-diagonal term in the one-electron Hamiltonian [4]:

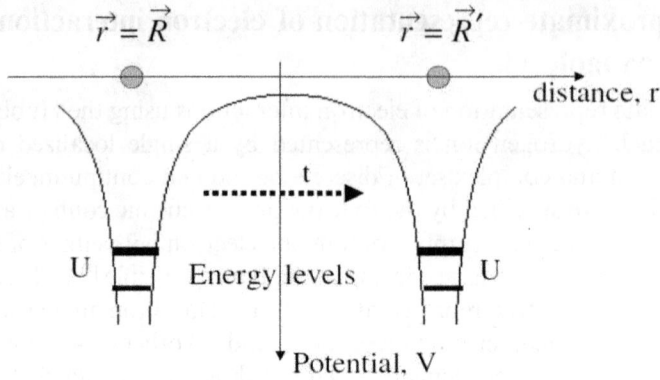

**Figure 3.1.** Schematic representation of the cross-section of the attractive potential due to the two protons at $r = R$ and $r = R'$ in the hydrogen molecule, the position of the energy levels indicating the possibility of tunneling from one atom to the other, and the Coulomb repulsion for electrons on the same atom. Reproduced from [4]. © IOP Publishing Ltd. All rights reserved.

$$\langle \phi(\mathbf{R})|h|\phi(\mathbf{R}') \rangle = \langle \phi(\mathbf{R}')|h|\phi(\mathbf{R}) \rangle = -t \tag{3.2}$$

Here the phases of $|\phi(\mathbf{R})\rangle$ and $|\phi(\mathbf{R}')\rangle$ are chosen to make the number $t$ real and positive. The diagonal terms in the Hamiltonian are expressed as:

$$\langle \phi(\mathbf{R})|h|\phi(\mathbf{R}) \rangle = \langle \phi(\mathbf{R}')|h|\phi(\mathbf{R}') \rangle = E_0 \tag{3.3}$$

The terms expressed in equations (3.2) and (3.3) together define the one-electron Hamiltonian problem for an ionized hydrogen molecule $H_2^+$. The stationary energy states of this Hamiltonian can be obtained from the diagonalization of the Hamiltonian, expressed in matrix representation as [4]:

$$\begin{bmatrix} E_0 & -t \\ -t & E_0 \end{bmatrix} \begin{bmatrix} a \\ b \end{bmatrix} = E \begin{bmatrix} a \\ b \end{bmatrix} \tag{3.4}$$

Here $a$ and $b$ are the components of the states $|\phi(\mathbf{R})\rangle$ and $|\phi(\mathbf{R}')\rangle$ with the normalization condition $a^2 + b^2 = 1$. The stationary energy states are $\frac{1}{\sqrt{2}}(|\phi(\mathbf{R})\rangle \mp |\phi(\mathbf{R}')\rangle)$ with the corresponding energies $E_0 \pm t$.

## 3.3 Hydrogen molecule $H_2$

To start with let us consider the two-electron problem of the hydrogen molecule using the independent electron approximation for the singlet spatially symmetric ground state. Within this approach, having both electrons into the one-electron level of lowest energy would cost a total energy of $2(E_0 - t)$. However, this approach entirely ignores the Coulomb interaction energy $U$, which becomes important when two electrons are found to be on the same hydrogen atom. The simplest way to improve upon this estimate of $2(E_0 - t)$ is the addition of the intra-atomic Coulomb repulsion $U$ multiplied by the probability of actually finding two electrons on the same hydrogen atom when the $H_2$ molecule is in the ground state within the independent electron approximation [3]. This probability is $2 \times 1/2 \times 1/2 = 1/2$ considering that both electrons are independent. The improved estimation of the ground state energy over the independent electron approximation is [4]:

$$E_{\text{IE}} = 2(E_0 - t) + \frac{1}{2}U \qquad (3.5)$$

Within the independent electron approximation, the approximate ground state is expressed as [4]:

$$\Psi_{\text{IE}} = \frac{1}{\sqrt{2}}\Psi_0 + \frac{1}{2}(\Psi_1 + \Psi_2) \qquad (3.6)$$

$\Psi_0$, $\Psi_1$, and $\Psi_2$ are the full set of singlet spatially symmetric states of the two-electron problem given by [4]:

$$\Psi_0 = \frac{1}{\sqrt{2}}[|\phi(\mathbf{R})\rangle\,|\phi(\mathbf{R'})\rangle + |\phi(\mathbf{R'})\rangle\,|\phi(\mathbf{R})\rangle]$$
$$\Psi_1 = |\phi(\mathbf{R})\rangle\,|\phi(\mathbf{R})\rangle \qquad (3.7)$$
$$\Psi_2 = |\phi(\mathbf{R'})\rangle\,|\phi(\mathbf{R'})\rangle$$

Here $|\phi(\mathbf{R})\rangle\,|\phi(\mathbf{R'})\rangle$ has electron 1 on the ion at $\mathbf{R}$ and electron 2 on the ion at $\mathbf{R'}$. The full two-electron Hubbard Hamiltonian $H$ in the matrix from $H_{ij} = (\Psi_i H \Psi_j)$ can be expressed in the space of the singlet states given in the equation (3.7), as [4]:

$$\begin{bmatrix} H_{00} & H_{01} & H_{02} \\ H_{10} & H_{11} & H12 \\ H_{20} & H_{21} & H_{22} \end{bmatrix} = \begin{bmatrix} 2E_0 & -\sqrt{2}t & -\sqrt{2}t \\ -\sqrt{2}t & 2E_0 + U & 0 \\ -\sqrt{2}t & 0 & 2E_0 + U \end{bmatrix} \qquad (3.8)$$

Here the off-diagonal terms for instance are expressed as:

$$\begin{aligned} H_{10} &= \langle \Psi_1 | H | \Psi_0 \rangle \\ &= \frac{1}{\sqrt{2}}[\langle\phi(\mathbf{R})|h_1|\phi(\mathbf{R})\rangle\langle\phi(\mathbf{R})|\phi(\mathbf{R'})\rangle + \langle\phi(\mathbf{R})|h_1|\phi(\mathbf{R'})\rangle\langle\phi(\mathbf{R})|\phi(\mathbf{R})\rangle] \\ &\quad + \frac{1}{\sqrt{2}}[\langle\phi(\mathbf{R})|h_2|\phi(\mathbf{R})\rangle\langle\phi(\mathbf{R})|\phi(\mathbf{R'})\rangle + \langle\phi(\mathbf{R})|h_2|\phi(\mathbf{R'})\rangle\langle\phi(\mathbf{R})|\phi(\mathbf{R})\rangle] \\ &= -\sqrt{2}t \end{aligned} \qquad (3.9)$$

In the matrix form of the Hubbard Hamiltonian equation (3.8), the diagonal terms in the states $\Psi_1$ and $\Psi_2$ that place two electrons on the same hydrogen atom contain the extra Coulomb repulsion energy $U$. This is a consequence of the electron–electron interaction $V_{12}$. This Coulomb repulsion energy contribution $U$ is not present in the diagonal element in the state $\Psi_0$ because the electrons here are on different atoms. In fact, this diagonal term in the state $\Psi_0$ agrees with the prediction of the Heitler–London (HL) treatment for the hydrogen molecule, and the ground state estimate is just $H_{00}$, i.e., $E_{\text{HL}} = 2E_0$.

It may be noted here that the one-electron tunneling amplitude $t$ only connects states where a single electron is transferred from one hydrogen atom to the other. It would take further two-body interactions to give nonvanishing matrix elements between states in which the positions of two electrons are changed [4]. The first set of diagonal terms leads to the formation of local magnetic moments in the absence of the second off-diagonal terms This is due to the fact that it would suppress the possibility of a second electron with oppositely aligned spin at singly occupied sites. On the other hand, in a solid material, the presence of the off-diagonal terms in the absence of the first set of diagonal terms can lead to a conventional electron band spectrum and one-electron Bloch levels where each electron is distributed through-out the entire solid crystal [2]. Finally, even the Hubbard model is too difficult for exact analysis when both sets of terms are present [1, 2].

The Hubbard Hamiltonian (equation (3.8)) then has a solution when [4]:

$$
\begin{vmatrix}
2E_0 & -\sqrt{2}\,t & -\sqrt{2}\,t \\
-\sqrt{2}\,t & 2E_0 + U & 0 \\
-\sqrt{2}\,t & 0 & 2E_0 + U
\end{vmatrix} = 0
\tag{3.10}
$$

The three eigenvalues obtained by solving equation (3.10) are:

$$
E_1 = 2E_0 + U
\tag{3.11}
$$

$$
E_{2,3} = E_\pm = 2E_0 + \frac{1}{2}U \pm \sqrt{4t^2 + \frac{1}{4}U^2}
\tag{3.12}
$$

The lowest of these three eigenvalues is the ground state energy of the Hubbard Hamiltonian:

$$
E_{\text{Hubbard}} = 2E_0 + \frac{1}{2}U - \sqrt{4t^2 + \frac{1}{4}U^2}
\tag{3.13}
$$

The ground state energy eigenfunction except for a normalization constant is given by:

$$
\Psi_{\text{Hubbard}} = \frac{1}{\sqrt{2}}\Psi_0 + \left(\sqrt{1 - \left(\frac{U}{4t^2}\right)} - \frac{U}{4t}\right)\frac{1}{2}(\Psi_1 + \Psi_2)
\tag{3.14}
$$

Figure 3.2 presents the ground state energy of the Hubbard model ($E_{\text{Hubbard}}$), the independent electron approximation to the ground state energy $E_{\text{IE}}$, and also the ground state energy within Heitler–London model $E_{\text{HL}}$ as a function of $U$ for the particular case of $E_0 = -10$ eV and $t = 0.5$ eV. The probability of finding two electrons on the same hydrogen atom in the ground state energy of the Hubbard model obtained from equation (3.14) is shown in figure 3.3. The molecular orbital theory and the kinetic approach of the bonding is the most appropriate treatment, when $U/t \ll 1$ both electrons are independent and delocalized, In contrast, when $U/t \gg 1$, the Hubbard model is in consonance with the Heitler–London prediction,

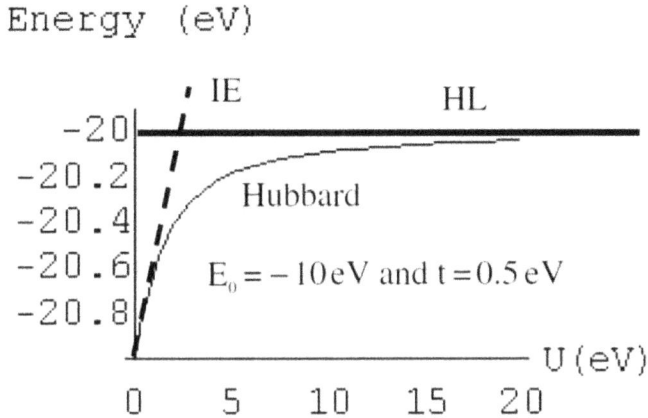

**Figure 3.2.** Ground state energy of the hydrogen molecule as a function of the Coulomb repulsion potential $U$ obtained from (i) Hubbard model (continuous line), (ii) the independent electron approximation (dashed line), and (iii) the Heitler–London model (bold straight line). The calculations are performed using the values of $E_0 = -10$ eV, and $t = 0.5$ eV. Reproduced from [4]. © IOP Publishing Ltd. All rights reserved.

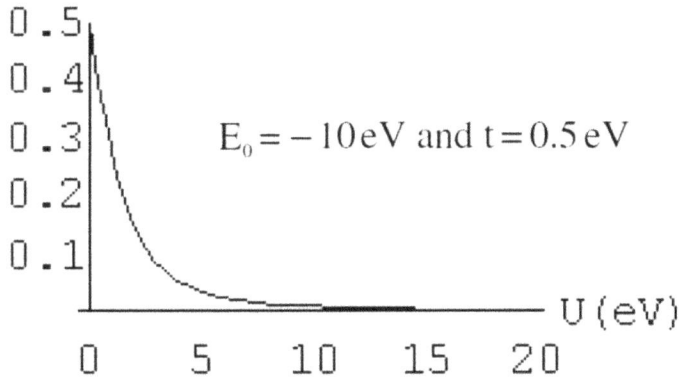

**Figure 3.3.** Probability of finding two electrons on the same atom in the hydrogen molecule as a function of the Coulomb repulsion potential $U$. The calculations are performed using the values of $E_0 = -10$ eV, and $t = 0.5$ eV. Reproduced from [4]. © IOP Publishing Ltd. All rights reserved.

in which the two electrons are well localized with each electron on a particular atom. Summarizing the discussion, one can say that the degree of localization (or delocalization) of electron is directly related to the competition between hopping kinetic energy $t$ and Coulomb repulsion energy $U$. Due to the relatively high energy cost of the charge fluctuations related to hopping, a large value of $U$ favors the localization of electrons.

Certain sections of text in this chapter have been reproduced with permission from [4] © IOP Publishing Ltd. All rights reserved.

# References

[1] Gebhard F 1997 *Metal Insulator Transition: Models and Methods* (Berlin: Springer)
[2] Roy S B 2019 *Mott Insulators: Physics and Applications* (Bristol: IOP Publishing)
[3] Ashcroft N W and Mermin N D 1976 *Solid State Physics* (Philadelphia, PA: Saunders College Publishing)
[4] Alvarez-Fernández B and Blanco J A 2002 *Eur. J. Phys.* **23** 11

**IOP** Publishing

Hydrogen
Physics and technology
**Sindhunil Barman Roy**

# Chapter 4

# Thermodynamic properties of fluid hydrogen

In the earlier chapters, we have seen how the understanding of the microscopic properties of hydrogen atoms and molecules played a fundamental role in the development of the quantum-mechanical theory of many-body systems. The properties of a single hydrogen atom led to the basis for atomic physics, and a hydrogen molecule is the simplest compound and an interesting platform for studying chemical binding. In this chapter, we will extend our study to fluid hydrogen, namely the macroscopic state of gaseous and liquid hydrogen.

Hydrogen has three recognized isotopes. The hydrogen isotopes have mass numbers of 1, 2, and 3. The most prevalent isotope has a mass of one and is commonly referred to as hydrogen. However, it is also occasionally termed as protium. The mass 2 isotope, deuterium, or heavy hydrogen, is designated with the symbol D and has a nucleus made up of one proton and one neutron. Deuterium makes up 0.0156% of the ordinary mixture of hydrogen. The mass 3 isotope is tritium, which has a nucleus made up of one proton and two neutrons. The properties of hydrogen isotopes differ significantly from one another. We will begin by discussing about the properties of hydrogen.

Figure 4.1 presents the phase diagram of hydrogen. There are three curves shown on this phase diagram. One curve represents the change of boiling temperature with pressure during the transition from liquid to a gaseous state. The same curve represents condensation when approached from the gaseous side. Another name for this condensation process is liquefaction. Another curve illustrates how melting (or freezing) temperature changes with pressure. A third curve represents temperatures and pressures for the sublimation process [1].

At ambient pressures and temperatures, hydrogen is a gas, and it becomes liquid or a solid at relatively low temperatures and/or high pressures. Figure 4.1 illustrates how the different phases of hydrogen respond to variations in pressure and temperature. Variation of the hydrogen boiling point with pressure is represented by the curve between the critical point and the triple point [1]. At the critical point in

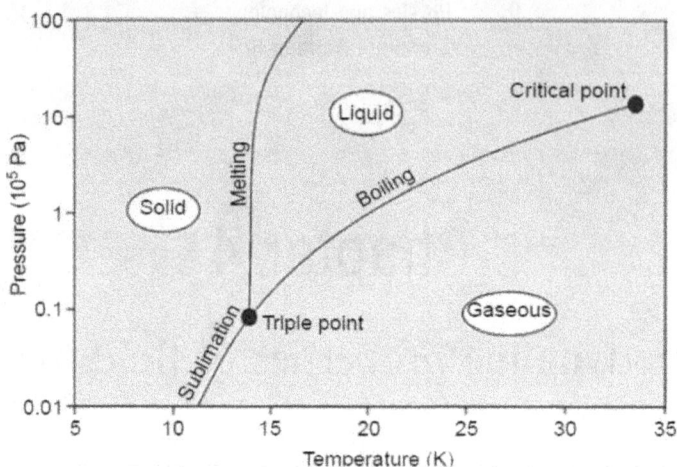

**Figure 4.1.** Phase diagram of hydrogen. Reprinted from [1], Copyright (2019), with permission from Elsevier.

this phase diagram, no change of state of hydrogen takes place when the pressure is increased or if heat is added. On the other hand, the triple point in the phase diagram denotes the temperature and pressure at which three different phases (i.e., gas, liquid, and solid) of hydrogen coexist while maintaining thermodynamic equilibrium.

Hydrogen remains in the liquid state in the temperature region of roughly 14 K to 33 K [2]. Hydrogen liquid is nearly the most volatile one among ordinary substances, and it is exceeded only by the isotopes of helium. At the same time, in most ways it is qualitatively quite normal in its behavior as a liquid. The low boiling point of liquid hydrogen is attributed to its relatively weak intermolecular attraction [2].

The properties of hydrogen are affected by two special characteristics: (i) the presence of quantum diffraction effects and (ii) the existence of the ortho and para forms of the hydrogen molecule.

The enhanced influence of quantum mechanics on intermolecular interactions causes a departure from the classical treatment of molecular interactions with a decrease in molar mass. The Lennard-Jones 6-12 potential is used in the traditional theoretical framework for treating molecular interactions [3]:

$$U = 4\varepsilon \left[ \left( \frac{\sigma}{r} \right)^{12} - \left( \frac{\sigma}{r} \right)^{6} \right] \tag{4.1}$$

The intermolecular distance is denoted by $r$, whereas $\varepsilon$ and $\sigma$ are the fluid-specific constants. The constant $\varepsilon$ is the maximum well depth (J mol$^{-1}$) and $\sigma$ is the intermolecular distance (m), where the intermolecular potential energy is zero [3]. In equation (4.1), the first term denotes molecular repulsion, whereas the second term denotes molecular attraction. In the 1940s, de Boer introduced a useful quantum parameter $\Lambda^*$ by using the interaction parameters of the Lennard-Jones potential

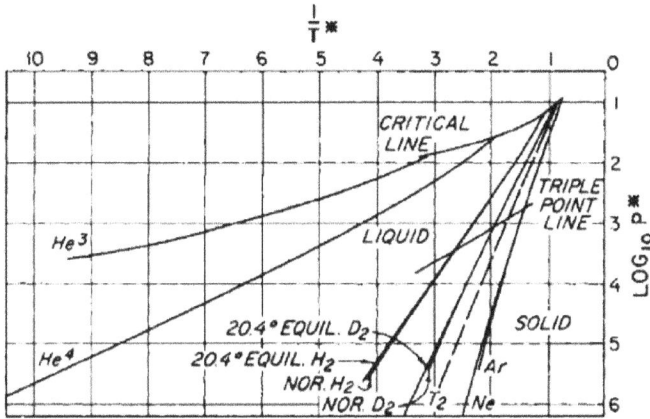

**Figure 4.2.** Reduced vapor-pressure data for various quantum substances. The curve for argon approximates the classical limit. Reprinted from [2], Copyright (1966), with permission from Elsevier.

together with Avogadro number $N_A$, the reduced Planck constant $\hbar$, and the molar mass $M$ [3]:

$$\Lambda^* = \frac{N_A 2\pi\hbar}{\sigma(M\varepsilon)^{1/2}} \tag{4.2}$$

This quantum effect is significant only when the de Broglie wavelength, $h/(M\varepsilon)^{1/2}$, for the relative motion of two hydrogen molecules each of mass $M$ with relative energy $\varepsilon$ is appreciable compared to the collision diameter $\sigma$ [2]. The relative magnitude of these effects is represented by their ratio $\Lambda^*$. In the classical limit $\Lambda^*=0$. The importance is manifested by larger-than-normal reduced volumes and vapor pressures, subnormal surface tensions, and heat of vaporization. We may recall here that due to their quantum-mechanical properties, the molecules retain vibrational energy, even at the absolute zero temperature, which is known as zero-point energy. In condensed states of light and weakly-interacting molecules, the zero-point energy plays a very important role by providing a repulsive force, which expands the lattice [2] (figure 4.2).

## 4.1 Nuclear spin: ortho-hydrogen and para-hydrogen

In chapter 2 we have seen that with the proton possessing a spin of 1/2 unit, the spins of the two nuclei in the hydrogen molecule may be parallel ($S = 1$) or antiparallel ($S = 0$). These two molecular species of hydrogen are called ortho-hydrogen and para-hydrogen, respectively. They are identified with the ortho form showing nuclear paramagnetism, while the para form is not. As a result, the parameters of the intermolecular potential are slightly different in the two forms of hydrogen, and thus the equation of state (EOS) and viscosity are slightly different for the two forms These differences in the mechanical properties are characteristically of the order of 1% or less [2]. The para-hydrogen has slightly higher volumes, thermal expansion coefficients, compressibilities, vapor pressures, and lower viscosities in the liquid state, and its second virial coefficient in the attractive region is smaller in absolute magnitude. These effects are mainly due to its force constant $\varepsilon/k$ being slightly smaller [2].

There is a large difference in the thermal properties of the two molecular species of hydrogen. This is attributed to the wave-mechanical symmetry requirements applied to the total wave function of the molecule. It is well known that electrons and protons have a magnetic quantum number or spin of 1/2 and an antisymmetric Schrödinger wave function. The net spin is zero when the spins are antisymmetric because they cancel each other out. In contrast, the net spin is 1 when the spins are symmetric. The significance of this is highlighted when the vibrational, rotational, and nuclear spin wave functions are combined into the total nuclear wave function for the hydrogen molecule [3]:

$$\psi_{\text{Total}} = \psi_{\text{Vib}} \psi_{\text{Rot}} \psi_{\text{Spin}} \tag{4.3}$$

The vibrational contribution is always symmetric due to the linear-diatomic nature of the hydrogen molecule and the lack of particle exchange between nuclei. Combining the two potential states of the nuclear spin, the one possible state of the vibration contribution, and the need for the whole nuclear wave function to be antisymmetric, two possible scenarios result [3]:

$$\psi_{\text{Antisym}} = \psi_{\text{Sym}} \psi_{\text{Sym}} \psi_{\text{Antisym}} \tag{4.4}$$

$$\psi_{\text{Antisym}} = \psi_{\text{Sym}} \psi_{\text{Antisym}} \psi_{\text{Sym}} \tag{4.5}$$

The rotational wave function $\psi_{\text{Rot}}$ is symmetric in equation (4.4) and antisymmetric in equation (4.5). The question remains whether the symmetric rotational contribution or the antisymmetric rotational contribution is associated with even rotational wave numbers ($\psi_{\text{rot}} = 0, 2, 4,...$) or odd wave numbers ($\psi_{\text{rot}} = 1, 3, 5,...$). It has been proven that only the symmetric rotational contribution can occupy the even levels because this was the only combination that could explain the experimental behavior observed in the heat capacities [3]. Thus, equation (4.4) represents the even ($\psi_{\text{rot}} = 0, 2, 4,...$) rotational energy levels associated with para-hydrogen, the form prevalent at low temperatures, and equation (4.5) represents the odd ($\psi_{\text{rot}} = 1, 3, 5,...$) rotational energy levels associated with ortho-hydrogen, the form prevalent at high temperatures.

In an isolated dihydrogen molecule, the rotational energy levels are given by $E_{\text{rot}} = BJ(J + 1)$, where $J = 0, 1, 2,...$ is the rotational quantum number and $B = 7.37$ meV. The rotational degeneracy is $g_r = 2J + 1$. The corresponding sublevels are characterized by the quantum number $m_J = -J, -J + 1,..., +J$. Then, it is found that only the even rotational states, $J = 0, 2, 4,...$, are accessible to the para-hydrogen molecules, and only the odd rotational states, $J = 1, 3, 5,...$, are accessible to the ortho-hydrogen molecules [2].

The complete system of rotational energy levels of hydrogen molecules can be described approximately by the formula for a rigid body [2]:

$$\varepsilon_j = J(J + 1)h^2/8\pi^2 I \tag{4.6}$$

Here $I$ is the moment of inertia about the center of mass. Equation (4.6) can be rewritten as:

$$\varepsilon_j = J(J + 1)k\theta_r \tag{4.7}$$

Here $\theta(\equiv h^2/8\pi^2 kI)$ is a characteristic temperature for rotation. The symmetry requirements lead to the selection rule, $\Delta J = \pm 2$, for rotational transitions.

In the gas phase, molecular hydrogen, deuterium, and tritium exist as ortho- and para-isomers because their rotational energy level separation corresponds to temperatures above their triple points $T_{tp}$ ($\theta_{rot} = 85.4$ K versus $T_{tp} = 13.80$ K for $H_2$, $\theta_{rot} = 43.0$ K versus $T_{tp} = 18.72$ K for $D_2$, and $\theta_{rot} = 29.1$ K versus $T_{tp} = 20.6$ K for $T_2$) [4]. With the decrease in temperature, ortho-hydrogen gets converted into para-hydrogen. This is because the lowest energy state of para-hydrogen has $J = 0$. The concentration of para-hydrogen increases with decreasing temperatures, and tends to reach 100% as the temperature approaches absolute zero. The equilibrium between ortho-hydrogen and para-hydrogen is expressed by [4]:

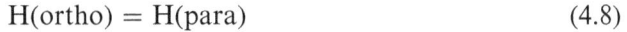

$$H(ortho) = H(para) \tag{4.8}$$

Here $H$ stands for enthalpy and the reaction is represented by equation (4.8), which does not depend on the total pressure. The standard enthalpy ($H$), entropy ($S$), and Gibbs free energy ($\Gamma$) of reaction for equation (4.8) are [4]:

$$\Delta_r H_T = (H_T - H_0)_{para} - (H_T - H_0)_{ortho} \tag{4.9}$$

$$\Delta_r S_T = S_T(para) - S_T(ortho) \tag{4.10}$$

$$\Delta_r G_T = \Delta_r H_T - T\Delta_r S_T \tag{4.11}$$

The equilibrium abundances of ortho-hydrogen and para-hydrogen are related to the Gibbs energy and are expressed as:

$$\Delta_r G_T = -RT \ln\left(\frac{P_{para}}{P_{ortho}}\right) = -RT \ln\left(\frac{\chi_{para}}{\chi_{ortho}}\right) \tag{4.12}$$

The partial pressures of ortho-hydrogen ($P_{ortho}$) and para-hydrogen ($P_{para}$) in equation (4.12) are equal to their mole fractions ($\chi_{ortho}$ or $\chi_{para}$) times the total pressure, which cancels out of the equation [4]. Figure 4.3 shows the equilibrium abundances of ortho-hydrogen and para-hydrogen in the temperature range of 10 K to 350 K. The figure shows that the molar ratio ortho–para approaches a value of 3:1 (i.e., 75% ortho-hydrogen and 25% para-hydrogen) with an increase in temperature. The term equilibrium hydrogen is used to describe hydrogen with an equilibrated ortho–para ratio at the ambient temperature. In the absence of paramagnetic catalysts, ortho-hydrogen and para-hydrogen equilibration is a rather slow process. Because the conversion reaction is so slow, hydrogen gas does not

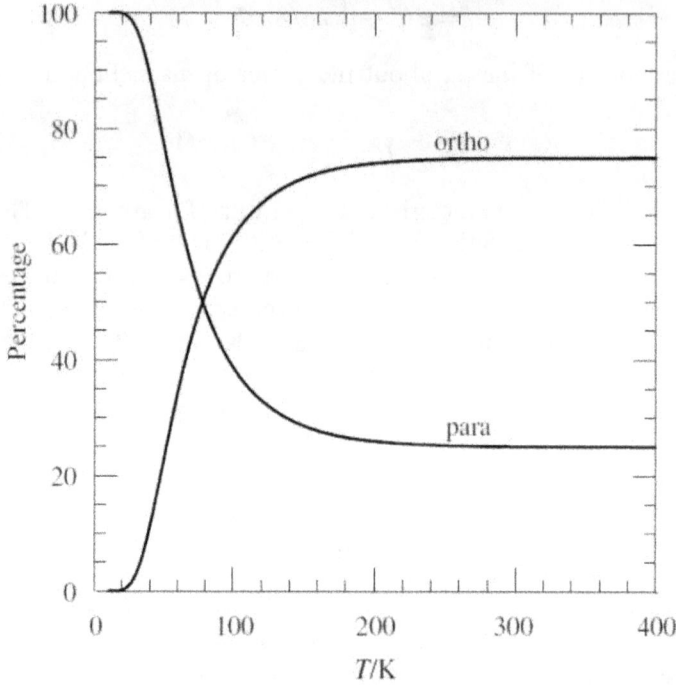

**Figure 4.3.** The temperature-dependent ratio of ortho–para hydrogen in equilibrium hydrogen. Reprinted from [4], Copyright (2013), with permission from Elsevier.

change from its high-temperature 3:1 ortho–para hydrogen ratio when cooled. Normal hydrogen (n-$H_2$) is defined as hydrogen having a 3:1 ortho–para ratio [4].

The nuclear-rotational partition functions of the two forms are expressed as [2]:

$$q_r(\text{para}) = \sum_{J=0,2,\ldots} (2J+1)e^{-J(J+1)\theta_r/T} \tag{4.13}$$

$$q_r(\text{ortho}) = 3 \sum_{J=1,3,\ldots} (2J+1)e^{-J(J+1)\theta_r/T} \tag{4.14}$$

The thermodynamic functions can be calculated from equations (4.13) and (4.14) by standard methods. The rotational specific heats (for example) are obtained as [2]:

$$\frac{C_r}{R} = \frac{\partial}{\partial T}\left(T^2 \frac{\mathrm{d}\ln q_r}{\mathrm{d}T}\right) \tag{4.15}$$

Here $R$ is the gas constant.

In a mixture of the ortho-hydrogen and para-hydrogen molecules at equilibrium, the concentrations stand in the same ratio as the rotational partition functions. Figure 4.4 presents the equilibrium concentration of the para-hydrogen molecule. The high and low-temperature limits are 25% and 100%, respectively, and the former results from the triple spin weight of the $S = 1$ state. The energy difference

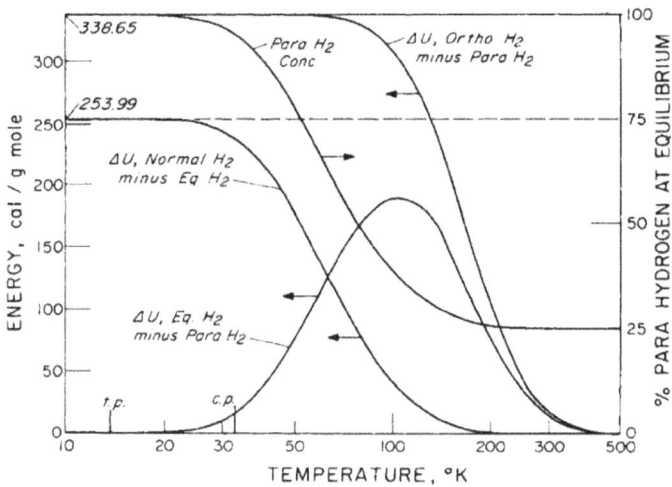

**Figure 4.4.** Equilibrium concentration of para-hydrogen and the energy differences between ortho-hydrogen and para-hydrogen, equilibrium and para-hydrogen, and normal and equilibrium hydrogen versus temperature. Reprinted from [2], Copyright (1966), with permission from Elsevier.

between ortho-hydrogen and para-hydrogen is also shown in figure 4.4 in the ideal-gas state. Below 40 K that value is 338.65 cal mole$^{-1}$, which is the difference in energy between the $J = 1$ and $J = 0$ rotational states [2]. The high-temperature normal hydrogen with an equilibrium mixture of 75% ortho-hydrogen and 25% para-hydrogen, has three-fourths this much energy or 253.99 cal mole$^{-1}$. This value exceeds the heat of vaporization, which is about 215 cal mole$^{-1}$ at the boiling point. Thus, the normal liquid hydrogen, which slowly releases the above energy by autocatalytic transformation to the equilibrium form of nearly pure para-hydrogen, would eventually be vaporized by the released energy, even if the liquid were contained in a perfectly insulated vessel. This is the reason why hydrogen is converted to the nearly pure para-hydrogen form for technological purposes.

Figure 4.5 presents the rotational specific heats of the ortho-hydrogen and para-hydrogen. At the high temperatures limit both species attain the classical value, $R$. The specific heat of the metastable mixtures of ortho-hydrogen and para-hydrogen can be calculated by a linear combination of these two contributions. The curve shown in figure 4.5 for normal hydrogen, i.e., the high-temperature equilibrium mixture of 75% ortho-hydrogen and 25% para-hydrogen, was obtained in this manner.

The equations of the state of ortho-hydrogen and para-hydrogen are nearly the same. The differences in the thermal properties arise mainly due to the internal energy states of the molecules. These differences are independent of density to a good approximation, and the differences calculated for the ideal-gas state should carry over to dense fluid states only with a slight change. This prediction has been checked by measuring the thermodynamic functions accurately for normal hydrogen and para-hydrogen at temperatures around 100 K and pressures up to 340 atm, and comparing the differences with the differences calculated for the ideal gases at the

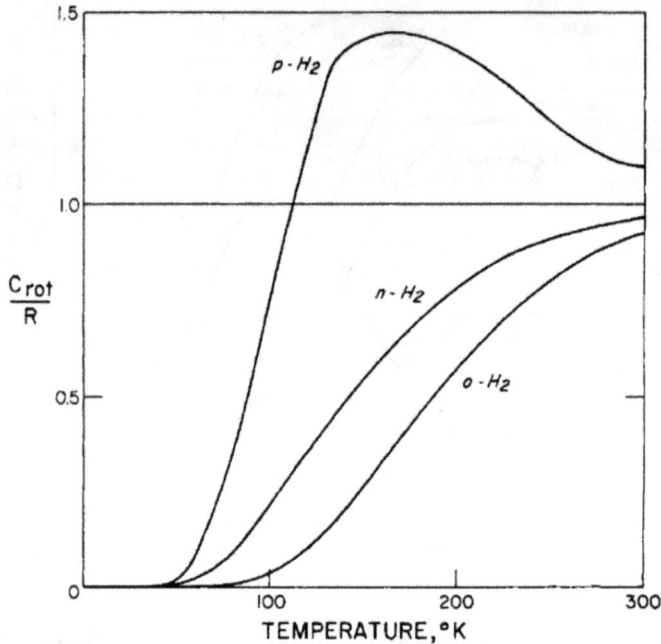

**Figure 4.5.** The rotational specific heats of ortho-hydrogen, para-hydrogen, and normal hydrogen. Reprinted from [2], Copyright (1966), with permission from Elsevier.

same temperature [2]. If energy, enthalpy, entropy, or specific heat in a particular state is known for any one species or ortho–para mixture of hydrogen, it can then be estimated for any other composition of ortho–para hydrogen in the same state by transposition of the corresponding differences in the ideal-gas state [2].

The thermal energy is insufficient to appreciably excite rotational transitions in the ortho-hydrogen and para-hydrogen molecules below 50 K. In this temperature regime, the ortho-hydrogen and para-hydrogen molecules remain in their respective ground states, $J = 0$ and $J = 1$, and the differences in their thermal and mechanical properties will be small. The distinction between ortho-hydrogen and para-hydrogen will be significant only in the nuclear magnetic susceptibilities and the optical absorption spectra [2].

## 4.2 The quantum law of corresponding states

A quantum-mechanical law of corresponding states needs to be used in hydrogen to properly understand its properties. This takes into account the variation of the quantum effect [2]:

$$p_r = f'(V_r, T_r, \Lambda^*) \tag{4.16}$$

Furthermore, the differences in experimentally obtained vapor pressures and critical-region properties cannot be reproduced without making a distinction between the potential parameters for ortho-hydrogen and para-hydrogen.

The differences in calculated polarizability of ortho-hydrogen and para-hydrogen have been used by Knaap and Beenakker [5] to deduce different intermolecular parameters applicable to data in the ground states. The difference in polarizability was based on differences in the calculated internuclear separation distance between ortho-hydrogen and para-hydrogen in their ground rotational states [3]. It was found that [3]:

$$\frac{\varepsilon_{ortho}}{\varepsilon_{para}} = 1.006 \tag{4.17}$$

$$\frac{\sigma_{ortho}}{\sigma_{para}} = 1.0003 \tag{4.18}$$

The values for normal hydrogen are then determined by linear interpolation [3]:

$$\frac{\varepsilon_{norm}}{\varepsilon_{para}} = 1.0045 \tag{4.19}$$

$$\frac{\sigma_{norm}}{\sigma_{para}} = 1.000\,23 \tag{4.20}$$

The parameters for the Lennard-Jones potential for normal hydrogen are $36.7\,\mathrm{J\,mol^{-1}}$ for the maximum well depth $\varepsilon$ and $2.959 \times 10^{-10}\,\mathrm{m}$ for the intermolecular distance $\sigma$ [3]. It was thus possible to distinguish between the properties of ortho-hydrogen and para-hydrogen in the ground states with respect to the quantum parameter with the help of these new parameters for the intermolecular potential. One can obtain the dimensionless quantities by rewriting $p$, $\rho$, and $T$ in terms of these constants:

$$p^* = \frac{N_A \sigma^3 p}{\varepsilon} \tag{4.21}$$

$$\rho^* = N_A \sigma^3 p \tag{4.22}$$

$$T^* = \frac{RT}{\varepsilon} \tag{4.23}$$

A plot of reduced thermodynamic properties versus the quantum parameter can be useful in determining the functional relationship between the quantum parameter and temperature, pressure, or density for light fluids including hydrogen. Figure 4.2 presents the reduced vapor pressures of several quantum liquids including hydrogen and its isotopes deuterium and tritium, converging towards the classical limit for $\Lambda = 0$ on the right-hand side of the figure represented by argon. The pressure and temperature in figure 4.2 have been reduced by the molecular parameters, $\varepsilon/\sigma^3$ and $\varepsilon/k$, respectively, rather than by the critical pressure and temperature and are designated $p^*$ and $T^*$.

A corresponding-states approach was used by Van Dael et al [6] to explain the change in critical temperatures between normal hydrogen and para-hydrogen. With the assumption of a linear relationship between the critical temperature and quantum parameter, the equation for the reduced critical temperature of para-hydrogen can be expressed as [3, 6]:

$$T^*_{c,\text{Para}} = T^*_{c,\text{ norm}} + (\Lambda^*_{\text{para}} - \Lambda^*_{\text{norm}})\frac{\Delta T^*_c}{\Delta \Lambda^*} \quad (4.24)$$

The subscripts c in the above equation denote critical point. A similar equation can also be obtained by using pressures or densities. More properties can be studied by using combinations of reduced temperature, pressure, and density with the help of dimensional analysis. The reduced speed of sound (for example) can be described as [3]:

$$w^* = \frac{w}{\left(\dfrac{N_A \varepsilon}{M}\right)^{1/2}} \quad (4.25)$$

If the data points include experimentally measured temperature and pressure, then that can be transformed while holding the value of the third experimental property, e.g., density or speed of sound, constant.

## 4.3 Some experimental results of the properties of liquid hydrogen

In this section, we will present some examples of the experimentally obtained thermodynamic and other physical properties of hydrogen. Technological application of liquid hydrogen usually involves hydrogen in its nearly pure para form, and hence the experimental information presented here will be mainly focused on para-hydrogen. Interested readers are referred to the articles by Corruccini for more exhaustive information [1, 2]. The temperatures are based on the definition of the triple point of water as 273.16 K. The molecular weight of hydrogen is 2.015 94 on the $C^{12} = 12.000$ scale.

### 4.3.1 Pressure–volume–temperature isotherms and thermodynamic properties

Experimental studies to obtain the pressure $(P)$–volume $(V)$–temperature $(T)$ isotherms of normal hydrogen began in the 1940s. There exists an early extensive correlation and compilation of properties of normal hydrogen by Woolley, Scott, and Brickwedde [7]. Subsequently, Johnston et al [8] determined $P$–$V$–$T$ isotherms of the liquid hydrogen from 20 K to 33 K at pressures up to 100 atm and gaseous isotherms from 34 K to 300 K at pressures up to 200 atm. Michels and collaborators determined the EOS thermodynamic properties of normal hydrogen in the temperature range between 98 K and 423 K and at pressures up to 2500 atm [2].

The thermodynamic and transport properties of para-hydrogen from the triple point (at 14 K) to 100 K and at pressures up to 340 atm have been determined by Goodwin and coworkers [9] in the 1960s. The studied properties included the $P$–$V$–$T$ surface, vapor pressure, specific heat at constant volume and saturation, the velocity of sound, viscosity, and dielectric constant [2]. These experimental studies led to the determination of the second and third virial coefficients, the liquid–vapor saturation densities heat of vaporization, critical constants, the melting pressure and the

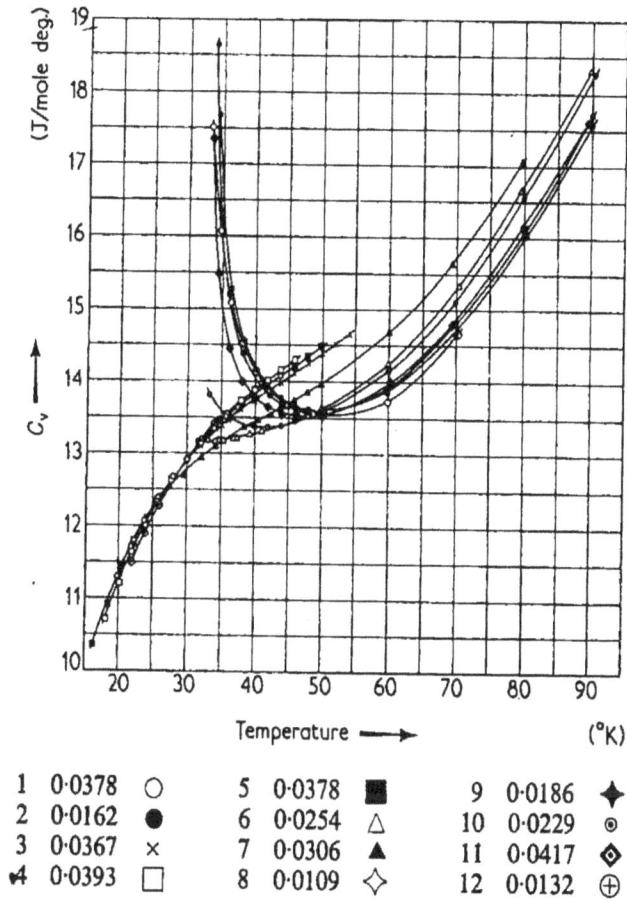

**Figure 4.6.** Specific heat as a function of temperature for several densities. Reprinted from [12], Copyright (1960), with permission from Elsevier.

volume of freezing liquid, and the internal energy, enthalpy, entropy, and specific heat at constant pressure. Younglove [10] measured the velocity of sound in compressed fluid para-hydrogen in the temperature range 15–100 K and pressures up to 300 atm. The velocity of sound in liquid normal and para-hydrogen were also measured by Van Itterbeek and Van Dael [11], and they combined those results with the available results for the density, vapor pressure, and saturation specific heats to calculate the compressibilities, thermal expansion coefficients, and the specific heats, $C_p$ and $C_v$ of both liquids from 14 K to 20 K. Some of these results are presented in figures 4.6–4.10, which are self-explanatory.

### 4.3.2 Thermal and electrical conductivity

The thermal conductivities of gaseous and liquid hydrogen have been measured by Roder and Diller [15] in the temperatures range 17 K–2000 K and at pressures up to

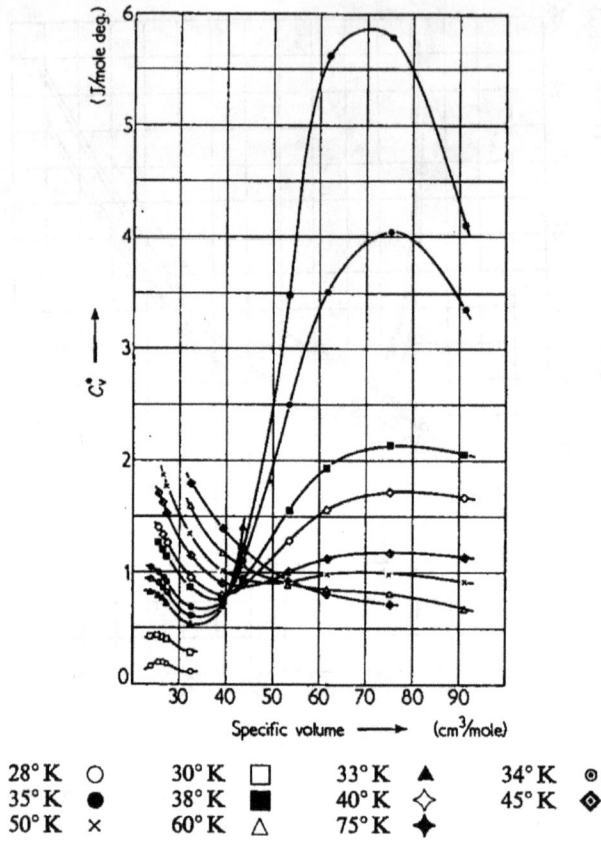

| 28° K | O | 30° K | □ | 33° K | ▲ | 34° K | ◉ |
| 35° K | ● | 38° K | ■ | 40° K | ◇ | 45° K | ◈ |
| 50° K | × | 60° K | △ | 75° K | ✦ | | |

**Figure 4.7.** Residual specific heat as a function of specific volume for several temperatures. Reprinted from [12], Copyright (1960), with permission from Elsevier.

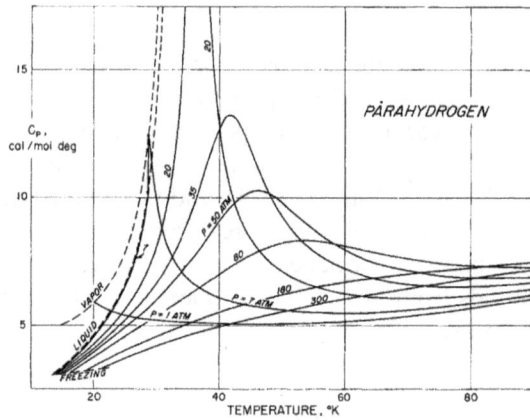

**Figure 4.8.** Specific heat at constant pressure for para-hydrogen. Reproduced from [13]. Image stated to be in public domain.

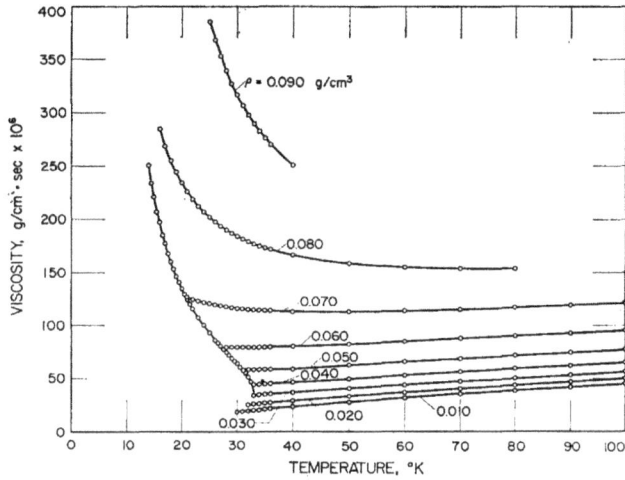

**Figure 4.9.** Viscosity of para-hydrogen on the lines of constant density and of the saturated liquid. Reprinted from [14], with the permission of AIP Publishing.

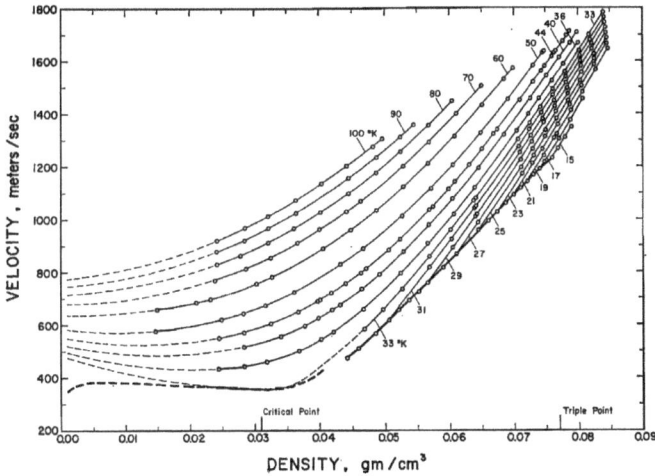

**Figure 4.10.** Velocity of sound in fluid para-hydrogen on isotherms The dashed lines indicate values calculated from $P$–$V$–$T$ data. The heavy lines represent the liquid–vapor coexistence boundary. Reprinted from [10], with the permission of AIP Publishing. © 1965 Acoustical Society of America.

15 MN m$^{-2}$. Figure 4.11 presents the thermal conductivity of dilute gaseous normal and para-hydrogen. The experimental results were analyzed as a function of density at fixed temperatures and as a function of temperature at fixed densities. Figure 4.11 (figure 4.12) presents the thermal conductivity of para-hydrogen as a function of density (temperature) at a fixed temperature (densities). Outside the critical region, the thermal conductivity of both the gas and the liquid increases continuously with

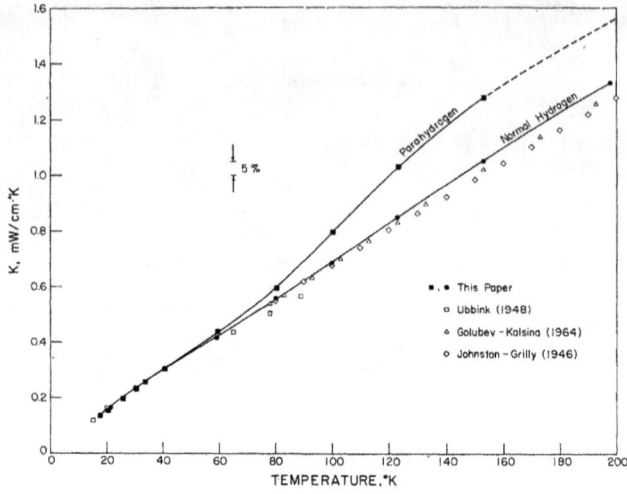

**Figure 4.11.** Thermal conductivity of dilute gaseous normal and para-hydrogen. Reprinted from [15], with the permission of AIP Publishing.

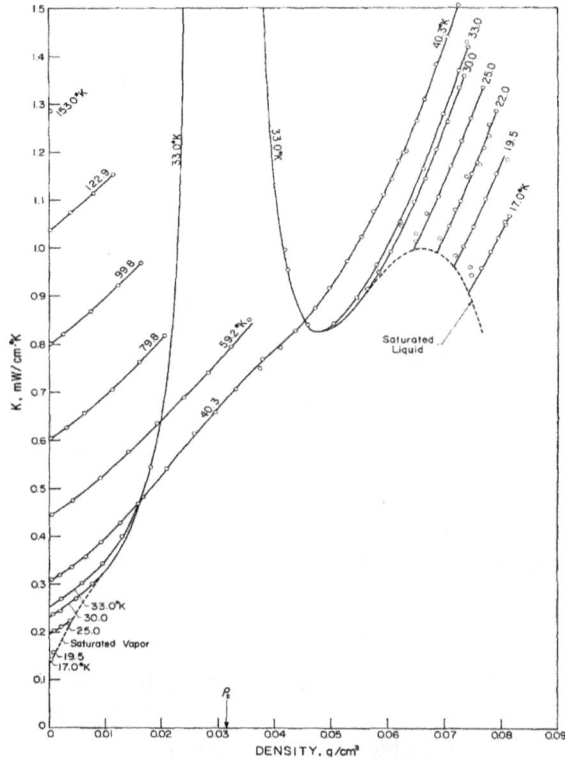

**Figure 4.12.** Thermal conductivity of para-hydrogen as a function of density at fixed temperatures. Reprinted from [15], with the permission of AIP Publishing.

temperature and density. In comparison to most other simple liquids, the temperature derivative of the thermal conductivity at fixed density in the compressed liquid hydrogen is positive and unusually large. In the critical region, the thermal conductivity increases rapidly with both temperature and density as these parameters approach their critical values [15].

Relatively less information is available on the electrical conduction in liquid hydrogen. The upper limit to the electrical conductivity mentioned in the literature is $10^{-17}$ ohm$^{-1}$ cm$^{-1}$ [2], and therefore the conductivity must be less than this. The reported breakdown strength is about $1.0 \times 10^6$ V cm$^{-1}$. The mobilities of both positive and negative ions have been determined to be $8.40 \times 10^{-3}$ cm$^2$ V$^{-1}$ s$^{-1}$, presumably at the boiling point [2].

For a detailed survey of experimental works on the thermophysical and transport properties of liquid hydrogen carried out starting from mid-1900, the readers are referred to the articles by Younglove [16], Jacobson [17], and Leachman *et al* [3].

## 4.4 The equations of state

B A Younglove at the National Bureau of Standards (NBS), in Boulder, Colorado, USA (now the National Institute of Standards and Technology (NIST)) introduced an EOS in the early-1980s for computing the thermophysical properties of several fluids,including hydrogen [3, 16]. This equation was used for the computation of pressure from density and temperature, and was also used with appropriate ancillary equations to calculate derived properties, including internal energy, entropy, enthalpy, heat capacity, and velocity of sound. Other relations used included the equation for vapor pressure, saturated liquid and saturated vapor densities, and melting pressure.

The Younglove equation [3, 16] was based on experimental data for the thermodynamic properties of pure para-hydrogen. It was appropriately modified to predict the thermodynamic properties of normal hydrogen by replacing the ideal-gas heat capacity equation and fixed-point properties of the para-hydrogen model with values for normal hydrogen. The 'International Practical Temperature Scale' of 1968 was employed in the derived models, and the upper limits of pressure and temperature in the EOSs were 121 MPa and 400 K [3] (figure 4.13).

Two surveys of the thermophysical properties of hydrogen were conducted in the first decade of 2000 [17, 18]. These surveys indicated the need for updating the EOSs for the thermodynamic properties of the various forms of hydrogen, as well as for new formulations for the transport properties, viscosity, and thermal conductivity. Subsequently, the new EOSs were developed for para-hydrogen and normal hydrogen to replace the earlier EOSs and an EOS was developed for ortho-hydrogen [3]. These new EOSs are valid for temperatures from the triple-point temperatures (13.8033 K for para-hydrogen, 13.957 K for normal hydrogen, and 14.008 K for ortho-hydrogen) to 1000 K and for pressures up to 2000 MPa, and the extrapolation was found to behave well, even at much higher temperatures and pressures [3].

The new EOSs are explicit in the Helmholtz free energy *a*, which can be expressed as [3]:

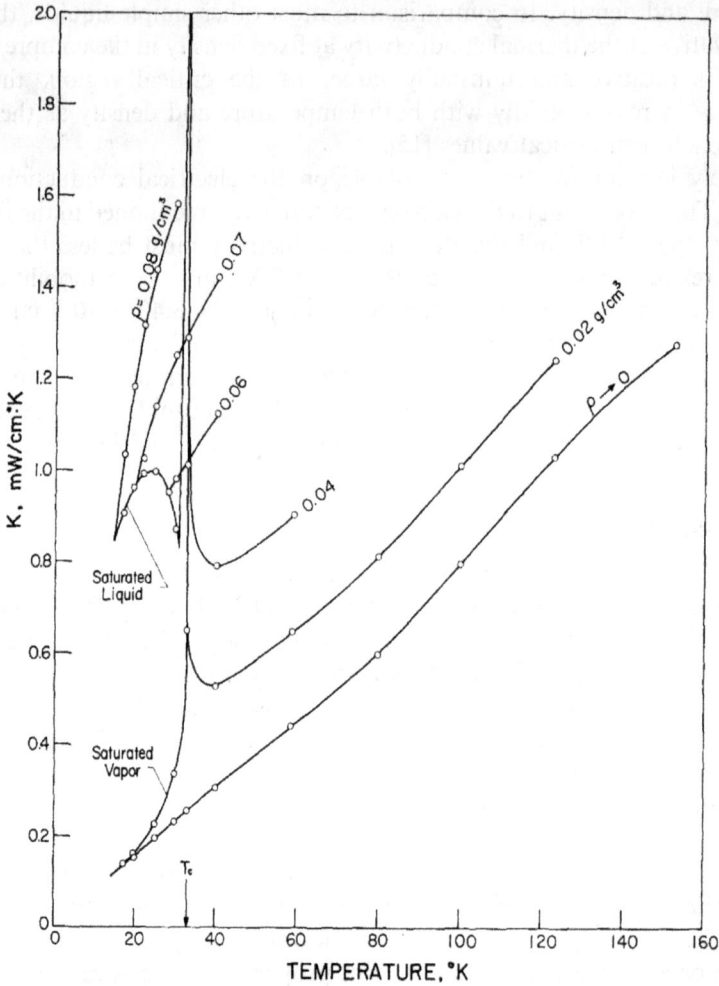

**Figure 4.13.** Thermal conductivity of para-hydrogen as a function of temperature at fixed densities. Reprinted from [15], with the permission of AIP Publishing.

$$\frac{a(T, \rho)}{RT} = \alpha(\tau, \delta) \tag{4.26}$$

Here $\alpha$ is the reduced Helmholtz free energy, and $\tau$ and $\delta$ are the reciprocal reduced temperature and reduced density, respectively.

$$\tau = \frac{T_c}{T} \tag{4.27}$$

$$\delta = \frac{\rho}{\rho_c} \tag{4.28}$$

The subscript c denotes a critical-point property. There are two parts in the reduced Helmholtz free energy: the ideal-gas contribution $\alpha_0$ and the residual contribution $\alpha_r$:

$$\alpha(\tau, \delta) = \alpha^0(\tau, \delta) + \alpha^r(\tau, \delta) \tag{4.29}$$

The ideal contribution to the Helmholtz free energy is based on nonlinear regression of calculated ideal-gas heat capacity data from the literature. The residual contribution to the Helmholtz free energy is based on nonlinear regression of experimental thermodynamic and caloric data [3]. The reduced Helmholtz free energy can then be used to compute further thermodynamic parameters.

### 4.4.1 The regression process

A nonlinear fitting algorithm based on the Levenberg–Marquardt method was used to optimize the ideal and residual contributions of equation (4.29). The ideal and residual contributions are composed of polynomial terms and the fitting algorithm minimizes the function [3]:

$$S = \sum W_\rho F_\rho^2 + \sum W_P F_P^2 + \sum W_{C_v} F_{C_v}^2 + \cdots \tag{4.30}$$

Here $W$ represents the weight assigned to a particular property and $F$ specifies the deviation of the property from the equation being optimized, and the two are summed for all of the points being fitted.

Each data point used in the fitting process corresponds to a different weight W. Equations (4.31)–(4.33) shown below present the results of some of these deviation calculations and similar equations exist for other properties included in the fit [3]:

$$F_P = \frac{P_{\text{data}} - P_{\text{calc}}}{P_{\text{data}}} \tag{4.31}$$

$$F_P = \frac{C_{v,\text{data}} - C_{v,\text{calc}}}{C_{v,\text{data}}} \tag{4.32}$$

$$F_\rho = \frac{\rho_{\text{data}} - \rho_{\text{calc}}}{\rho_{\text{data}}} \left( \frac{\partial \rho}{\partial P} \right)_T \tag{4.33}$$

A comparison of calculated deviations of individual data points and the total deviations of specific data sets is used to determine the quality of the resulting EOS. After assessment of the quality of the resulting EOS representation of the experimental data, new data points and numerical weights are assigned to improve deficiencies. Numerous iterations of this process result in the final EOS.

### 4.4.2 Ideal-gas contributions to the reduced Helmholtz free energy

The ideal-gas heat capacity equation is expressed as [3]:

$$\frac{c_P^0}{R} = 2.5 + \sum_{k=1}^{N} u_k \left(\frac{v_k}{T}\right)^2 \frac{\exp(v_k/T)}{[\exp(v_k/T) - 1]^2} \tag{4.34}$$

The parameters and coefficients used in equation (4.34) are available in reference [3]. These were regressed to the most accurate predictions available in the literature on the ideal-gas isobaric heat capacity. The correlations are valid in the range from nearly 0 to 1000 K. The estimated uncertainty of the formulations is less than 0.04% for para-hydrogen, less than 0.1% for normal hydrogen, and less than 0.1% for ortho-hydrogen.

The ideal-gas Helmholtz free energy is expressed as [3]:

$$a^0 = h^0 - RT - Ts^0 \tag{4.35}$$

Here the ideal-gas enthalpy $h^0$ is expressed as:

$$h^0 = h_0^0 + \int_{T_0}^{T} c_P^0 dT \tag{4.36}$$

And the ideal-gas entropy $s^0$ is expressed as:

$$s^0 = s_0^0 + \int_{T_0}^{T} c_P^0 dT - R \ln\left(\frac{\rho T}{\rho_0 T_0}\right) \tag{4.37}$$

By combining equations (4.35)–(4.37), one gets:

$$a^0 = h_0^0 + \int_{T_0}^{T} c_P^0 dT - RT - T\left[s_0^0 + \int_{T_0}^{T} c_P^0 dT - R \ln\left(\frac{\rho T}{\rho_0 T_0}\right)\right] \tag{4.38}$$

In reduced form, it is expressed as:

$$\alpha^0 = \frac{h_0 \tau}{RT_C} - \frac{s_0^0}{R} - 1 + \ln\left(\frac{\delta \tau_0}{\delta_0 \tau}\right) - \frac{\tau}{R}\int_{\tau_0}^{\tau} \frac{c_p^0}{\tau^2} d\tau + \frac{1}{R}\int_{\tau_0}^{\tau} \frac{c_p^0}{\tau} d\tau \tag{4.39}$$

Here $\delta_0 = \rho_0/\rho_c$ and $\tau_0 = T_C/T_0$. Equation (4.39) can be expressed in a computationally convenient parameterized form as [3]:

$$\alpha^0 = \ln \delta + 1.5 \ln \tau + a_1 + a_2 \tau + \sum_{k=3}^{N} a_k ln[1 - \exp(b_k \tau)] \tag{4.40}$$

The subscript $k$ is the index of each term in the ideal-gas heat capacity equation. The ideal-gas heat capacity coefficients $a_k$ and $b_k$ used in the formulations for para-hydrogen, normal hydrogen, and ortho-hydrogen are available in a tabular form in reference [3].

### 4.4.3 Residual contribution to the reduced Helmholtz free energy

The residual contribution to the reduced Helmholtz free energy is expressed as [3]:

$$\alpha^r(\tau, \delta) = \sum_{i=1}^{l} N_i(\delta^{d_i}\tau^{t_i}) + \sum_{i=l+1}^{m} N_i(\delta^{d_i}\tau^{t_i}) \exp(-\delta^{p_i})$$
$$+ \sum_{i=m+1}^{n} N_i(\delta^{d_i}\tau^{t_i}) \exp[\phi_i(\delta - D_i)^2 + \beta_i(\tau - \gamma_i)^2] \tag{4.41}$$

In the new EOSs, the first summation is a simple polynomial. This consists of seven terms, with exponents $d_i$ and $t_i$ on the reduced density and temperature, respectively. The second summation has an exponential density component to aid in the calculations of liquid and critical-region property calculations and consists of two terms There are five modified Gaussian bell-shaped terms in the third summation. These terms improve the modeling of the critical region. Spanning the critical region experimentally is the most difficult task, and it is also the most difficult to model theoretically [3].

The values of $N_i$, $d_i$, $t_i$, $p_i$, $\phi_i$, $\beta_i$, $\gamma_i$, and $D_i$ are somewhat arbitrary. The bounds on some of their values were, however, set to obtain a physically correct form of the equation. The values of $d_i$, $p_i$, $t_i$, $\gamma_i$, and $D_i$ are positive and those of $d_i$ and $p_i$ are integers. The values of $\phi_i$ and $\beta_i$ are negative. For these EOSs, $l = 7$, $m = 9$, and $n = 14$ [3]. Interested readers can find the simple polynomial terms for the para-hydrogen, normal hydrogen, and ortho-hydrogen EOSs and the modified Gaussian terms for each fluid in tabular forms in reference [3].

Plots of certain characteristic properties were created to test the thermodynamic consistency of the final functional form of the EOS. The monotonic characteristic of the Boyle curve, ideal curve, Joule–Thomson inversion curve, and Joule inversion curve are represented well the new para-hydrogen, normal hydrogen, and ortho-hydrogen EOSs.

### 4.4.4 Fixed-point properties and vapor pressures

Nonlinear regression techniques were used by Leach *et al* [3] to determine the critical-point properties based on other critical-region data as part of the optimization process. A particular selected experimental critical point of the actual fluid therefore may not exactly match with the calculated critical point for the EOS. However, it is generally expected that the final value may lie within the uncertainties of published experimental values. There can be some instances where the predicted critical state may be considered to be more accurate than experimentally measured values [3].

Leach *et al* [3] used the critical properties from the Younglove EOSs [16] as initial starting points in the nonlinear regression. In this fitting technique, the critical-point values were considered as adjustable parameters. Hence, the critical-point values determined by the regression for the new formulations changed slightly from those of the Younglove models [16].

It is difficult to reproduce the critical density experimentally. This is due to the small change in slope of the critical isotherm very near the critical pressure [3]. The critical densities of normal hydrogen and ortho-hydrogen have never been measured, and the value for normal hydrogen was based on an approximation from the available saturation data [3]. The rectilinear diameter method was used as an estimate along with the experimental data to increase the accuracy of the critical density for para-hydrogen. On the other hand, the quantum law of corresponding states was used to predict normal hydrogen values by transforming para-hydrogen experimental data near the critical point to the normal hydrogen surface [3].

Leach *et al* [3] used an ancillary equation to approximate the saturation line predicted by the Maxwell criterion with equation (4.41). The vapor pressure at saturation can then be expressed as [3]:

$$\ln\left(\frac{p_\sigma}{p_c}\right) = \frac{T_C}{T}\sum_{i=1}^{4} N_i \theta^{k_i} \tag{4.42}$$

where $\theta = (1 - T/T_C)$, $p_\sigma$ is the saturated vapor pressure, and the values of the coefficients $N_i$ and the exponents $k_i$ are given in [3].

### 4.4.5 Comparison of calculated data to experimental data

Leachman *et al* [3] made an excellent compilation of experimental data sets available on hydrogen until about the early-2000s and compared those with the data calculated using the new EOSs for para-hydrogen, normal hydrogen, and ortho-hydrogen. Furthermore, they extended these comparisons with calculations performed with the Younglove EOS [16]. The absolute average deviation (AAD) for each data set was expressed as:

$$AAD = \frac{1}{n}\sum_{i=1}^{n} |\%\Delta X_i| \tag{4.43}$$

$$\%\Delta X_i = 100 \times \left(\frac{X_{Data} - X_{Calculated}}{X_{Calculated}}\right) \tag{4.44}$$

Different caloric and near-critical region properties are predicted by the equations for para-hydrogen, ortho-hydrogen, and the 3:1 mixture of normal hydrogen. In the temperatures range from the triple point to 250 K and pressures up to 40 MPa, the estimated uncertainty with a coverage factor of 2 for primary data sets in the density of the EOS for para-hydrogen is 0.1%. In the critical region, generally an uncertainty of 0.2% in pressure is observed. In the temperature range between 250 and 450 K and at pressures up to 300 MPa, the uncertainty in density is 0.04%. This increases to 1% in the temperature range 450–1000 K. At pressures above 300 MPa, uncertainties of calculated densities increase from 0.04% to a maximum of 1% near the extreme of 2000 MPa. The speed-of-sound data is calculated to within 0.5% below 100 MPa, and the estimated uncertainty for heat capacities is 1.0%. Vapor pressures and saturated liquid densities were calculated using the Maxwell criterion, and the estimated uncertainties for each property are 0.1% [3].

For normal hydrogen and ortho-hydrogen, the vapor-pressure near-critical region single-phase density and speed-of-sound data were predicted using the quantum laws of correspondence [3]. The projected values for normal hydrogen were found to be within the uncertainty of the available data sets after a comparison with the available experimental data. Both an ortho-hydrogen EOS and an improved version of the new standard hydrogen EOS were created using these data. These results were used to improve the new normal hydrogen EOS and also for the formation of an ortho-hydrogen EOS.

The estimated uncertainties in the normal hydrogen EOS and the para-hydrogen EOS are quite similar [3]. The deviations in density are 0.1% at temperatures from the triple point to 250 K and at pressures up to 40 MPa. There is a 0.2% pressure uncertainty in the crucial region. In the temperature region between 250 and 450 K and at pressures up to 300 MPa, the uncertainty in density is 0.04%. This increases to 1% in the temperatures range 450–1000 K. The uncertainties in calculated densities above pressures of 300 MPa increase from 0.04% to a maximum of 1% near the extreme of 2000 MPa. The data of speed-of-sound are represented within 0.5% below 100 MPa, except near the critical region [3]. For heat capacities, the estimated uncertainty is 1.0%. The estimated uncertainties of vapor pressures and saturated liquid densities calculated using the Maxwell criterion are 0.2% for each property [3].

# References

[1] Kecebaşa A and Kayfeci M 2019 *Solar Hydrogen Production* (Amsterdam: Elsevier) ch 1
[2] Robert J 1966 *Corruccini Liquid Hydrogen: Properties, Production and Applications* (Amsterdam: Elsevier) ch 2
[3] Leachman J W, Jacobsen R T, Penoncello S G and Lemmon E W 2009 *J. Phys. Chem. Ref. Data* **38** 721
[4] Fegley B 2013 *Practical Chemical Thermodynamics for Geoscientists* (Amsterdam: Elsevier)
[5] Knaap H F and Beenakker J J 1961 *Physica (Amsterdam)* **27** 523
[6] van Dael W, van Itterbeek A, Cops A and Thoen J 1965 *Cryogenics* **5** 207
[7] Woolley H W, Scott R B and Brickwedde F G 1948 *J. Res. Natl. Bur. Stand.* **41** 379
[8] Johnston H L, Keller W E and Friedman A S 1954 *J. Am. Chem. Soc.* **76** 1482
[9] Goodwin R D, Diller D E, Roder H M and Weber L A 1963 *J. Res. Natl. Bur. Stand.* **61A** 173
[10] Younglove B A 1965 *Acoust. Soc. Am.* **38** 433
[11] Van Itterbeek A and Van Dael W 1961 *Physica* **27** 1202
[12] Younglove B A and Diller D E 1960 *Cryogenics* **2** 348
[13] Roder H M, Weber L A and Goodwin R D 1965 NBS Monograph No. 94, National Bureau of Standards https://www.osti.gov/biblio/5035661
[14] Diller D E 1965 *J. Chem. Phys.* **42** 2089
[15] Roder H M and Diller D E 1970 *J. Chem. Phys.* **52** 5928
[16] Younglove B A 1982 *J. Phys. Chem. Ref. Data* **11** Suppl. 1
1985 *J. Phys. Chem. Ref. Data* **14** 619
[17] Jacobsen R T, Leachman J W, Penoncello S G and Lemmon E W 2007 *Int. J. Thermophys.* **28** 758
[18] Leachman J W, Jacobsen R T, Penoncello S G and Huber M L 2007 *Int. J. Thermophys.* **28** 773

**IOP** Publishing

Hydrogen
Physics and technology
**Sindhunil Barman Roy**

# Chapter 5

## Exotic properties of dense hydrogen

We have studied in the previous chapters that hydrogen is a molecular gas at room temperature and under atmospheric pressure. The interaction between hydrogen atoms is a purely quantum mechanical effect, which leads to the formation of one of the strongest bonds in chemistry, namely the covalent H–H bond. Due to this strong bonding, hydrogen exists in molecular form, with atoms separated by approximately 0.74 Å and a bond dissociation energy of approximately 4.52 eV under ambient conditions. We also learned that the behavior of hydrogen is strongly influenced by quantum mechanical effects, with the nuclear quantum effects being larger for hydrogen than any other atom. However, the intermolecular bonding is very weak, and it requires extreme conditions to bring the molecules together and bind them into a solid state. Solid hydrogen has a large quantum zero-point energy, far greater than its latent heat of melting, and has a Debye temperature well above melting point. The solidification of hydrogen was first achieved in 1899 by Dewar at a relatively low temperature of 19 K. We shall now study the behavior of hydrogen in its dense states obtained under applying external pressure at various temperatures.

### 5.1 Hydrogen under pressure

Eugene Wigner was one of the pioneers of modern condensed matter physics. Back in 1935, Wigner along with his colleague Hillard Huntington first studied what would happen to hydrogen if it was compressed to very high densities [1]. With the help of a nearly free-electron model, they predicted that above 25 GPa ($\approx$ 250000 atm) of pressure, hydrogen would enter a metallic state [1, 2]. However, this kind of pressure was unimaginable to be achieved in the laboratory at that time. In any case, Wigner and Huntington were quite far off in their estimate of the actual pressure required because they did not take into account of the compressibility of hydrogen,

Over the next eight decades, high-pressure physics developed and matured significantly. As a result, it was possible to subject hydrogen to pressures of the order of 400 GPa. This was an almost 16-fold increase compared with the original

doi:10.1088/978-0-7503-5172-0ch5
© IOP Publishing Ltd 2024

prediction of Wigner and Huntington, and varieties of exciting and interesting phenomena have been revealed in dense hydrogen [2]. It was established that hydrogen forms a quantum molecular solid at low temperatures and pressures [3, 4]. This solid is characterized by large zero-point motion, rotational disorder, and the distinguishability of molecules into para-hydrogen and ortho-hydrogen states. The interactions between the hydrogen molecules increase with increasing pressure, which give rise to phase transitions that depend on pressure and temperature, the ortho–para hydrogen state, and the isotope of hydrogen. The interactions increase to such an extent at higher pressures that the hydrogen solid can no longer be considered a system of weakly interacting molecules. Instead, the condensed system evolves toward a network solid. It is predicted by the theory that at sufficiently high pressure, the hydrogen molecules break down to form a dense plasma of high-density atomic metal [5]. As a quantum metal, it is predicted that the strong dynamical character associated with the low mass in such a quantum metal leads to the formation of a quantum fluid ground state. This may give rise to very high-temperature superconductivity and/or cause the materials to remain in a liquid state down to the lowest possible temperatures.

At present, five solid phases of hydrogen are known (see figure 5.1). Hydrogen is, however, unique among the stable elements in that full structural information (e.g., the locations of the atomic centers and the shapes of the molecules) is absent for all of these phases [2]. We shall first focus on the three molecular phases, which are relatively well-studied. These are phase I, which is a high-temperature, low-pressure phase; phase II, which is a low-temperature, high-pressure phase; and phase III, which is a high-pressure molecular phase existing above 150 GPa [4].

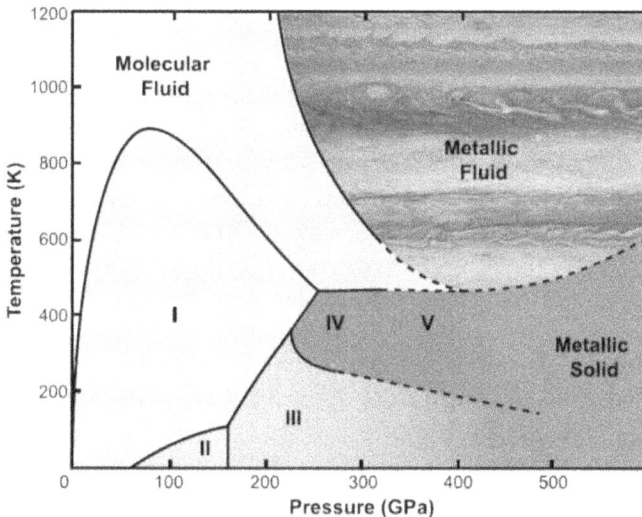

**Figure 5.1.** Schematic pressure-temperature phase diagram of hydrogen. Solid phase lines are a combination of static compression studies of solid hydrogen and dynamic compression studies of fluid deuterium. Dashed lines represent extrapolations of these combined results. The dark brown color of phases III and V at higher pressures suggests the closing of the bandgap. Reproduced from [2]. CC BY 4.0.

The high-temperature phase I comprises a hexagonal close-packed (hcp) lattice of freely rotating and, on average, spherically symmetric molecules [6]. The two low-temperature phases—Phase II and Phase III—possess lower symmetry, which suggests some kind of ordering of the hydrogen molecules. The properties of the two phases are, however, very distinct. These necessitate more general concepts, going beyond the mere crystallographic dissimilarities, to understand the origin of various phenomena observed in the low-temperature phases.

Molecules in solid hydrogen have almost free vibrational and rotational quantum states for the following reasons [7]:

1. The hcp structure has high symmetry and half of the intermolecular distance (1.90 Å) is well above both the inter-nuclear distance (0.74 Å) and the van der Waals radius (1.57 Å) of $H_2$.

2. The electronic charge distribution in $H_2$ is nearly spherical, which is experimentally manifested by the fact that the g-value of the rotational magnetic moment is nearly equal to +1. This minimizes the angular-dependent intermolecular interactions resulting from electron overlap and dispersion. The largest of the orientation-dependent interactions is the electric quadrupole-quadrupole (EQQ) interaction. This is of the order of magnitude of $Q^2/R^5 \sim 2.8$ cm$^{-1}$, where $Q_2$ is the electric quadrupole of $H_2$ and $R$ is intermolecular distance.

3. Separation of the rotational energy levels is large due to the very large rotational constant of $H_2$ (B = 59.3 cm$^{-1}$). Since ortho- and para-$H_2$ may be regarded as separate species, the smallest energy interval is between $J = 2$ and 0 (355.87 cm$^{-1}$), and therefore the mixing of these levels by the EQQ interaction is less than 1%.

Since the earliest investigations, vibrational spectroscopy has been instrumental in elucidating the crystal structures, bonding properties, and orientational state in the solid hydrogens [8]. The Raman excitations of these solids consist of vibrons, rotons, and phonons. Raman and infrared vibrons and rotons characterize the intra-molecular vibration and rotation of molecules, respectively. In combination, this experimentally determined information on various excitations leads to a wealth of knowledge because of selection rules derived from weak interactions between molecules in the solid, the persistence of $J$ as a good quantum number to moderate pressures, coupling of nuclear spin and angular momentum, and translational lattice symmetry [8].

Measurements of various excitations have been useful for structural studies in hydrogen as a function of temperature, pressure, and ortho–para state. McLennan and McLeod [9] used the Raman effect to study liquid hydrogen and observed a vibrational Raman transition $Q(O)$ and two rotational transitions, $S(O)$ and $S(1)$. The results of their study clearly showed that $H_2$ has well-defined rotational quantization, even in the condensed phase, and that $H_z$ at low temperature must be regarded as a mixture of effectively distinct sets of molecules, ortho-hydrogen ($J = 1$) and para-hydrogen ($J = 0$) [7]. At low densities of hydrogen, the Raman vibrons form a multiplet associated with the splitting of the $Q_1(J)$ bands, and the rotons are split into a triplet and a single phonon. These are characteristic of the hcp structure, which is found to be stable for the $J = 0$ solids and ortho/para mixtures at high temperatures near the melting line. On the other hand, the crystal structure of the low-temperature ordered state for $J = 1$ enriched solids at low densities is an ordered face-centered cubic (fcc) structure (space group Pa3), which is the lowest energy structure for electric quadrupoles [8]. For details of high-resolution spectroscopic measurements in the context of solid hydrogen, the readers are referred to the review article by T Oka [7].

If phase III is compressed at 300 K, it transforms into phase IV at around 230 GPa (see figure 5.1). Phase IV is thought to be driven by entropy and it is one of the most unusual phases of hydrogen [2]. If phase IV is further compressed at 300 K, it gradually transforms into phase V of hydrogen (see figure 5.1). This transformation takes place over a range of 50–60 GPa, starting at 275 GPa and effectively finishing at above 325 GPa [2]. It is speculated that phase V is a partially purely atomic state and it is a precursor to a fully metallic and atomic state.

We will discuss now some of these phenomena, concentrating on the properties associated with order–disorder of the rotational, vibrational, and electronic degrees of freedom.

Experimental studies of changes in the physical properties of materials at high density, employ a device called a diamond anvil cell to achieve ultrahigh compression. In a diamond anvil cell, a sample is confined to a microscopic chamber in a thin metal foil and squeezed between two diamond anvils. The device is based on a simple physical concept: pressure is inversely proportional to the area of a surface over which a force is applied. This simplicity, however, has an inherent drawback in that achieving extreme pressures inevitably means working with very small sample volumes.

In conventional techniques, applying extreme pressures to highly compressible materials such as hydrogen is not quite so straight forward. However, research groups around the world have expanded the boundaries of pressure generation over the past few decades. The tools and methods needed to accurately estimate pressures applied to microscopic samples of compressed gas have also been refined.

## 5.2 Phase I

Phase I of hydrogen is characterized by quantum spherically disordered molecules arranged in a hcp structure. This is shown schematically in figure 5.2. Hydrogen is a very good molecular insulator at room temperature and pressure above 5.5 GPa, with a bandgap of 9.5 eV [2]. This phase I occupies a very prominent part of the hydrogen phase diagram, reaching up to 190 GPa at 300 K. This phase shows remarkable pressure stability and to the best of our knowledge extends over the second largest pressure range for any molecular system, second only to molecular chlorine, whose phase I exists over a pressure interval of 230 GPa [2].

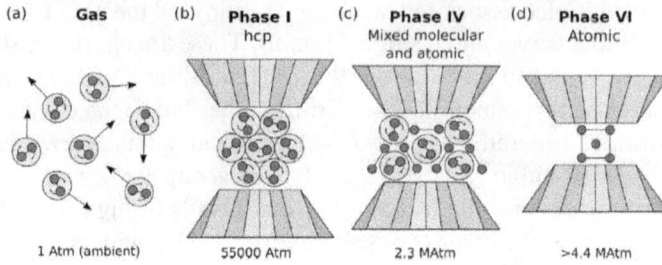

**Figure 5.2.** Schematic representation of the gaseous and solid states of hydrogen under different pressures at room temperature (300 K): (a) gaseous molecular state; (b) phase I, with hcp structure; (c) phase IV, with the mixed molecular and atomic state; and (d) purely atomic and metallic state. Reproduced from [2]. CC BY 4.0.

### 5.2.1 Rotational disorder in phase I

As mentioned above, this high-temperature phase of hydrogen consists of a close-packed lattice of freely rotating and, on average, spherically symmetric molecules. The excitations of the freely rotating molecules in phase I can be described by the rotational quantum number $J$ [6]. This behavior is in contrast with heavier molecular crystals, where molecular rotation is substantially hindered, even at low pressures. The properties of $H_2$ and $D_2$ crystals greatly depend on the ground state of the constituent molecules ($J = 0$ or 1), which in turn is determined by their total nuclear spin [6]. The hydrogen solids considered here are assumed to have a substantial fraction of spherically symmetric ($J = 0$)-molecules (e.g., para-$H_2$ or ortho-$D_2$).

The rotational disorder and orientational state in hydrogen as a function of pressure, temperature, and the ortho–para state have been studied extensively through vibrational spectroscopy [4]. Rotational Raman spectroscopy study of megabar pressures has revealed that rotational disorder persists throughout phase I. The presence of the rotational disorder is characterized by the persistence of rotational excitations over this $P$–$T$ range encompassing phase I. This was further confirmed by infrared measurements of the vibron–roton combination bands. The rotational disorder is remarkable, considering the large increase in intermolecular coupling ($\epsilon'$) over this pressure range, increasing from 3 to 510 cm$^{-1}$ from zero-pressure to 180 GPa [4]. The lattice symmetry of hydrogen molecular centers for phase I has been investigated by x-ray diffraction, first to the 40 GPa range and subsequently to a record pressure of 120 GPa at 295 K [4]. These measurements were performed on single crystals of hydrogen and deuterium within a surrounding medium of solid helium to help preserve the integrity of the crystals to very high pressures. These studies showed that the structure remained hcp throughout the range studied, thus confirming the results of Raman and infrared experiments.

### 5.2.2 Vbrational localization transitions

The bivibron bands corresponding to the double vibron excitation in a single hydrogen molecule and the simultaneous excitation of vibrons on two different molecules have been studied with Raman spectroscopy. This investigation aimed to explore the possibility of a novel transition in which the double excitation transforms from being localized on a single molecule to being delocalized throughout the crystal to give a bound–unbound transition [4]. The detailed Raman-scattering measurements were also performed on solid deuterium, where the intensity of the bands corresponding to the double excitations was a factor of $10^{-4}$ to $10^{-5}$ times weaker than those of the single excitation vibron bands described above. The increase in intermolecular coupling with pressure causes a shift in the frequency of the sharp localized bivibron excitation to reach that of the broad continuum of delocalized states [4]. The sharp bivibron band vanishes at this point. The experimentally observed transition is reasonably supported by the results of theoretical calculations, which are free of adjustable parameters. Localization effects are also observed in the intramolecular stretching mode, in both hydrogen and deuterium [4].

## 5.3 Symmetry breaking and phase II of hydrogen

We have seen above that in the ground state at low temperature and low pressure, hydrogen molecules are in the $J = 0$ spherical rotational states. With the application of pressure, the molecules are brought closer together, and at high enough pressure a new equilibrium state of hydrogen is reached with a trade-off between going higher in kinetic energy (i.e., $J \neq 0$ rotational levels) and gain in negative potential energy through an orientational ordering, which minimizes EQQ energy [6, 10]. It has been shown by first-principles calculations that the EQQ interactions dominate over other interactions in the pressure range up to 100 GPa [11].

A first-order phase transition, which breaks the rotational symmetry of the orientationally disordered phase of hydrogen to the orientationally ordered phase, was first predicted in para-$H_2$ and ortho-$D_2$ in the pressure range of ∼15–30 GPa [10]. This was accompanied by a structural transition from the hcp to the fcc phase. Subsequently, it was observed that the broken symmetry phase transition occurred with a strong isotopic shift in the transition pressure, 28 GPa for $D_2$, 69 GPa for HD, and 110 GPa for $H_2$ [10]. The orientationally ordered phase of hydrogen is named phase II.

Spontaneous Raman spectra demonstrated the presence of a structural phase transformation in solid $H_2$ beginning at 145 GPa at 77 K [12]. This transition was characterized by a discontinuous drop in frequency ($\sim100$ cm$^{-1}$ at 77 K) of the Raman-active vibron. This was confirmed in several subsequent spectroscopic studies. The magnitude of the vibron discontinuity diminished as the temperature increased from 77 K, and the discontinuity disappeared altogether at a critical point near 140 K. The Raman vibron frequency decreased continuously with increasing pressures above this temperature. The magnitude of the vibron-frequency shift and the pressure-temperature conditions of the phase transformation were consistent with the theoretically predicted pressure-induced orientational ordering of the molecular solid [12].

Experimental investigation of hydrogen at megabar pressures with infrared spectroscopy revealed a new absorption line in the vibron-frequency range appearing at 110 GPa. This was followed by a discontinuity in the infrared vibron at 150 GPa, and also a striking three-order of magnitude increase in absorptivity at this pressure. Together, these results indicate that there is a large electronic change at the 150 GPa transition.

The lower pressure portions of the phase diagram of both ortho–para mixtures and ortho-rich samples of deuterium have been investigated using Raman spectroscopy. Significant changes take place in low-frequency rotational excitations, along with a huge increase in the intensity of high-frequency vibron sidebands at the transition to phase II. The transition was characterized by the appearance of a new roton and vibron band, changes in their intensities, and narrowing and splitting of the bands, which point towards an increase of molecular ordering [4]. A detailed study of the multiple vibrons and roton suggested a structure for deuterium, in which the molecules oriented themselves in different directions in successive planes of the original hcp solid to form an orientational superstructure. It was also revealed that there is an intermediate phase between the high-temperature hcp structure and this orientational superstructure. This intermediate phase may be an orientational glass phase [4].

### 5.3.1 Direct investigation of phase II crystal structure

While infrared and Raman spectroscopies have revealed various interesting aspects of phase II of hydrogen, the only direct experimental methods available to study crystal structure in the high-pressure phases of $H_2$ and $D_2$ are x-ray or neutron scattering experiments. There are no inner electronic shells in $D_2$ or $H_2$, while x-rays are scattered by electrons in molecular orbitals. Therefore, the x-rays are mostly insensitive to the individual positions of H(D) atoms in the molecules and to orientational ordering. In contrast, neutrons are scattered by individual nuclei, and are therefore useful to determine the individual positions of H(D) in the structure, as well as the orientations of the molecules [10]. Due to the very small scattering power of a hydrogen crystal in a diamond anvil cell, the use of third-generation synchrotron-based x-ray sources and the growth of a single crystal in helium are required to measure the equation of state of $H_2$ and $D_2$ to 120 GPa at 300 K.

The neutron scattering study of the hydrogen in a high-pressure environment is in general more challenging because of the low intensity of neutron sources. However, the significant improvement in neutron instrumentation and pressure techniques at various neutron scattering laboratories across the world has pushed the pressure-related limitations for single-crystal neutron diffraction experiments.

The structure of the broken symmetry phase in solid $D_2$ has been studied with a combination of neutron and x-ray diffraction up to about 65 GPa [10]. In the x-ray diffraction studies, while cooling the single crystal of $D_2$ at a constant pressure of 63.4 GPa, a small positive discontinuity in the lattice parameter $c$ ($\Delta C/c = 3 \times 10^{-4}$) and a negative discontinuity in the parameter $a$ ($\Delta a/a = -6 \times 10^{-4}$) were observed at 70 K. This result gives a transition point to phase II, which is in excellent agreement with previous spectroscopic studies. The corresponding volume discontinuity, $\Delta V/V = 10^{-3}$ supports a previous estimate from the Clapeyron equation and the slope of the I–II boundary line [10]. From this x-ray measurement, it was found that the phase I–phase II transition did not affect the diffraction peaks of the hcp lattice, except for a very small volume discontinuity [10].

The neutron experiments were performed on the same $D_2$ sample at 38 GPa in the temperature range 1.5–70 K, above and below the I–II transition (I–II transition temperature $T_{I-II} \approx 44$ K at 38GPa). The theoretical model that visualizes the I–II transition as purely rotational with a Pa3-type local order was used to explain the experimentally obtained neutron data. The Pa3 structure is the structure observed in pure ortho-$D_2$ or para-$H_2$ (metastable in the $J = 1$ rotational state), which orders at low temperature in a cubic close-packed structure with neighboring molecules directed along the different cubic diagonals. This structure minimizes the EQQ energy for a three-dimensional compact structure. To have a similar type of order on a hexagonal lattice, the molecules need to form four sublattices with molecular axes directed perpendicular to the four different faces of a tetrahedron [10]. The molecules are then arranged so that every molecule is surrounded by molecules belonging to the other three sublattices. The unit cell has a $P\bar{3}$ symmetry and consists of eight molecules. The neutron intensities calculated with this structure were found to be in good agreement with the experimental data [10].

These structural studies suggest a purely orientational transition on the hcp lattice of Pa3-type. The results also explain the incommensurate modulation of the structure. This incommensurate modulation lowers the symmetry of the hcp lattice, and hence could explain the multiplet structure of the roton bands and the additional vibron bands observed in the Raman and infrared spectra of phase II [8].

## 5.4 Phase III: symmetry breaking at higher pressure

Phase III is obtained by compressing phase II with pressure above ~155 GPa and temperature below 100 K or at around 190 GPa at 300 K [2]. It has been shown to possess an hcp lattice [13, 14] with unusually intense infrared activity [15].

Raman-scattering experiments have demonstrated that solid hydrogen undergoes a structural phase transition beginning at 145 GPa at 77 K [12]. This transition is characterized by a discontinuous drop in frequency (~100 cm$^{-1}$ at 77 K) of the Raman-active vibron. The magnitude of the vibron discontinuity diminishes with the increase in temperature from 77 K and disappears altogether at a critical point near 140 K. The Raman vibron frequency decreases continuously with increasing pressures above this temperature.

Synchrotron infrared (IR) spectroscopy studies at ultrahigh pressures have provided further evidence of this high-pressure symmetry-breaking transition [15]. A striking three orders of magnitude increase in vibron absorbance was observed at the pressure of 150 GPa at 85 K. Furthermore, a discontinuity in the frequency of the infrared vibron was observed, which was identical to that observed in earlier Raman spectroscopy studies at the same temperature [12].

The transition from phase II to phase III has been investigated through x-ray powder-diffraction experiments with pressures up to 183 GPa and a low temperature of 100 K [13]. In that study, a transition into phase III was confirmed by in situ

Raman-scattering measurements. Two diffraction peaks corresponding to the 100 and 101 lines of the hcp structure were observed above the phase II–phase III transition pressure of 160 GPa, which indicated that the hydrogen molecules in phases II and III are still in the vicinity of the hcp lattice point up to 183 GPa. Assuming a hexagonal structure for these phases, the lattice parameter $c$ was found to be decreased by 1.5% at phase II–phase III phase transition [13].

Theoretically, the structure of phase III has been considered to be a superlattice of hcp phase I with six hydrogen molecules in the primitive cell [16]. This is because a large number of molecular vibration modes have been reported for phase III from Raman-scattering and IR absorption experiments. While no superlattice reflection was detected in the x-ray diffraction study due to the low signal-to-noise ratio, overall the experimental results are consistent with a theoretically proposed picture of the classical orientational ordering of rotating hydrogen molecules for phase III of solid hydrogen. A further synchrotron-based x-ray diffraction study has confirmed that the high-pressure transitions in $H_2$ were not caused by major crystallographic changes of the hcp structure, and the phases remained isostructural except for an increase in anisotropy, which was reflected in a severe distortion in the c/a ratio [14].

## 5.5 Phase IV and phase V of solid hydrogen

From the discovery of phase III, it took about 25 years to find the existence of phase IV of hydrogen and deuterium in the early-2010s [2, 17, 18]. If phase III is compressed at 300 K, then it transforms into phase IV at around 230 GPa. Erements *et al* [17] reported that on compression of normal molecular hydrogen at room temperature (295 K), at 200 GPa the Raman frequency of the molecular vibron strongly decreased and the spectral width increased. These results suggested a strong interaction between hydrogen molecules. Deuterium behaved similarly. Above 220 GPa, hydrogen turned opaque and electrically conductive. At 260–270 GPa, the conductance of hydrogen sharply increased and changed little on further pressurizing up to 300 GPa or cooling down to at least 30 K [17]. The sample also reflected light well.

Howie *et al* used Raman and visible optical transmission spectroscopy to investigate dense hydrogen (deuterium) up to 315 (275) GPa at 300 K [18]. They observed the phase III transformation at around 200 GPa. This transition to phase III was succeeded at 220 GPa by a reversible transformation to a new phase. This new phase IV was characterized by the simultaneous appearance of the second vibrational fundamental and new low-frequency phonon excitations, and a dramatic softening and broadening of the first vibrational fundamental mode. The optical transmission spectra of phase IV showed an overall increase of absorption and a closing of the bandgap, which reaches 1.8eV at a pressure of 315 GPa [18].

Phase IV is thought to be entropy-driven and along with phase V (described below) is possibly the most unusual phase of hydrogen. With the results of Raman spectroscopy experiments in combination with theoretical structural investigations, it has been speculated that phase IV of hydrogen is made up of alternating layers consisting of six hydrogen atom rings and unbound hydrogen molecules (see figure 5.2). The interatomic distances in the ring are around 0.82 Å, leading to a vibrational frequency of around 2700cm$^{-1}$, which is significantly reduced compared with that under ambient conditions, while the atoms in the unbound molecules have a vibrational frequency close to 4200cm$^{-1}$. An x-ray diffraction study has revealed that the transition to phase IV of $H_2$ is not accompanied by major crystallographic changes of the hcp structure, and it remains isostructural except for a severe distortion in the $c/a$ ratio, i.e., an increase in anisotropy [14].

Using in situ high-pressure Raman spectroscopy, Dalladay *et al* [19] presented evidence that at pressures greater than 325 GPa at 300 K, hydrogen and hydrogen deuteride transform to a new phase V. This new phase of hydrogen is characterized by the substantial weakening of the vibrational Raman activity, a change in pressure dependence of the fundamental vibrational frequency, and a partial loss of the low-frequency excitations. The transformation takes place over a range of 50–60 GPa, starting at 275 GPa and effectively finishing at above 325 GPa. Due to the differences in quantum mechanical properties between hydrogen and deuterium, phase V has not been observed in the latter [2]. It has been speculated that phase V is a partially purely atomic state, and a precursor to a fully metallic and atomic state [19].

## 5.6 Phase VI: metallic hydrogen

In the high-pressure physics community, metallic hydrogen is known as the Holy Grail [20]. Following the earlier prediction by Wigner and Huntington [1], in his seminal paper Ashcroft [5] had theorized that if the hydrogen molecule in $H_2$ solid is dissociated and a purely atomic alkali-metal-like solid is formed, then this solid could be a room temperature superconductor. These predictions, together with the possibility that metallic hydrogen could be a strong explosive with ultrahigh energy density, make this a wonder material attracting the attention of experimentalists in the high-pressure community worldwide. Modern theory predicts that the transition into the metallic state of hydrogen takes place at a pressure of about 500 GPa.

The initial study of solid hydrogen in the diamond anvil cell to around 66 GPa using Raman spectroscopy revealed that the intramolecular vibrational frequency of hydrogen increased with pressure up to 33 GPa, and it then started to decrease as more pressure was applied [22]. Vibrational frequency is a measure of H–H bonding strength in the $H_2$ molecule, and one can easily extrapolate that at some very high pressure the bond will be broken and molecular hydrogen can transform into an alkali-like free-electron metal similar to Li or Na [2]. The decrease in the intramolecular vibrational frequency of hydrogen indicated that the molecular bonds are weakening. The molecular bonds will be broken eventually when molecular hydrogen transforms to the atomic state originally envisaged by Wigner and Huntington [1]. Such an atomic solid hydrogen state is presented schematically in figure 5.2(d).

The sample environment of the diamond anvil cell is quite restricting. Still, several probes can be used to evaluate the degree of metallicity in solid hydrogen. However, all such probes have their limitations. This along with the small linear size (2–3 $\mu$m) of the hydrogen samples required to reach pressures above 350 GPa can easily lead to misinterpretation of the data, and in turn to erroneous claims of metallization [2]

Mao and Hemley [23] made the first claim on the metallization of hydrogen in 1989. The claim was made based on the reduced Raman signal and increased absorption by the sample, and it was concluded that the metallic state had reached somewhere above 200 GPa. Soon after, Lorenzana *et al* [24] made another claim on achieving hydrogen metallization. With the subsequent improvements in experimental methods, it soon became apparent that the observed experimental features supporting the metallization of hydrogen could be explained by the loss of hydrogen at high pressures and by increased fluorescence of the diamonds being mistaken for the closing of the electron energy bandgap [2].

In 2011, combining Raman spectroscopy with direct electrical measurements of sample resistance, Erement and Troyan claimed that they had observed 'liquid atomic metallic hydrogen' at pressures above 260 GPa [17]. The claim was based on the disappearance of the Raman signal and an abrupt drop in sample resistivity at 260 GPa. However, it was soon shown that hydrogen remains in a mixed molecular and atomic semiconducting solid (phase IV) phase to at least 315 GPa at 300 K, transforming to phase III at lower temperatures [2]. The loss of the Raman signal and the drop in sample resistance were explained by the loss of hydrogen and the collapse of the sample chamber.

With the discovery of phase V above 325 GPa, and the suggestion that this phase could represent the onset of dissociation and the first step toward a completely metallic state [19], the experimental efforts towards achieving metallization in hydrogen started with a new vigor. To this end, a paper entitled 'Observation of the Wigner–Huntington transition to metallic hydrogen' was published from Harvard University with the claim of metallization of hydrogen at 500 GPa [25]. However, this work immediately drew several criticisms [2], and even generated a public debate on metallic hydrogen.

Dense hydrogen becomes more and more opaque to visible light under increasingly extreme pressures. In recent times, Loubeyre *et al* [26] have studied the optical transparency of solid hydrogen at extreme pressure and low temperature using a diamond anvil cell in a synchrotron radiation source. In the initial pressure range of tens of GPa, the dense hydrogen sample was transparent to both infrared and visible light (see figure 5.3(a)). The dense hydrogen lost its transparency to visible light when the pressure was raised to roughly 300 GPa (see figure 5.3(b)). Finally, the sample became reflective to both infrared and visible light when the pressure was raised above 425 GPa [26, 27]. Loubeyre *et al* [26] attributed the change in optical reflectivity to a probable pressure-induced phase transition in which electrons in the dense hydrogen sample become free to move like those in metal.

This work is also not free from controversy because the results of infrared absorption measurement by themselves are not enough proof for claiming the existence of a metallic state [2]. A definite proof for metallic hydrogen would only come from a measurement of the electrical conductivity of hydrogen sample at high pressure as a function of temperature. The sample needs to show a high level of electrical conduction, which should decrease as the sample temperature is raised. Despite the significant development of experimental techniques to study condensed

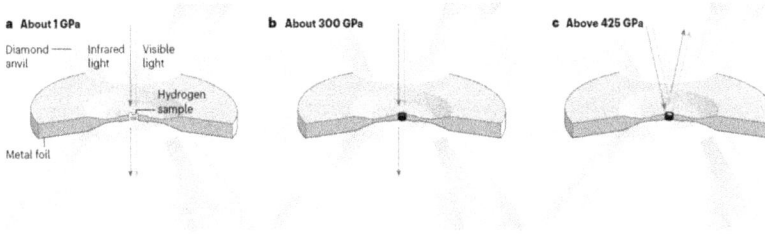

**Figure 5.3.** Effect of increasing pressure on cold solid hydrogen. Reprinted from [27], Copyright (2020), with permission from Nature Publishing.

matter in extreme conditions in the past few decades, electrical-transport measurements of hydrogen still remain a huge challenge [27].

# References

[1] Wigner E and Huntington H B 1935 *J. Chem. Phys.* **3** 764

[2] Gregoyyanz E *et al* 2020 *Matter Radiat. Extremes* **5** 038101

[3] van Kmnendonk J 1983 *Solid Hydrogen* (New York: Plenum)

[4] Hemley R J and Mao H 1996 *J. Non-Cryst. Solids* **205–7** 282

[5] Ashcroft N W 1993 *J. Non-Cryst. Solids* **156-8** 621

[6] Mazin I I *et al* 1997 *Phys. Rev. Let* **78** 1066

[7] Oka T 1993 *Annu. Rev. Phys. Chem.* **44** 299

[8] Goncharov A F *et al* 1996 *Phys. Rev.* B **54** R15590

[9] McLennan J C and McLeod J H 1929 *Nature* **23** 60

[10] Goncharenko I and Loubeyre P 2005 *Nature* **435** 1206

[11] Kaxiras E and Guo Z 1994 *Phys. Rev.* B **49** 11822

[12] Hemley R J and Mao H K 1988 *Phys. Rev. Lett.* **61**

[13] Akahama Y *et al* 2010 *Phys. Rev.* B **82** 060101(R)

[14] Ji C *et al* 2019 *Nature* **573** 558

[15] Hanfland M, Hemley R J and Mao H K 1993 *Phys. Rev. Lett.* **70** 3760

[16] Tse J S *et al* 2008 *Solid State Commun.* **145** 5

[17] Eremets M I and Troyan I A 2011 *Nat. Mater.* **10** 927

[18] Howie R T *et al* 2012 *Phys. Rev. Lett.* **108** 125501

[19] Dalladay-Simpson P, Howie R T and Gregoryanz E 2016 *Nature* **529** 63

[20] Geng H Y 2017 *Matter and Radiation at Extremes* **2** 275

[21] Ashcroft N W 1968 *Phys. Rev. Lett.* **21** 1748

[22] Sharma S K *et al* 1980 *Phys. Rev. Lett.* **44** 886

[23] Mao H K and Hemley R J 1989 *Science* **244** 1462

[24] Lorenzana H E *et al* 1990 *Phys. Rev. Lett.* **64** 1939

[25] Dias R P and Silvera I F 2017 *Science* **355** 6326

[26] Loubeyre P, Occelli P and Dumas P 2020 *Nature* **577** 631

[27] Desgreniers S 2020 *Nature* **577** 626

# Chapter 6

## Hydrogen in various solid matrix

It is possible to combine hydrogen with solid materials, either chemically or physically. Thus, chemisorption and physisorption are the names given to the two solid-state hydrogen-storage processes, respectively. In chemisorptions hydrogen chemically reacts with the solid to form a hydride, whereas in physisorptions the hydrogen molecule adheres to the solid surface. In this chapter, we will focus on the scientific aspects related to hydrogen in solids. In part II (section 12.3) of this book, we will discuss various classes of materials that are important for hydrogen energy technology.

### 6.1 Physically bound hydrogen

Similar to other gases, hydrogen physisorbs onto a solid surface. Weak van der Waals interactions keep the diatomic hydrogen molecule attached to the surface, preventing it from dissociating. With the enthalpy of adsorption ($H_a$) ranging from 4 to 10 KJ mol$^{-1}$, the strength of these interactions for hydrogen is quite modest [1]. These weak interactions necessitate low-temperature ambiance for a significant amount of hydrogen gas absorption. Physisorption is a surface phenomenon, thus high surface area materials have been the focus of research. It is important for mobile applications that there should not be a loss of hydrogen during adsorption and desorption, and also very fast kinetics because no activation energy is involved. To this end, porous nanostructured materials have drawn considerable attention. Such materials with a high surface area are advantageous to increase the hydrogen-storage density due to their low density and high porosity [2]. Figure 6.1 provides a schematic representation of such materials, which include nanoporous carbon materials, metal–organic frameworks (MOFs), covalent organic frameworks (COFs), porous aromatic frameworks (PAFs), and nanoporous organic polymers.

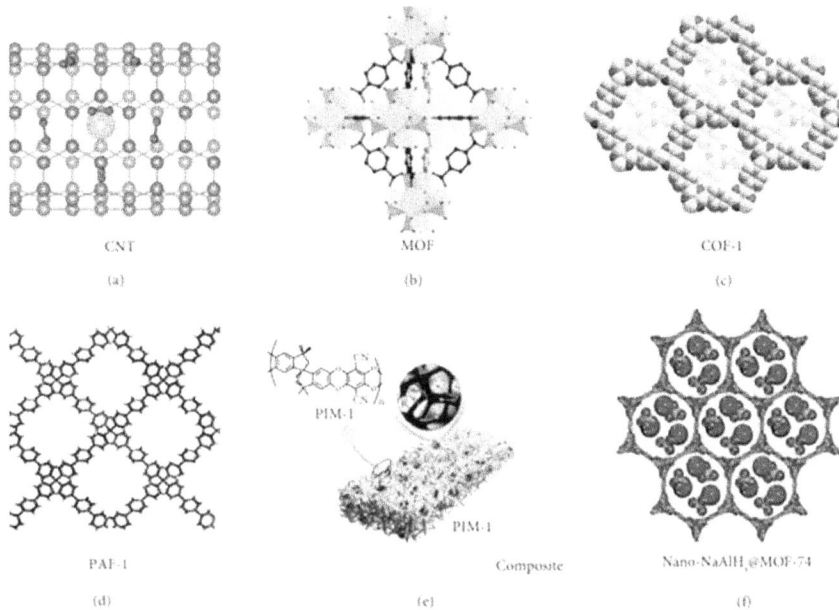

**Figure 6.1.** Schematic representation of various physically bound hydrogen systems: (a) nanoporous carbon materials (carbon nanotube (CNT)), (b) metal–organic frameworks (MOF), (c) COFs, (d) PAFs, (e) the structure of a nanoporous organic polymer (PIM-1) and the composite with a nanoporous filler, and (f) nano-hydrides (sodium alanate ($NaAlH_4$)) confined in the nanopores of a MOF. Reproduced from [2]. CC BY 4.0.

## 6.1.1 Nanoporous carbon materials

Nanoporous carbon materials are considered to be promising materials for hydrogen storage because of their high porosity, low density, and cost efficiency. Such carbon materials are available in nature in a variety of forms: single-wall carbon nanotubes (SWNTs), multi-wall nanotubes (MWNTs), and other carbon structures, e.g., carbon nanofibers (CNFs), carbon aerogels (CAs), etc. All these carbon materials have a benzene-like carbon hexagonal structure. However, they differ from each other in the way in which these carbon hexagons are arranged. These carbon-based materials can be roughly divided into two groups: (i) those possessing long-range order of carbon hexagons, namely carbon nanotubes and carbon nanofibers; and (ii) those having irregular structures, such as activated carbon [1].

### 6.1.1.1 Carbon nanotubes

Carbon nanotubes with diameters in the range of a nanometer are referred to as single-wall carbon nanotubes (SWCNTs) [2]. It can be assumed that SWCNTs are formed by rolling a single graphene sheet. Multi-wall carbon nanotubes (MWCNTs) consist of nested single-wall carbon nanotubes.

Carbon nanotubes are quite inert to surface impurities, and therefore they need severe activation conditions. In SWCNTs, all carbon atoms are exposed to the surface. Thus, they have the highest surface-to-bulk atom ratio. This property enables the SWCNTs to become highly surface-active. The densities of CNTs are considerably lower than that of metals, intermetallic alloys, COFs, and MOFs [2]. Furthermore, these nanomaterials have a large number of void spaces in the form of pores, which can accommodate a large amount of hydrogen. In SWCNT bundles, hydrogen can be physisorbed at a number of locations, including grooves, exterior wall surfaces, and interstitial channels [2]. Experimental results showed that the hydrogen-storage capacity of SWCNTs and MWCNTs can reach up to 4.5–8 wt% at 77 K, and a moderate capacity of approximate 1 wt% at ambient temperature and pressure [3]. The hydrogen-storage capacity of MWCNTs could be significantly improved at room temperature under high-pressure environments, e.g., 2.0 wt% at 40 bar, 4.0 wt% at 100 bar, and 6.3 wt% at 148 bar [3].

In an attempt to increase the amount of hydrogen that pristine nanotubes can store, efforts have been made to tailor their surface characteristics. High-energy atomic bombardment, reactive ball milling, high-temperature annealing, and transition-metal doping are some of the methods of surface alterations [2]. The structure of carbon nanotubes is partially destroyed by reactive ball billing and high-energy atomic bombardment, which produce a high density of defects. Defect-enriched carbon nanotubes are able to store a larger amount of hydrogen due to the additional hydrogen binding sites that these defects provide. But in this instance, hydrogen binds irreversibly via chemisorption, and therefore the samples need to be heated over 500 °C in order to release the stored hydrogen.

The increased hydrogen-storage capacities of transition-metal doped nanotubes can be understood in terms of the spillover phenomenon. Figure 6.2 presents a schematic of the spillover phenomenon. The hydrogen molecules are initially adsorbed by the transition-metal atoms and are then subsequently spilled into different adsorption sites of carbon nanotubes. The readsorption data of Pd- and V-doped carbon nanotubes have indicated that more than 70% of hydrogen can be spilled into the low-energy binding sites such as external wall or groove sites of nanotubes [3]. Such sites are associated with a considerably low desorption barrier ($\sim k_B T$) and therefore explain why the observed kinetics of desorption are not affected by the doping.

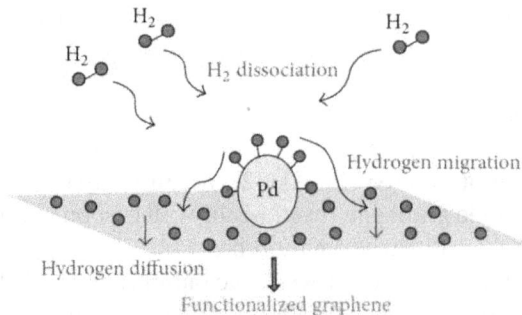

**Figure 6.2.** A schematic representation of the spillover phenomenon. Reproduced from [3]. CC BY 4.0.

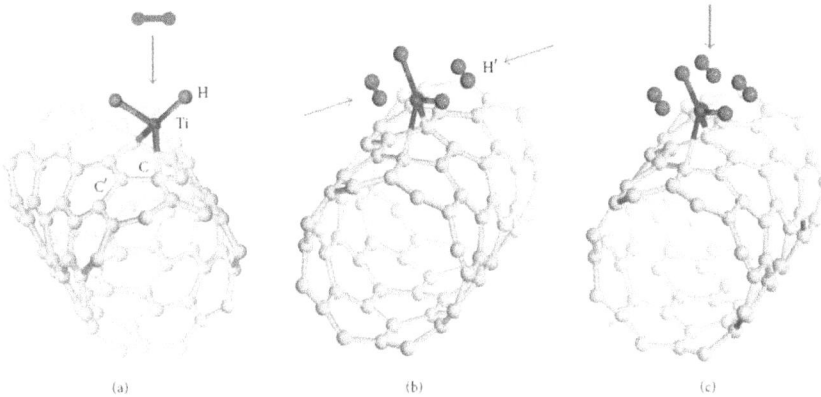

**Figure 6.3.** The configurations of hydrogen on Ti-doped carbon nanotubes. Reproduced from [3]. CC BY 4.0.

First principle calculations have been used to study the hydrogen-storage behavior of Ti-doped CNTs [3]. Four hydrogen molecules can bind to each Ti atom in the doped carbon nanotube, which is equivalent to a gravimetric storage capacity of 8.0 wt%. Figure 6.3 shows how the bonding of hydrogen molecules in Ti-doped carbon nanotubes takes place in stepwise processes. In the first step, a hydrogen molecule undergoes dissociative chemisorption and binds to a single titanium atom (see figure 6.3(a)). This event happens without an energy barrier. The second step involves physisorption of two hydrogen molecules (see figure 6.3(b)). Then, in the final step the fourth hydrogen molecule is physisorbed to the Ti atom in a direction parallel to the C-Ti plane (see figure 6.3(c)). A unique hybridization between Ti-$d$, hydrogen $\sigma^*$ antibonding, and SWCNT C-$p$ orbitals explains unexpected bonding. However, because of the wide range of storage capacities reported in different tests, the experimental studies of hydrogen storage continue to be a contentious topic [3]. The main cause of this is the uncertainty around the purity of the sample utilized in different trials. It is well known that some metallic contaminants from their fabrication method can be found in carbon nanotubes. These metallic contaminants take up hydrogen. As a result, the contributions that metals provide to hydrogen storage are occasionally mistakenly linked to the storage capacity of carbon nanotubes.

### 6.1.1.2 Carbon nanofibers

The mechanical properties of carbon nanofibers (CNFs) are outstanding and they have a huge surface area. Chemical vapor deposition, electrospinning, and templating techniques are among the synthesis methods used to create CNFs [2]. Such methods are simple and suitable for mass production. This makes CNFs a potential candidate for hydrogen storage due to their low cost and commercial availability. According to initial experimental research, CNFs may store hydrogen for up to 6.54 wt% to 0.7 wt% at room temperature and pressure of about 100bar [2]. CNFs produced through chemical activation treatment have gained a lot of attention recently because of their controllable pore sizes and larger surface area. At 298 K and 100 bar pressure, Ni-doped CNFs produced by metal doping demonstrated an improved hydrogen absorption of 2.2 wt%.

### 6.1.2 Metal–organic frameworks

Metal–organic frameworks (MOFs) are crystalline porous materials formed by metallic polyhedra and organic ligands (see figure 6.4(a)). With micropores smaller than 2 nm and a continuous skeleton structure, MOFs exhibit great porosity. Quite a few synthetic methods have been developed to produce two-dimensional and three-dimensional MOFs by coupling inorganic building units, metal ions, or clusters with organic units of carboxylates or other organic anions, such as phosphonate, sulfonate, and heterocyclic compounds [2, 3]. The selection of metal ions and organic building blocks enables one to control the framework topology, pore size, and surface area. MOFs have large cubic cavities of uniform sizes [3]. Because of the various cavity structures that can be created inside them, MOFs are highly porous materials and are suitable for hydrogen gas storage.

The first use of MOFs for hydrogen-storage applications was reported in the early-2000s [2, 3, 8]. A framework MOF-5 was synthesized from zinc salt and 1,4-benzenedicarboxylic acid (BDC) to give $Zn_4O(BDC)_3$, and resulted in a hydrogen intake of 4.5 wt% at 78 K and 1.0 wt% at room temperature and 20 bar pressure, and thus opening a new avenue for hydrogen-storage materials [2]. Figure 6.4(b) shows

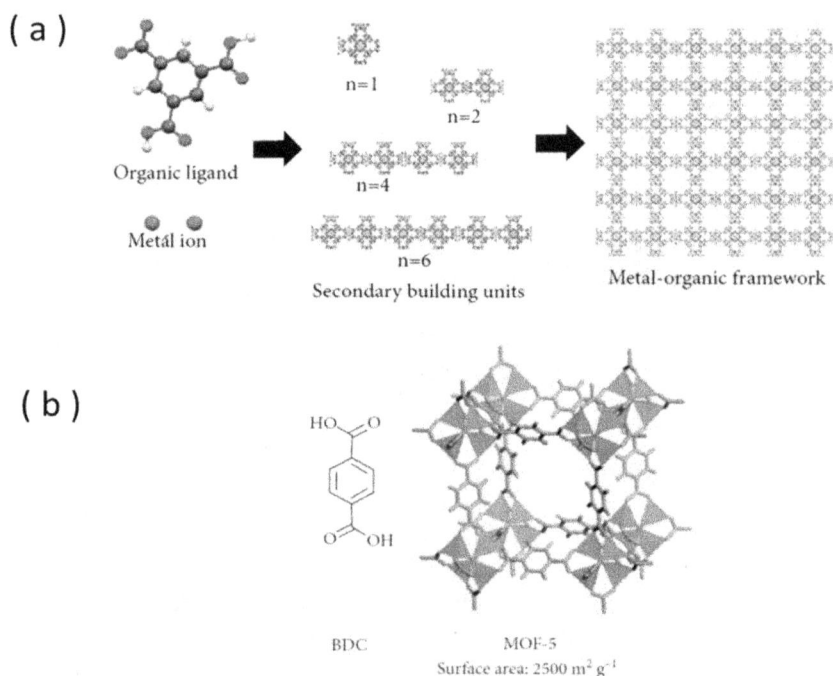

**Figure 6.4.** (a) Schematic representation of the mechanism and formation of MOFs. Reproduced from [4]. CC BY 4.0. (b) Chemical structure of BDC and MOF-5. Reprinted with permission from [5]. Copyright (2010) American Chemical Society.

schematically the chemical structure of BDC and MOF-5. However, the poor moisture-stability of MOF-5 caused limited applicable environments and unstable performance [2]. The altered chemical linkage and structural topology subsequently led to improved porosity, moisture, mechanical, and acid/base stabilities. A range of MOFs are prepared by several synthesis processes, including conventional heating, electrochemistry, microwave-assisted heating, mechanochemistry, and sonochemistry [2, 3].

### 6.1.3 Covalent organic framework

COFs are carbon-based crystalline nanoporous organic polymers. They are constructed with strong covalent linkages, e.g., B-O, C-O, B-C, C-C, and C-N, to give two-dimensional and three-dimensional structures. COFs have high porosity, well-order pores, and superior chemical and thermal stability. These properties make them quite suitable for hydrogen storage.

Figure 6.5 presents the molecular structures of building units and crystal structures of two COFs, namely COF-1 and COF-5. COF-1 is synthesized by using a self-condensation reaction of 1,4-benzenediboronic acid, and COF-5 was prepared from 1,4-benzenediboronic acid and 2,3,6,7,10,11-hexahydroxytriphenylene. COF-5

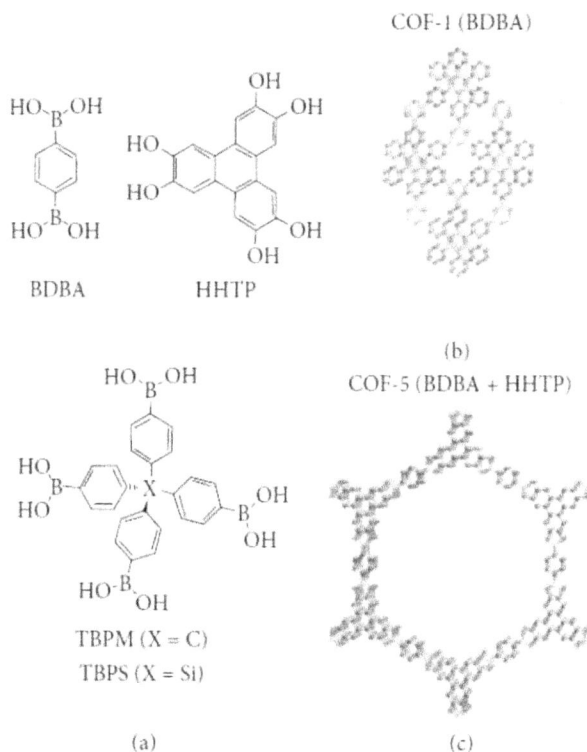

**Figure 6.5.** (a) Molecular structures of building units, (b) crystal structure of COF-1, and (c) crystal structure of COF-5. Reprinted with permission from [6]. Copyright (2008) American Chemical Society.

exhibits a high BET surface area as 1590 $m^2 g^{-1}$ and a 3.5 wt% hydrogen intake at 77 K and 80 bar pressure.

The capacity of COF deteriorates sharply when operating at an increased temperature. Metal doping with metals such as Li, Mg, Ti, and Pd, has been utilized in COFs to improve the hydrogen-storage capacity [2]. The metal incorporation on COF skeletons enhances the interactions between the COF skeleton and hydrogen atoms. It has been reported that lithium-doped COFs show an improved hydrogen-storage capacity in the range of 6.73 wt% and 6.84 wt% at 298 K and 100 bar pressure [2]. This enhancement of hydrogen uptakes originates from the bonds between positively charged lithium ions and hydrogen molecules.

### 6.1.4 Porous aromatic framework

Porous organic materials with a tetrahedral diamond-like structure are known as PAFs [2]. High porosity, a large surface area, low mass densities, and strong mechanical and thermal stability are among the beneficial characteristics of PAFs. To further enhance the capabilities, the organic frameworks allow for pore size modulation and post-synthetic functionalization. PAFs are often synthesized by irreversible cross-coupling processes and have many phenyl rings [2]. PAFs can be utilized in a variety of processes, including molecular separation, hydrogen storage, catalysis, and molecular sensing because of their distinct structure and characteristics.

The first synthesis of PAFs took place in the early-2000s [2], which was termed PAF-1 and exhibited a surface area of 5600 $m^2 g^{-1}$. Figure 6.6 presents a schematic of the synthesis of PAF-1. High uptakes of hydrogen (7.0 wt% at 77 K and 48 bar) was possible due to a high surface of PAF-1. However, PAF-1 exhibits a low heat of adsorption, indicating a weak interaction between hydrogen molecules and its surface. This would therefore suggest that hydrogen intake capacity of PAF-1 is low at high temperatures or ambient pressure. In order to get around this issue, a number of PAFs were synthesized [2] that had quadrivalent atom silicon (Si) or germanium (Ge) in place of the carbon center. At 77 K and 60 bar pressure, PAF-3 with Si centers shows a surface area of 2932 $m^2 g^{-1}$, 6.6 kJ $mol^{-1}$ heat of adsorption, and a hydrogen uptake of 5.5 wt%. PAF-4 with Ge centers exhibits surface area of 2246 $m^2 g^{-1}$, 6.3 kJ $mol^{-1}$ heat of adsorption, and has a hydrogen uptake of 4.2 wt% at 77 K and 60 bar pressure. It was also reported that the hydrogen-storage capabilities of PAF-4 with lithium tetrazolide doping were increased to 20.7 wt% (at 77 K and 100 bar) and 4.9 wt% (at 233 K and 100 bar), respectively. The increase

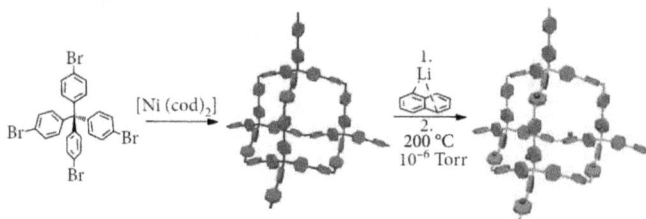

**Figure 6.6.** Schematic illustration of the synthesis of porous aromatic frameworks PAF-1 and Li-doped PAF-1. [7] John Wiley & Sons. Copyright © 2012 WILEY-VCH Verlag GmbH & Co. KGaA, Weinheim.

in hydrogen binding energy in PAFs on the lithium doping sites may be the cause of the capacity boost (figure 6.6). In addition to lithium, doping with magnesium and calcium has also been effectively used to boost hydrogen binding strength in PAFs. At 233 K and 100 bar pressure, the hydrogen-storage capacities for PAF-Mg and PAF-Ca are 6.8 wt% and 6.4 wt%, respectively. With the use of appropriate building units and doping materials, it might be possible to further increase the volumetric capacities and heat of adsorption.

### 6.1.5 Nanoporous organic polymers

While the surface area and capability of the COFs and PAFs are promising, they are typically not cost-effective and deliver powder-type polymers with comparatively poor mechanical and processability features. Their commercial applications are thus restricted. Other organic polymers with nanoporous properties include conjugated microporous polymers (CMPs), hyper crosslinked polymers (HCPs), and polymers of intrinsic microporosity (PIMs). These polymers have also found extensive application as adsorbents, catalyst carriers, separation materials, and gas storage materials [2]. Figure 6.7 presents schematic illustrations of the structures of HCPs, CMPs, and PIMs.

HCPs, CMPs, and PIMs possess several advantages and unique properties for hydrogen-storage applications [2]:
1. Variable polymer backbones and facile functionalization.
2. Tunable pose size and crosslinking density.
3. Lightweight and high surface area.
4. High processability for bulk, coating, and composite materials.
5. Low cost and accessibility for mass production.

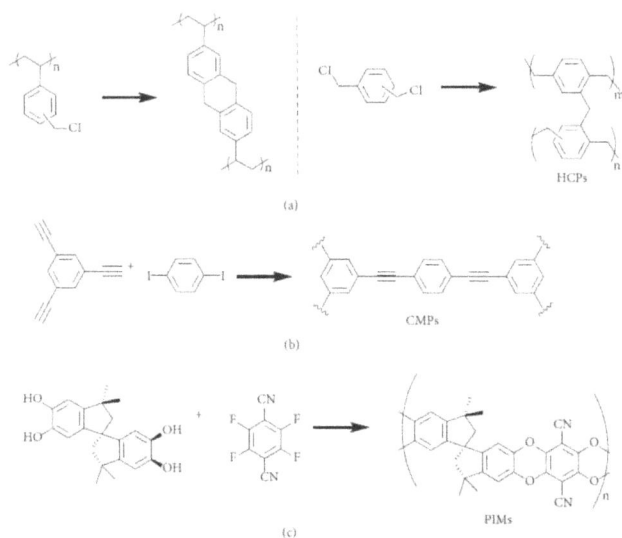

**Figure 6.7.** Synthetic routes of (a) HCPs, (b) CMPs, and (c) PIMs. Reproduced from [2]. CC BY 4.0.

The primary drawback is that the surface area ($< 2000$ m$^2$ g$^{-1}$) of these nanoporous organic polymers is very low compared to COFs and PAFs, which restricts their ability to absorb hydrogen.

## 6.2 Chemically bound hydrogen

When a hydrogen molecule comes into contact with the surface of a solid, it separates into two hydrogen atoms, which then diffuse in the solid and form a chemical bond with the solid material. This phenomenon is known as chemisorptions. The first information on the reaction of hydrogen with a solid dates from 1866, when Graham observed the absorption of hydrogen in palladium up to 935 times its volume [9]. Today it is known that hydrogen forms chemical compounds known as hydrides with all elements except inert gases (see figure 6.8).

Metal hydrides are a family of hydrogen compounds that are formed when hydrogen and metals combine. Metal hydrides have unique properties due to the nature of the hydrogen–metal bonding, which puts these compounds at the forefront of cutting-edge research and industrial developments [11]. Metal hydrides find great uses in chemical processing (reducing agents, strong bases, strong reductants, catalysts), physical separation processing (desiccants, isotope separation, gas separation, and hydrogen purification), nuclear engineering (neutron moderators, reflectors, and shields), thermal applications (heat pumps), and energy storage (hydrogen fuel tanks and secondary batteries) [11]. Table 6.1 presents a list of the historical events related to the development of metal hydrides and their major applications. In the rest of the section, we will discuss the interesting scientific aspects of metal hydrides in general in connection with hydrogen technology. There is an interesting monograph [10] and several review chapters published in specialized edited books [7–9], and we will follow these closely in the discussions here.

Metal hydrides are the classic chemically bonded hydrogen-storage materials. Under reasonable pressure and temperature conditions, metal hydrides—whether the host metal is an elemental metal, an alloy, or a metal—can function as solid-state

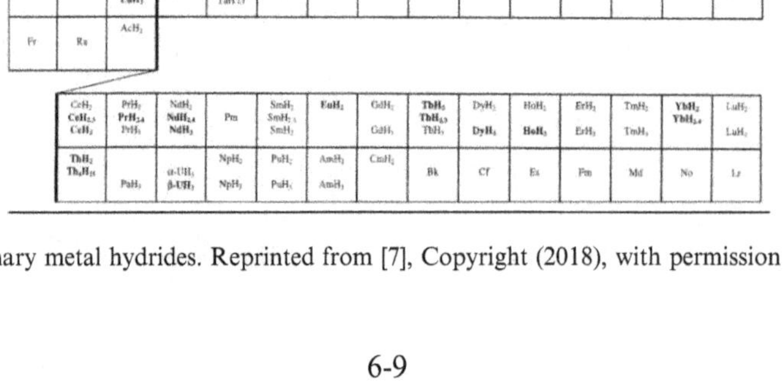

**Figure 6.8.** Binary metal hydrides. Reprinted from [7], Copyright (2018), with permission from Elsevier.

**Table 6.1.** Historical events related to discoveries and applications of metal hydrides. Reprinted from [11], Copyright (2018), with permission from Elsevier.

| Year | Contributor | Event |
|---|---|---|
| 1912 | Jurisch | Studies of hydrogen absorption of Pt and Pt–Au alloys |
| 1923 | Sieverts et al. | Studies of hydrogen absorption of rare-earth metals |
| 1928 | Tammann et al. | Studies of hydrogen absorption of Pd |
| 1936 | de Boer et al. | Studies of hydrogen absorption of Zr |
| 1946 | Burke | Studies of hydrogen absorption of U |
| 1956 | Trzeciak et al. | Hydrides of $AB_2$, $A_2B$ intermetallic alloys |
| 1958 | Libowitz | Reversible hydrogen storage of AB intermetallic alloys |
| 1967 | Pebler et al. | Reversible hydrogen storage of $AB_2$ intermetallic alloys |
| 1967 | Battelle | First NiMH battery using $TiNi+Ti_2Ni$ |
| 1969 | Zijlstra et al. | Reversible hydrogen storage of $AB_5$ intermetallic alloys |
| 1975 | Benz | Metal hydride-based bus (TiFe) |
| 1975 | Argonne National Laboratory | Heat pump as air conditioning unit with metal hydride |
| 1982 | Air Product and Ergenics | Industrial-scale hydrogen purifier with metal hydride |
| 1984 | Willems | AB5 HSA used in NiMH battery |
| 1986 | Tanaka et al. | Demonstration of metal hydride alkaline fuel cell |
| 1989 | Ovonic, Matsushita, Sanyo | NiMH battery commercially available |
| 1991 | Mazda | Monohydride (MH) tank used in a rotary engine powered car |
| 1996 | Toyota | MH tank used in solid-oxide fuel cell powered car |
| 1997 | Kadir et al. | AB3 HSA used in NiMH battery |
| 1999 | General Motor, Ovonic | First commercially available electrical vehicle equipped with NiMH battery |
| 2003 | Sanyo | Commercially available super-lattice HSA-based NiMH battery |

hydrogen-storage material. Therefore, they have a significant safety advantage over the ways of storing hydrogen that are gaseous and liquid. Compared to hydrogen gas (0.99 H atoms $cm^{-3}$) or liquid hydrogen (4.2 H atoms $cm^{-3}$), metal hydrides have a higher hydrogen-storage density (up to 6.5 H atoms $cm^{-3}$ for some materials). For onboard vehicle applications, metal hydride storage is therefore a safe, volume-efficient storage solution [1]. In part II of this book (see chapter 12), we will go into greater detail about the technological features of metal hydrides in relation to hydrogen energy technology.

### 6.2.1 Hydrogen–metal systems

Systems containing hydrogen and metals can exist in a wide variety of phases, each having its own stoichiometry, crystal structure, and electrical characteristics. Typical ionic compounds (e.g., NaH with NaCl type crystal structure), covalent compounds (e.g., discrete tetrahedral $SnH_4$ molecules), or metallic compounds with hydrogen in the interstices of the closest metal atom packing can all be considered as metal hydrides within the classical framework of chemical bonding [9]. In many systems, hydrides have features of different bonding types. As a result, the conventional classification as metallic (interstitial), covalent, and ionic hydrides cannot be applied in this context. Instead, a more natural order based on the location of metals in the periodic table is used.

The following is a summary of the appealing properties of hydrogen in metals [12]:

1. With the lowest atomic mass of all the elements in the periodic table, the behavior of hydrogen in metals can be rigorously studied using quantum mechanics.

2. In metals, the diffusivity of hydrogen is very high. Thus, it is possible to study a variety of kinetic properties associated with hydrogen down to ultra-low temperatures.

Hydrogen has the following options when it comes to bonding with metals: (i) obtaining an additional electron, (ii) giving up its electron, and (iii) sharing the electron with the metal. There are three possible phases for the metal hydride: solid, liquid, and gas. Depending on the type of hydrogen–metal bond, in the solid phase the metal hydride can be a metal, insulator, semimetal, or semiconductor [11]. Table 6.2 provides examples of metal hydrides. The examples are classified according to the electronegativity (EN) of the host metal system, which establishes

Table 6.2. Classification of metal hydrides according to metal type and EN. Reprinted from [11], Copyright (2018), with permission from Elsevier.

| Metal type | Nature of chemical bond | General appearance | Example | Hydrogen dissociation temperature (K) | Heat of formation (kJ mol $H_2^{-1}$) |
|---|---|---|---|---|---|
| | | | | temperature (K) | |
| Element | | | | | |
| EN<1.0 | Ionic | White solid | NaH | 693 | ~113.0 |
| EN 1.1–1.3 | Partial ionic, partial metallic | White solid | $MgH_2$ | 550 ~74.5 | |
| EN 1.3–1.6 | Metallic | Metal | VH | Room temperature ~33.5 | |
| EN > 1.6 | Partial metallic, partial covalent | Colorless solid | $AlH_3$ | 400 ~11.3 | |

(Continued)

**Table 6.2.** (*Continued*)

| Metal type | Nature of chemical bond | General appearance | Example | Hydrogen dissociation temperature (K) | Heat of formation |
|---|---|---|---|---|---|
| EN>1.6 | Covalent | Gas or liquid | GeH$_4$ | 488 | ~45.3 |
| Intermetallic compound | Metallic | Metal powder | LaNi$_5$H$_6$ | Room temperature | ~30.1 |
| Solid solution | Metallic | Metal powder | (TiCrV) H$_2$ | Room temperature | ~39 |
| Complex metal hydride | Partial ionic, partial metallic | White solid | LiBH$_4$ | 643 | ~95.4 |

the nature of the chemical bond and in turn determines the type of metal hydride [11]. Electrons from hydrogen are unlikely to be accepted by metal hydrides including alkali (Group 1a) or alkaline earth metals (Group 2a) with low EN. Ionic solids with a very negative heat of formation are formed by such metal hydrides. They do not conduct electricity and are stable. The EN of the host metal rises as it takes an electron from hydrogen. The metal–hydrogen bond becomes increasingly metallic in this scenario, leading to the emergence of metal-like characteristics. The bond tends to become more like covalent with even higher EN metals [11]. The electron from hydrogen is shared but remains localized, and the hydride acquires the characteristics of insulators similar to other covalent compounds.

The heat of hydride formation ($\Delta H_h$) is one method of measuring the strength of the metal–hydrogen bond [11]. A more stable hydride with a greater bond strength is indicated by a lower (i.e., more negative) $\Delta H_h$ value. LaH$_2$, for instance, has a $\Delta H_h$ of ~208 kJ mol$^{-1}$. Hydrogen can also combine with alloys and complex metals other than elemental metals to form hydrides; these combinations usually result in metallic and insulator-type hydrides, respectively [11].

### 6.2.1.1 Thermodynamics

The majority of binary metal hydrides are produced by a solid-gas reaction between the metal and hydrogen [9]. An idealized pressure-composition isotherm and van't Hoff plot for reaction of hydrogen with a metal or intermetallic complex are shown in figure 6.9. A solid solution of hydrogen in metal ($\alpha$-phase) forms at low hydrogen pressure $p_{H_2}$, where hydrogen occupies interstitial places in the metal host lattice. When the equilibrium pressure $p_{eq}$ is attained, a metal hydride $\beta$-phase (also called $\alpha'$) emerges. The pressure stays constant until the $\alpha$-phase is completely transformed into the $\beta$-phase at the end of the plateau region. The temperature and pressure of the plateau zone determine the conditions for the synthesis of metal hydrides and reversible hydrogen storage in these materials. The maximal hydrogen absorption

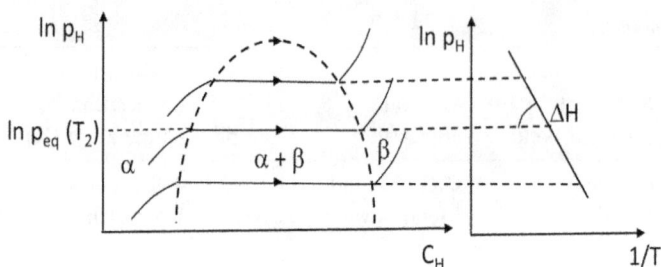

**Figure 6.9.** (a) Pressure-composition isotherms (left-hand), and (b) van't Hoff plot (right-hand) for the reaction of hydrogen with a metal or intermetallic compound. Reprinted from [9], Copyright (2003), with permission from Elsevier.

and the capacity suitable for reversible storage are determined, respectively, by the hydrogen concentration ($c_H$) at the end of the plateau and the length of the plateau region [9]. In the single-phase region $\beta$ beyond the plateau, the pressure increases sharply again with very minor compositional changes. Higher pressure plateau regions may follow this region. Above a critical temperature $T_c$, the two-phase region disappears and the hydrogen concentration in the metal hydride can vary continuously.

From a van't Hoff plot of $\ln p_{eq}$ versus $1/T$, the enthalpy $\Delta H$ of metal hydride production (which is often negative) and the reaction entropy $\Delta S$ can be obtained by applying the formula $\ln p_{eq} = \Delta H/RT - \Delta S/R$, where $R$ is the gas constant (see right-hand side of figure 6.9). Thermodynamical data for a few metal hydrides are presented in table 6.3. The liquid and gaseous hydrogen densities in this table are 70.9 and 0.084 g l$^{-1}$, respectively, and can be used for comparison.

It is sufficient to discuss $\Delta H$ values for the considerations of stability. This is due to the fact that the $\Delta S$, which represents the entropy of hydrogen gas (130 J K mol$^{-1}$) lost during absorption by the metal or intermetallic compound, is generally constant for the majority of systems. Therefore, in general, a $\Delta H$ value of less than $-38$ kJ mol$^{-1}$ is required to meet the necessary condition $\Delta G = \Delta H - T\Delta S < 0$ for hydride formation at room temperature [9]. Here, the free enthalpy of hydride production is denoted by $\Delta G$. However, this idealized depiction of thermodynamics ignores kinetic effects. In real systems, the plateaus are not perfectly flat, and hysteresis is often observed in the absorption and desorption of hydrogen. This is indicative of a nonequilibrium situation. In this situation, purity and homogeneity of the systems play an important role.

### 6.2.1.2 Site occupancy of hydrogen in metals
The x-ray diffraction technique cannot be used for atomistic structure determination of hydrogen–metal systems because the coherent scattering amplitude of x-rays from the hydrogen atom is quite small due to its small atomic number. Ion channeling technique in which the yield of a nuclear reaction between a probing ion beam and hydrogen or deuterium atoms is measured as a function of crystal orientation, and neutron scattering experiments including inelastic neutron scattering which

**Table 6.3.** Thermodynamic and hydrogen-storage properties for selected metal hydrides. Reprinted from [9], Copyright (2003), with permission from Elsevier.

| Hydride | $\Delta H$ (kJ mol$^{-1}$ H$_2$) | Weight fraction of H (%) | H density (g L$^{-1}$) |
|---|---|---|---|
| LiH | −180 | 12.7 | 98 |
| NaH | −112 | 4.2 | 58 |
| MgH$_2$ | −74 | 7.7 | 109 |
| CaH$_2$ | −188 | 4.8 | 93 |
| AlH$_3$ | −8 | 10.1 | 149 |
| TiH$_2$ | −136 | 4.0 | 152 |
| MnH$_{0.5}$ | −16 | 0.9 | 62 |
| FeH$_{0.5}$ | +20 | 0.9 | 59 |
| PdH$_{0.7}$ | −41 | 0.7 | 72 |
| LaH$_2$ | −208 | 1.4 | 73 |
| UH$_3$ | −127 | 1.3 | 137 |
| Mg$_2$FeH$_6$ | −98 | 5.5 | 150 |
| Mg$_2$Co$_H$5 | −86 | 4.5 | 125 |
| Mg$_2$NiH$_4$ | −64 | 3.6 | 97 |
| FeTiH$_2$ | −30 | 1.9 | 96 |
| LaNi$_5$H$_6$ | −31 | 1.4 | 92 |
| ZrCr$_2$H$_{3.8}$ | −96 | 2.0 | 111 |

measures the vibrational mode of hydrogen in crystals, are used to investigate the site occupancy of hydrogen in metals [12].

According to experimental studies, hydrogen atoms are found to reside octahedral (O) or tetrahedral (T) sites in the host metals [12]. Figure 6.10 is an illustration of this. Table 6.4 provides an overview of the hydrogen site occupancies in various hydride and solid-solution phases. In solid-solution phases, the hydrogen atoms occupy the interstitial positions at random. However, in hydride stages, they consistently occupy specific interstitial locations.

Hydrogen atoms generally occupy the T sites in body-centered cubic (BCC) metals. On the other hand, hydrogen atoms prefer to occupy the O sites when the lattice parameter is small. One example of such a hydrogen–metal system is $\beta$-V$_2$H. This is so that the closest neighboring host metal atoms can be pushed away, creating a sizable hole. In the BCC structure, for both O and T site occupancies the strain field formed around the hydrogen atom exhibits tetragonal symmetry [12]. Compared to the T site, where the distances between the four metal atoms surrounding the core hydrogen atom are equal, the O site has a significantly higher tetragonality [12]. Conversely, the T- and O sites in the face-centered cubic (FCC) structure exhibit cubic symmetry. When analyzing the structure of hydrides, the diffusion process, and the internal friction brought on by stress-induced hydrogen jumps, the symmetry surrounding the interstitial site plays an important role.

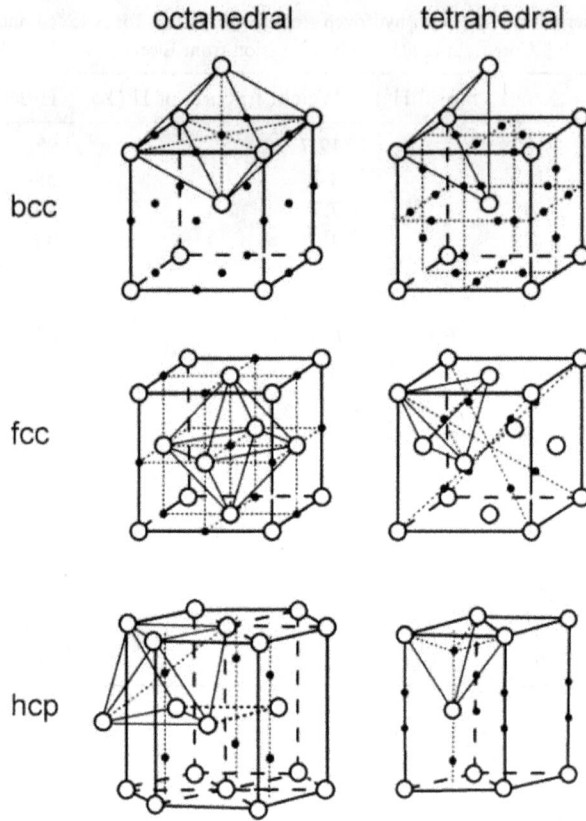

**Figure 6.10.** The octahedral and the tetrahedral sites in BCC, FCC, and HCP metals. Reprinted from [12], Copyright (2001), with permission from Elsevier.

**Table 6.4.** The site occupancy and atomic volume of hydrogen in metals. Reprinted from [10], Copyright (1993), with permission from Springer.

| Structure | Site | Metal/hydride | $\Omega_H$ ($1 \times 10^{-3}$ nm$^3$) |
|-----------|------|---------------|------------------------------------|
| BCC | T | V | 2.64 |
| | | Nb | 3.13 |
| | | Ta | 2.8 |
| | O | $V_2H$ | 2.23 |
| HCP | T | Y | 3 |
| FCC | T | $TiH_2$ | 2.78 |
| | | $VH_2$ | 3 |
| | | $NbH_2$ | 2.67 |
| | | $CeH_2$ | 4.48 |
| | O | Ni | 2.98 |
| | | Pd | 2.8 |
| | | RhH | 2.42 |

Atoms of hydrogen cannot approach one another within a certain atomic distance. Empirical evidence indicates that the minimal approach distance is 0.21 nm [12]. This results in the situation where hydrogen atoms cannot occupy all T-sites in BCC metals at once. All metals and hydrides frequently exhibit this phenomenon, which is known as blocking [12].

### 6.2.1.3 Lattice expansion

A hydrogen atom placed in an interstitial site of a metallic solid pushes away metal atoms. This gives rise to a dilatational strain field around the hydrogen atom, leading to an expansion of the metal lattice as a whole. Let there be a change in the lattice parameter $a$ of the metallic solid by $\Delta a$ after the introduction of hydrogen by a concentration increment of $\Delta c$. Then, $\lambda = \Delta a/(a\Delta c)$ is the expression for the corresponding strain $\lambda$ per unit concentration. In this case, the hydrogen-to-metal atom ratio $c = [H]/[M]$ gives the concentration $c$. The strain field is isotropic in the case of O- and T-occupancies in the FCC lattice, and $\lambda$ is independent of the crystal orientation [12]. The volume change resulting from hydrogen in this case is given by the following expression: $\Delta V/V = 3\,\Delta a/a = 3\,\lambda\Delta c = (\Omega_H/\Omega)\Delta C$, where $\Omega_H$ and $\Omega$ denote the respective atomic volumes of hydrogen and the metal.

The strain field surrounding a hydrogen atom occupying the T- or O-site in BCC metals is not isotropic. An example of O-site occupancy is provided in figure 6.11. If the hydrogen atom is located at site 3 in figure 6.11, then the two nearest-neighbor metal atoms in the $z$ direction are pushed away from the hydrogen atom. On the other hand, the four next-nearest-neighbor atoms in the (x, y), (x,-y), (-x, y), and (-x,-y) directions are slightly pulled in. The amount of displacement in the $V_H$ hydride is 19.5% in the $z$ direction, but $-2.56\%$ along the $x$ and $y$ directions. In figure 6.11, the length is extended along $x$ and $y$ directions by the hydrogen atoms at sites 1 and 2, respectively. As a result, there are three types of O-sites in the BCC lattice of a hydrogen–metal system [12]. Therefore, when the strain surrounding hydrogen is anisotropic, the strain parameter $\lambda$ needs to be regarded as a tensor quantity. If the metal crystal contains hydrogen with concentration $c$, the total strain $\epsilon_{ij}$ can be expressed as [12]:

$$\epsilon_{ij} = \sum_{p-1}^{n_d} \lambda_{ij}^{p} c_P \tag{6.1}$$

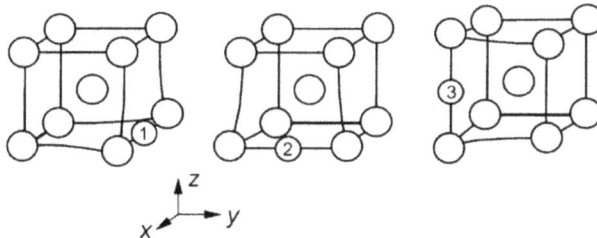

**Figure 6.11.** The strain field due to hydrogen in the O-site. Reprinted from [12], Copyright (2001), with permission from Elsevier.

$$c = \sum_{p-1}^{n_d} c_P \qquad (6.2)$$

Here $n_d$ is the number of types of interstitial sites, $p$ is the index to distinguish the types, and $c_p$ is the hydrogen concentration at the sites indexed with $p$. In the BCC lattice, $n_d = 3$ for the O-sites, and the $\lambda$ tensors are expressed as:

$$\lambda^{(1)} = \begin{pmatrix} \lambda_1 & 0 & 0 \\ 0 & \lambda_2 & 0 \\ 0 & 0 & \lambda_2 \end{pmatrix} \qquad (6.3)$$

$$\lambda^{(2)} = \begin{pmatrix} \lambda_2 & 0 & 0 \\ 0 & \lambda_1 & 0 \\ 0 & 0 & \lambda_2 \end{pmatrix} \qquad (6.4)$$

$$\lambda^{(3)} = \begin{pmatrix} \lambda_2 & 0 & 0 \\ 0 & \lambda_2 & 0 \\ 0 & 0 & \lambda_1 \end{pmatrix} \qquad (6.5)$$

The volume expansion is then expressed as the trace of the $\lambda$ tensor [12]:

$$\Omega_H/\Omega = \lambda_{11} + \lambda_{22} + \lambda_{22} = \lambda_1 + 2\lambda_2 \qquad (6.6)$$

The volume expansions for various metal–hydrogen systems are summarized by Fukai and are presented in table 6.4. The magnitude of $\Omega_H$ is approximately constant in ordinary metals (except for the rare-earth metals), i.e., $\Omega_H = (2.6 \pm 0.5) \times 10^{-3}$ $nm^{-3}$. The $\Omega_H$ for the rare-earth metals is considerably larger, i.e., $\Omega_H = (4.0 \pm 0.5) \times 10^{-3}$ $nm^{-3}$. This large difference in $\Omega_H$ has been attributed to the difference in the electronic states [12].

### 6.2.1.4 Vibrational state of hydrogen

The hydrogen atom at its interstitial site in a metal performs a vibrational motion in a potential well having a minimum at the particular O or T-site. Inelastic neutron scattering can be used experimentally to study this vibrating state of hydrogen. Energy transfers between neutrons and hydrogen atoms take place when a neutron beam passes through a crystal containing hydrogen [12]. The energy of scattered neutrons is subsequently changed from their initial energy state as a result. The vibrational states of hydrogen are quantized, and hydrogen interacts with neutrons to get excited to higher energy levels. As a result, the energy spectrum of the scattered neutrons displays a sequence of peaks that are almost evenly spaced apart. The vibrational excitation energy of interstitial hydrogen in various metals is summarized in table 6.5. The BCC structure has two vibration energies. In addition, when compared to the FCC structure, its energy spectra are somewhat more complicated. This is correlated to the fact that the hydrogen potential energy strongly depends on the orientation of the crystal. It is observed that vibrational

**Table 6.5.** Vibrational excitation energy of interstitial hydrogen in metals. Reprinted from [10], Copyright (1993), with permission from Springer.

| Crystal structure | Metal/ hydride | Composition [H]/ [M] | Cite occupancy | Vibrational energies (meV) |
|---|---|---|---|---|
| FCC | Ni | 0.75 | O | 88 |
| FCC | Pd | 0.02-0.014 | O | 66 |
| FCC | PdH | 0.068 | O | 56 |
| FCC | $TiH_2$ | 2 | T | 147.6 |
| BCC | V | 0.012 | T | 170 |
| BCT | $V_2H$ | 0.51 | O | 56 |
| BCC | Nb | 0.03 | T | 163 |
| BCC | NbH | 0.75 | T | 167 |

energies of the O-site are lower than that of the T-site [12]. This is as a result of larger interstitial volume and milder potential slope of the O-site than the T-site. Thus, the analysis of vibration energy can be used to determine the site occupancy.

### 6.2.1.5 Hydrogen as a lattice gas and order–disorder transitions

Interstitial hydrides can be modeled as a host-guest systems as a first approximation, where hydrogen can be treated by a lattice gas model [9]. In the metal hydride, hydrogen can migrate nearly freely. Similar to the behavior of pair potential for gas particles, the primary H–H interactions are long-range attractive forces and repulsive in short-range. At room temperature, self-diffusion constants of hydrogen in the different metallic mediums are high; for example, $D = 4 \times 10^{-4}$ mm$^2$ s$^{-1}$ for hydrogen in PdH $_{0.7}$, which is comparable to that of protons in water. It is possible to vary hydrogen concentration continuously, i.e., a solid-solution $MH_x$ is formed with $x$ covering a broad stoichiometric range [9]. Two distinct phases of different density can exist in equilibrium in the real hydride systems below a critical temperature $T_c$: two phases $MH_x$ and $MH_y$ with nonoverlapping solid-solution regions $x$ and $y$.

As an example, figures 6.12 and 6.13 present the phase diagrams of the systems Pd–H and $ZrCr_2$–D, respectively, to illustrate these phenomena. The uniform Pd–H phase separates into two phases, $\alpha$, and $\beta$ below $T_c = 570$ K, with a miscibility gap in between [9]. The same features are also observed in $ZrCr_2$–D with $T_c = 350$ K. Short-range order effects drive this lattice gas to liquid transition. In the lattice liquid or gas phases, low occupancies of hydrogen in interstices are often observed, i.e., hydrogen is disordered statistically. The interstices are, however, not filled in a random fashion. The H–H repulsion introduces a short-range order, which blocks the nearest-neighbor sites around each hydrogen atom within a radius of 210 pm. This is revealed in neutron diffraction studies of disordered metal hydrides in the form of a very broad 'liquid-like' peak at $d = 210$ pm. This kind of hydrogen substructure has a configurational entropy different from zero. They are likely to

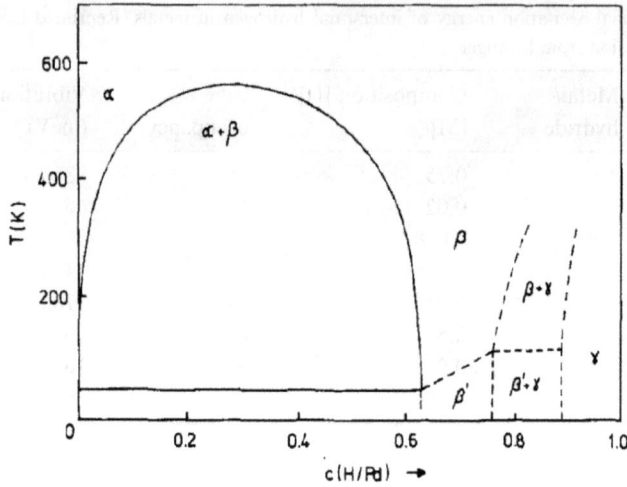

**Figure 6.12.** Phase diagrams of the systems Pd–H. $T$ stands for temperature and $c$(H/Pd), for concentration of hydrogen as atom ratio H/Pd; $\alpha$, $\beta$, and $\gamma$ are palladium hydride phases with disordered H distribution, and $\beta'$ is palladium hydride phase possibly with ordered H distribution. Reprinted from [13], Copyright (1976), with permission from Elsevier.

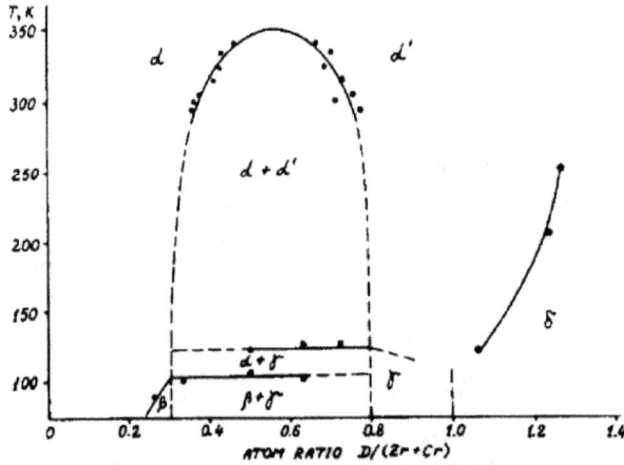

**Figure 6.13.** Phase diagrams of the systems ZrCr$_2$–D. $T$ stands for temperature; $\alpha$ and $\alpha'$ for zirconium dichromium hydride phases with disordered H distribution; $\beta'$, $\gamma$, and $\delta$ are zirconium di=chromium hydride phases with ordered H distribution. Reprinted from [14], Copyright (1984), with permission from Elsevier.

remain unstable at low temperatures. As a consequence, a further transition occurs, which can be described as a crystallization of the lattice liquid [9]. Here, the repulsive interaction of neighboring hydrogen start playing a dominant role, and a long-range order in the hydrogen sublattice is introduced. The 'liquid-like' peak revealed in the neutron diffraction studies vanishes on the ordering of the hydrogen substructure.

This ordering of hydrogen is accompanied by a reduction in symmetry and often a drastic drop in the mobility of hydrogen.

Within specific bounds, this lattice gas approach for hydrogen holds good for typical metallic hydrides, both binaries and ternaries [9]. Any departure from this idealized frame suggests that the metallic hydrides are actually chemical compounds rather than pure host-guest systems. Phase transitions provide a key distinction between real metal hydrides and the model of hydrogen as a lattice gas, liquid, or solid. Within the framework of Landau theory of phase transition [15], the crystallization of a material is a first-order transition phase transition. The order–disorder transition in a hydride may, however, be first- or second-order transition. The structural connections between ordered and disordered phases of metal hydrides have been proven in many cases by the crystallographic group–subgroup relationships. This indicates the possibility of continuous second-order phase transition [9].

### 6.2.2 Physical properties of hydrogen–metal systems

The electronic, magnetic, and thermal properties of a metal or an intermetallic compound are affected when hydrogen penetrates its crystal lattice structure. The main effects are that of the generally observed lattice expansion (which we have discussed earlier in this section), the electronic interaction between hydrogen and the neighboring metal atoms (metal–hydrogen bonding), and the hydrogen-hydrogen interactions. Additionally, the addition of hydrogen can modify electrical and magnetic properties in a number of ways. Metal-semiconductor transitions, ferromagnetic or antiferromagnetic order, the diamagnetic transformation of paramagnetic metals, the ferromagnetic semiconductor transformation of antiferromagnetic metals, and the change of metal valences can all be brought about by hydrogenation [9].

*6.2.2.1 Mechanical properties*

The mechanical properties of materials are degraded by the formation of metal hydrides. This is a significant issue in the field of engineering. High internal pressure results from the precipitation of $H_2$ in the gaps and cracks inside a material structure, and the development of hydrides in high-stress regions reduces a cohesion of a material.

Internal strain and stress are frequently introduced by inhomogeneities during hydrogenation. On the other hand, inhomogeneities in hydrogen concentration may be introduced by mechanical stress [9]. The latter effect is used to study long-range hydrogen diffusion, where a sample is mechanically deformed and the hydrogen redistribution is examined afterwards. The metal or intermetallic combination is cracked into a fine powder by the lattice expansion that occurs during the formation of a metal hydride. Compared to $LaNi_5$, the volume effect following hydrogenation for the complete hydride $LaNi_5H_7$ can reach up to 25%. In the system $CeRu_2H_5$, volume change can be up to 37% due to a valence change for Ce. Generally, for every atom of hydrogen absorbed, there is a volume increase of $2 - 3 \times 10^6 \text{ pm}^3$ [9].

Repeated cycles of hydrogen absorption and desorption may result in phase segregation. Magnetic hyperfine splitting in $^{63}$Ni Mössbauer spectroscopy revealed the creation of very tiny Ni clusters in $LaNi_5$ powder cycled 1500 times. Cycled materials have a high surface area, which makes them more sensitive to contaminants such as water or oxygen.

### 6.2.2.2 Electrical properties

Throughout the hydrogenation process, the resistivity of metal hydrides may increase or decrease. For instance, in $LaNi_5$, hydrogenation results in the formation of the $H^-$ ion, which lowers the conduction electron count in metal hydride and raises electrical resistivity. The M–H band in Pd is located close to the lower edge of the conduction band [11]. As a result, when Pd is hydrogenated, the conduction electron density decreases and the resistivity is increased. Other important influencing factors on electrical resistivity may include lattice expansion, the randomization of hydrogen in the hydride, and changes in the lattice structure that reconstruct the Brillouin zone, as in the case of Ti and V.

### 6.2.2.3 Magnetic properties

Numerous changes in bulk magnetic properties on hydride formation might result from the complex interplay between volume expansion, or the changing of interatomic distances, and the electron density of states at the Fermi surface. The same cubic structure is crystallized in ferrimagnetic $Y_6Mn_{23}$ and Pauli-paramagnetic $Th_6Mn_{23}$. Interestingly, $Th_6Mn_{23}$ turns into a ferromagnet upon hydrogenation, whereas $Y_6Mn_{23}$ loses its magnetic character. A structural transformation (cubic-tetragonal) that occurs in $Y_6Mn_{23}H_{30}$, but not in the analogous Th compound, was used to explain this distinct behavior [9]. When cerium forms hydrides, it frequently shifts its valency from IV+ to a III+ in compounds formed. This valence shift leads to the emergence of ferromagnetism in Pauli-paramagnetic $CeNi_3$ by hydrogenation.

### 6.2.2.4 Optical properties

The hydrogen content of metal hydrides can drastically alter their optical properties. For instance, the trihydride $MH_3$ ($M^{3+}(H^-)_3$) made of a metal with a 3p oxidation state differs in transparency from the dihydride $MH_2$ ($M^{3+} e^- (H^-)_2$) of the same metal [11]. Incoming light is reflected from $MH_2$ because of the extra electron that makes it more metal-like. $MH_3$; on the other hand, it has a huge optical band gap and acts like an insulator, allowing visible light to pass through. Utilizing this shift in optical characteristics, switchable mirrors can be created [11].

### 6.2.2.5 Electronic structure

It can be assumed within the rigid band approximation that the energy bands in the metal or intermetallic complex stay unaltered upon the introduction of hydrogen [16]. This results in: (i) the protonic model for hydrogen if the hydrogen $1s$ band lies above the conduction band in the metal or intermetallic compound with hydrogen being an electron donor; (ii) the anionic model if it lies below with hydrogen as electron acceptor. As a result, the concentration of free electron carriers ought to increase or decrease, accordingly.

A more realistic approach takes the hydrogen-induced changes of the conduction band into account [9]. Metal hydride research is increasingly employing quantum mechanical computations. The nature of the chemical bonding between metal hydrides, the preferred hydrogen sites, and the reasons limiting the hydrogen capacity of storage materials are among the topics of interest. Semiempirical self-consistent methods such as Linearized Muffin-Tin Orbital (LMTO) and Linearized Augmented Plane-Wave (LAPW) are often employed based on the complexity of the crystal structure [11].

Photoelectron spectroscopy experiments together with quantum mechanical calculations have revealed that hydrogen indeed strongly influences the metal conduction band and induces new low-lying energy states a few electron volts (eV) below the Fermi level $E_F$. The difficulty of synthesizing good quality single crystals and the limited thermodynamic stability under experimental conditions such as ultrahigh vacuum make it challenging to use experimental techniques like x-ray absorption near-edge structure spectroscopy and photoelectron spectroscopy [11]. Diffraction techniques yield a fairly detailed image of the crystal structure averaged over time and space, but they are unable to reveal details about local structures and coordinated dynamics in materials. The local structure and dynamics of hydrogen can be studied with nuclear magnetic resonance, infrared, Raman, Mössbauer, and muon spin rotation spectroscopy. Metal–hydrogen and hydrogen-hydrogen interactions in metal hydrides are best studied with inelastic neutron scattering (INS). Because of the strong incoherent scattering of $^1H$, INS can also be used to locate hydrogen sites in hydrides with very low hydrogen concentrations, where diffraction techniques are not effective.

Electronic structural considerations frequently provide an explanation for the maximum hydrogen content of hydrides. The prediction that Pd can accept 0.76 additional electrons fits well with the observation that Pd readily takes up hydrogen to form $PdH_{0.7}$ [9]. The metal–hydrogen bonding in actinide base hydrides appears to be more influenced by the interaction between hydrogen-1 $s$ and actinide metal-5$f$ electrons, in comparison to the hydrogen-$s$, metal-$d$ overlap in transition-metal and rare-earth hydrides.

### 6.2.2.6 Characterization of metal hydrides

Transition metal-based metal hydrides typically form as a metal powder. The hydrogen diffusion constant, surface area, and particle size distribution are crucial physical characteristics to describe these powders. These qualities can be measured using a variety of techniques, which can be cross-checked and compared. For instance, particle size distribution (PSD) can be carried out by laser diffraction, graduated sieving, air elutriation analysis, or scanning electron microscope (SEM) micrograph/optical counting methods [11]. A variety of surface area analyzers are available for use in measuring surface area [11]. Measures of membrane permeability, ion implantation with secondary ion mass spectrometry, electrochemical techniques, internal friction measures, Gorsky-effect relaxation tests, and direct absorption methods can all be used to determine the hydrogen diffusion constant in a metal.

Other significant properties are hydrogen kinetics and reaction heat. Differential scanning calorimetry can be used to measure the reaction heat directly or to calculate it thermodynamically [11]. Thermo-gravimetric analysis can be used to compare the hydrogen absorption/desorption kinetics in more detail. Another significant physical property is hardness, whose measurement is helpful in determining how defects form in metal hydrides.

## 6.3 Different classes of metal hydrides

Summarizing the discussion so far, we can say that hydrogen with its electronic configuration $1s^1$ can form bonds to metals in different ways. Receiving a further electron leads to a stable helium $1s^1$ configuration. This is possible while combining with very electropositive metals. Because of its charge, the resultant hydride anion $H^-$ is extremely deformable and less stable than helium. In order to create a covalent $\sigma$-bond with less electropositive metals, hydrogen can provide its $s$-electron to them. Finally, hydrogen has the ability to create interstitial metal hydrides of varying compositions with a metallic character [11]. These different bonding mechanisms may also be combined in many ternary hydrides. The crystal structure and other characteristics of metal hydrides are mostly determined by the type of chemical bond present in the compounds. We shall briefly introduce below the main classes of metal hydrides.

### 6.3.1 Metal hydrides from elemental metals

Hydrogen may combine with a wide range of elements to create binary metal hydrides. Figure 6.8 presents examples of the wide variety of such metal hydrides. While metals from Group 3a form trihydrides ($MH_3$) [11], elements of alkali metals and alkaline earth metals form monohydrides (MH) and dihydrides ($MH_2$), respectively. Transition metal-based hydrides range in hydrogen-to-metal ratio (H/M) from 0.2 to 3.75, with $MH_2$ and $MH_3$ being the prominent members. As we have discussed in section 6.2.1 and shown in table 6.2, these metal hydrides can be classified into ionic, metallic, and covalent metal hydrides, according to the EN of the host metal.

#### 6.3.1.1 Ionic metal hydrides

When hydrogen combines with alkali and alkaline earth metals, atomic hydrogen will receive one electron from the host metal in order to produce the $H^-$-ion, which has a complete outer shell of electrons (1s), and an ionic bond is formed [11]. The crystal structures of the ionic metal hydrides are identical to NaCl, and their chemical and physical properties are quite similar. The melting and boiling points are high for ionic metal hydrides. Under normal circumstances, they typically appear as a white powder or as a brittle solid. One example of an ionic metal hydride that is a potent reductant is sodium hydride, which releases hydrogen gas and forms hydroxide when it reacts irreversibly with water [11]:

$$NaH + H_2O \rightarrow NaOH + H_2 \tag{6.7}$$

In moist air, the powdered ionic metal hydride burns on its own and produces a mixture of products, some of which are nitrogenous compounds. In addition, they react readily with low-humidity air to generate hydroxide, oxide, and carbonate.

### 6.3.1.2 Metal hydrides with metallic character

Metal hydrides originating from actinides, lanthanides, and transition metals generally exist in the form of conductive black powder. At the microscopic rather atomic level, electron energy levels from hydrogen and the host metal overlap. The hydrogen and metal are said to be alloyed, and they share delocalized electrons characterized by a metallic bonding. As shown in figure 6.10, the host metal matrix contains hydrogen atoms at the occupation sites in either an octahedral or a tetrahedral configuration, formed by six and four host metal atoms, respectively.

The quantum tunneling process, in which protons can hop between neighboring occupation sites, allows hydrogen to diffuse into the host metal [11]. When the host metals in a single row of elements (left-hand side of the periodic table) have a lower atomic number, metal hydrides from the transition-metal series have stronger metal–hydrogen bonds [11]. Because of the increased conduction electron density, metal hydrides from heavier transition metals in the same row (going towards the right-hand side) have weaker metal–hydrogen bonds. Elements such as Ti, V, Zr, Nb, Hf, and Ta on the left-hand side of the transition-metal rows are referred to as hydride formers, whereas the remaining elements are called hydride modifiers. The $\Delta H_h$ of hydrides containing hydride formers are strongly negative. On the other hand, when hydrides have hydride modifiers, $\Delta H_h$ is either near zero or positive.

### 6.3.1.3 Covalent metal hydrides

This class of metal hydrides usually consists of covalent compounds with low melting points. They can exist at standard pressures and temperatures as a gas, liquid, or solid. Boron and aluminum-containing metal hydrides are in this group. At room temperature, diborane ($B_2H_6$) is a colorless gas with a sweet yet repulsive smell. Diborane, when combined with air, produces an explosive mixture that spontaneously ignites in moist air at ambient temperature. Additionally, it interacts with water to produce hydrogen and boric acid [11]:

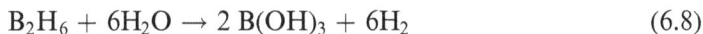

$$B_2H_6 + 6H_2O \rightarrow 2\,B(OH)_3 + 6H_2 \qquad (6.8)$$

Other boron hydrides with varying stoichiometry can also be produced. Some are stable ($B_5H_9$, $B_6H_{10}$, $B_6H_{12}$ $B_6H_{15}$, $B_6H_{16}$, $B_{20}H_{16}$), some are unstable ($B_4H_{10}$), and some very unstable ($B_5H_{11}$). The hydride of Al, $AlH_3$ is a colorless pyrophoric solid. It is known as alane, and can be produced by the following reaction [11]:

$$2LiAlH_4 + BeCl_2 \rightarrow 2AlH_3 + LiBeH_2Cl_2 \qquad (6.9)$$

$\alpha$-alane, $\alpha'$-alane, $\beta$-alane, $\delta$-alane, $\epsilon$-alane, $\theta$-alane, and $\gamma$-alane are among the several known polymorphs of aluminum hydride. While $\gamma$-alane forms a bundle of fused needles, $\alpha'$-alane forms needle-like crystals, and $\alpha$-alane has a cubic or rhombohedral shape [11].

### 6.3.2 Metal hydrides from alloys

Alloy-derived metal hydrides are more versatile in terms of their composition and structure. Binary alloys, which are composed of two metallic elements, are the most basic combination. The nature of the alloy depends on its heat of formation ($\Delta H$) [11]. When $\Delta H$ is very positive, the two metallic elements in the alloys are immiscibly when melted and stay separated during cooling. An eutectic reaction takes place during cooling if the alloy's $\Delta H$ value decreases to zero. An alloy with mixtures of each element is then formed with limited solubility. It is possible to create a solid solution with a broad range of compositions between the two elements by further decreasing $\Delta H$ into the negative range. Ultimately, an intermetallic compound with a fixed stoichiometry is created when $\Delta H$ becomes highly negative.

The number of constituent elements can be used to categorize intermetallic compounds (IMC). They include binary, ternary, quaternary, and more complex systems. The two constituent elements in binary IMCs typically differ greatly in terms of electronegativity and size. The size of the bonding atoms will typically change toward an optimum ratio when electron transfer occurs. The reduction of the unit cell size in IMCs is caused by the electron transfer from the less electronegative element (usually the larger one) to the more electronegative element [11]. This charge transfer alters the local electron density in the host alloy, which is important for hydrogen storage.

IMCs can have a stoichiometry that is narrow or broad. For instance, every IMC that forms between La and Ni in the La–Ni binary phase diagram has a unique stoichiometry. All IMCs in this phase diagram exist at vertical lines [11]. Conversely, in the Zr–Ni binary phase diagram, the IMCs $ZrNi_5$, $ZrNi_3$, and $Zr_5Ni_{10}$ exist at ranges of composition. Its solubility, which is associated with defect formation and lattice-adjusted contraction, is determined by the width of the composition range within a given IMC [11].

Any excess A atoms added to an IMC containing A and B elements of an ideal stoichiometry $A_mB_n$ migrate to sites where B atoms occur or generate vacancies at the B-sites, while preserving the same crystal structure. An anti-site defect is created by the former event, and a vacancy defect by the latter. Examples of defects originating from compositions with off-stoichiometry include vacancy defects and anti-site defects. An off-stoichiometric structure may not have the ideal radii ratio ($r_A/r_B$), but the defect formations will modify the A-B distance, leading to a lattice that is almost ideal. For instance, the Laves phase IMC $TiCr_2$ has a higher solubility for the A-rich side of the $AB_2$ stoichiometry, while having a radii ratio of 1.225, which is less than the ideal ratio of $\sqrt{1.5}$.

### 6.3.2.1 Hydrides from binary intermetallic compounds

The binary phase diagram of the constituent metals is the starting point for the search for binary IMCs. For example, in the La–Ni binary phase diagram, there exist eight IMCs: $La_3Ni$, $La_7Ni_3$, $LaNi$, $La_2Ni_3$, $LaNi_2$, $LaNi_3$, $La_2Ni_7$, and $LaNi_5$. As discussed earlier, the heat of hydride formation ($\Delta H$) of an IMC is a measure of metal–hydrogen bond strength. It can be estimated from the $\Delta H$ of the constituent elements using the following expression [11]:

$$\Delta H(A_m B_n) \approx \frac{m}{m + n}\Delta H(A) + \frac{n}{m + n}\Delta H(B) \qquad (6.10)$$

Here A is generally a larger atom from the alkaline earth, rare-earth, or transition-metal elements, and B is a smaller one, usually a transition-metal element. Useful binary IMC systems for hydrogen storage include $A_3B$ ($Nb_3Sn$), $A_2B$ ($Mg_2Ni$), AB (TiFe), $AB_2$ ($TiMn_2$), $AB_3$ ($LaNi_3$), $A_2B_7$ ($Nd_2Ni_7$), $A_6B_{23}$ ($Ho_6Fe_{23}$), and $AB_5$ ($LaNi_5$). There are possibilities of more than one stable crystal structure in each of these systems. The IMC $AB_2$, also known as Laves phases, can exist in hexagonal C14, FCC C15, and hexagonal C36 structures. The electron density near to the edge of the Brillouin zone, which is influenced by the atomic size and number of outer-shell electrons in the component elements, is the primary determinant of the stable crystal structure that forms at room temperature [11]. The lattice structure of metal hydride may change during hydride formation in order to minimize the electron energy near to the Fermi level. One such instance is the transition of the crystal structure from BCC to body-centered-tetragonal (BCT) to FCC in the hydride of TiCr alloy.

### 6.3.2.2 Hydrides from pseudo-binary intermetallic compounds
Intermetallic compounds suitable for a particular application may not always be found on the binary phase diagram. In such situations, the specifically targeted IMC can be synthesized by partial substitution of A, B, or both elements for elements containing the same or similar number of outer-shell electrons. The heat of formation of such pseudo-binary IMCs can be estimated by the molar ratio and $\Delta H$ of the constituent elements. The commercially available $AB_2$ compounds for NiMH batteries are examples of pseudo-binary IMCs. According to the theoretical estimations, the stoichiometry of $TiNi_2$ is suitable for room-temperature NiMH battery applications [11]. However, this phase does not exist on the Ti–Ni phase diagram. There is no $ZrNi_2$ phase either in the Zr–Ni phase diagram. $TiNi_2$ and $ZrNi_2$ have too large an electron density to form one of the stable $AB_2$ Laves phases, which are excellent materials for storing hydrogen. When Ni is substituted with transition metals such as Cr and Mn that have fewer electrons, a Laves-phase compound called $AB_2$ is formed. Another example is compound $A_2B_7$, where Mg is utilized to modify the metal–hydrogen bond strength for a more appropriate $\Delta H$.

### 6.3.2.3 Hydrides from ternary intermetallic compounds
Progress in the search for pseudo-ternary IMC is relatively slow since there are not enough complete ternary phase diagrams for metal systems, even though some ternary alloy systems are appropriate for hydrogen-storage applications. On the other hand, ternary IMCs are usually present in multiphase IMCs as a secondary phase. $MgAlNi_4$, $MgLaNi_4$, $LaMg_2Ni_9$, LaNiSn, and YNiSn are a few examples [11].

### 6.3.2.4 Hydrides from metal solid solutions

Metal solid solutions can have a wide range of compositions. This is in contrast to the rather narrow range of intermetallic compounds. Most of the metal solid solutions capable of storing hydrogen contain vanadium (V). An example of such a metal hydride originating from a metal solid solution is CrVTi. The ternary solid solution derived from the elements Cr and Ti has a broad composition range, and V and TiCr share the same BCC structure [11]. However, in the composition range of interest to the present context (about Cr:V: Ti = 1:1:1), the stable crystal structure at room temperature is an FCC C15 Laves phase. It is required to anneal the alloy at 1650 K and a subsequent fast quench to obtain the ternary solid solution in the BCC phase. In the reaction between hydrogen and this BCC alloy, the crystal structure changes from BCC to BCT to FCC at saturation. The pressure plateau of the BCT-FCC transition is around 0.1 MPa. In this pressure range, hydrogen storage is reversible. At room temperature, however, the pressure plateau of the BCC-BCT transition is relatively low. This makes it difficult to get the stored hydrogen released from the alloy.

### 6.3.3 Transition metal hydride complexes

Metal hydrides originating from transition metals do not always form metallic M-H bonds. Transition metal hydride complexes can also have other bond structures. Examples are the ionic bonding in $HCo(CO)_4$ and $H_2Fe(CO)_4$, and the covalent bonds in molecular homoleptic metal hydrides containing $[ReH_9]^{2-}$ and $[FeH_6]^{4-}$. A complex transition-metal hydride can be formed by combining an anionic homoleptic transition-metal (M) hydrido complex $[MmHh]^{x-}$ with a cation, $A^+$ or $A^{2+}$ (A = Li–Cs, Mg–Ba, Eu, Yb) [11]. Hydrogen is covalently bonded to the metal M in the $[Mm Hh]^{x-}$ complex, and the attractive electrostatic Coulomb interactions between the cations and complex anions promote the formation of an extended solid. In complex transition-metal hydrides, the hydride fluoride analogs of ionic hydrides are less common. However, certain hydrides exhibit structural similarities with their corresponding halides.

The hydride complexes adhere to the 18-electron rule from coordination chemistry. For instance, $[ReH_9]^{2-}$ in $K_2ReH_9$ or $[NiH_4]^{4-}$ in $CaMgNiH_4$. They may also contain free hydride anions $H^-$ that are not part of the complex anions, as in $K_3PtH_5$ ($= (K^+)_3[PtH_4]^{2-} H^-$). Depending on their bonding properties, complex transition-metal hydrides are stoichiometric, electron-precise compounds that are frequently colored, nonmetallic solids with ordered hydrogen distribution. The high mobility of hydrogen as a ligand in some of these compounds may cause a transition to a disordered high-temperature phase. Metal-metal interactions can happen in compounds that are rich in metals and have less electropositive components for A (= Li, Mg). As a result, metal hydrides form that are bordering more on the metallic side. Conversely, the metal hydrides have a more salt-like character with heavier, electropositive metals for A (= Rb, Cs, Ba).

### 6.3.4 Amorphous metal hydrides

Depending on the range of their ordered structures, solids are classified as crystalline and amorphous materials [15]. Typical amorphous materials are produced through a quick-quench method and/or the addition of glassifying elements, such as boron and silicon [11]. Window glass is a common example of an amorphous material. An amorphous material does not have long-range order in both composition and structure. Crystalline and amorphous materials are placed at opposite ends of the degree of disorder spectrum. Somewhere between these two extremes lie the multi-element and multiphase metal hydrides. As the degree of disorder in the metal hydride increases, the following trends are observed [11]:

1. The hydrogen-storage capacity increases.
2. The plateau region in the pressure-concentration isotherm shortens, and the curve becomes more slanted.
3. Less lattice expansion due to hydrogenation occurs and results in less pulverization during hydride/dehydride cycling.
4. More surface-active sites are available to facilitate chemical/electrochemical reactions.

In addition to hydrides originating from amorphous host metals, hydrogen-induced amorphization can also result in the production of amorphous metal hydrides during hydrogenation. The atoms in this later mechanism are moved out of their places in the lattice by the heat produced by the exothermic hydrogen absorption process. The C15 phase $AB_2$ alloy with an atomic radii ratio of ($r_A/r_B$) >1.37, is an example of an alloy that experiences hydrogen-induced amorphization. The enthalpy difference between the occupancies of hydrogen in crystalline and amorphous states drives this process.

# References

[1] Prabhukhot R P, Wagh R M, Mahesh M and Gangal C A 2016 *Adv. Energy Power* **4** 11
[2] Zheng J *et al* 2021 *Research* **2021** 3750689
[3] Zacharia R and Rather S 2015 *J. Nanomater.* **2015** 914845
[4] Dighe A V, Nemade R Y and Singh M R 2019 Modeling and Simulation of Crystallization of Metal–Organic Frameworks *Processes* **7** 527
[5] Omar K F and Joseph T H 2010 *Accounts of Chemical Research* **43** 1166–75
[6] Sang S H, Hiroyasu F, Omar M Y and William A G III 2008 *Journal of the American Chemical Society* **130** 11580–1
[7] Konstas K *et al* 2012 Lithiated Porous Aromatic Frameworks with Exceptional Gas Storage Capacity *Angew. Chem. Int. Ed.* **51** 6639–42
[8] Rosi N L *et al* 2003 *Science* **300** 1127
[9] Kohlmann H 2003 *Encyclopedia of Physical Science and Technology* 3rd edn (Amsterdam: Elsevier) 441 p
[10] Fukai Y 1993 *The Metal Hydrogen Systems* (Berlin: Springer)
[11] Young K 2018 *Module in Chemistry, Molecular Sciences and Chemical Engineering* (Amsterdam: Elsevier)

[12] Tanaka K and Yoshinari O 2001 *Encyclopedia of Materials: Science and Technology* 2nd edn (Amsterdam: Elsevier) 3905 p

[13] Jacobs J K and Manchester F D 1976 *L. Less. Com. Met.* **49** 67

[14] Somenkov V A and Irodova A V 1984 *L. Less. Com. Met.* **101** 481

[15] Chaikin P M and Lubensky T C 1995 *Principle of Condensed Matter Physics* (Cambridge: Cambridge University Press)

[16] Singleton J 2001 *Band Theory and Electronic Properties of Solids* (Oxford: Oxford University Press)

**IOP** Publishing

Hydrogen
Physics and technology
**Sindhunil Barman Roy**

# Chapter 7

## Solid proton conductor

We will see in Part II of this book that the core of hydrogen technology lies in the generation of large quantities of $H_2$, and utilizing it safely and efficiently. We will see that there are several possible ways to achieve both of these goals. However, probably the most promising and the most efficient in terms of utilization is electrochemical conversion [1].

Let us consider the reaction:

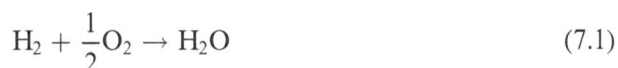

$$H_2 + \frac{1}{2}O_2 \rightarrow H_2O \tag{7.1}$$

This simple reaction releases a significant quantity of energy $\delta G = -237$ kJ mol$^{-1}$ and the only byproduct is pure water. Relation (7.1) thus tells that hydrogen is a source of clean energy, and it is also possible to generate large quantities of hydrogen from water, which is one of the most abundant compounds on Earth. From the angle of cost and development, both the generation and utilization devices can be built around largely the same materials and technologies [1]. The core of these electrochemical devices is an electrolyte.

### 7.1 Electrolytes and fuel cells

The basic property of an electrolyte is that it conducts electrical charge in the form of ions (charged atoms) but not in the form of electrons [1]. An electrolyte conducts through ions, but electronically it is insulating. Such a material will also show a high degree of specificity in the transported ions. that are transported. In applications, an electrolyte separates the electrodes of an electrochemical device by blocking electronic conduction but allows charge to pass in the form of selected ions (e.g., protons) [1].

There are numerous examples of electrolytes in various applications, and the most common ones are liquids such as molten or aqueous metal salt. However, these electrolytes are not of interest in the present context. In addition, the function of the electrolyte varies between applications. For example, in electroplating, the electrolyte may perform the dual function of being both an ionic conductor and a reactant. On the other hand, the electrolyte functions solely as a transport medium in a rechargeable lithium battery, while being inert with respect to the occurence of overall reaction [1]. Solid proton conductors fall into the latter category. The electrolyte material itself is not consumed in the electrochemical cell reactions. Figure 7.1 shows an example of this function with a generic fuel cell based around a proton-conducting electrolyte.

A fuel cell is an electrochemical device, essentially composed of an anode, an electrolyte, and a cathode, in which chemical energy is converted to electrical energy [2].

The basic operation of the fuel cell is shown in figure 7.1. Here the fuel gasses $H_2$ and $O_2$ are continuously fed to their respective electrodes, and the reaction product $H_2O$ formed at one electrode only is removed continuously. The oxidation of hydrogen gas at the anode produces protons plus electrons. The protons then travel through the electrolyte to recombine with the electrons, which travel via an external circuit and react with oxygen to produce water [1]. The useful energy is extracted from the fuel cell directly via the electrical circuit.

The driving force for this process in the fuel cell is the difference in chemical potential or activity between hydrogen at the anode and that at the cathode. In an

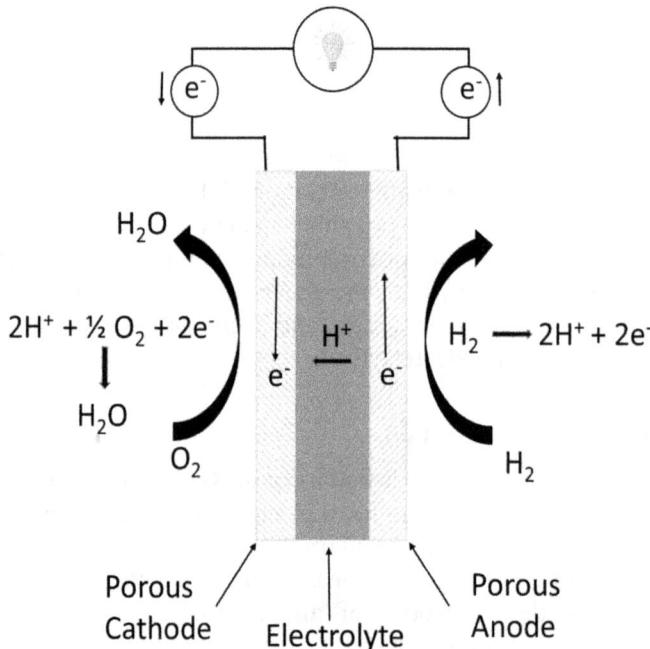

**Figure 7.1.** Schematic representation of a fuel cell. Hydrogen gas is reduced to protons plus electrons at the anode. The protons traverse the electrolyte to the cathode. The electrons travel via an external circuit. The protons and electrons react with oxygen at the cathode to form water. Adapted from [1], Copyright (2009), with permission from Elsevier.

ideal electrochemical system with no loss this difference in chemical potential can be related to the cell voltage via the Nernst equation [1]:

$$E = E^0 + \frac{RT}{nF} \ln \left( \frac{a_{\text{red}}}{a_{\text{ox}}} \right) \qquad (7.2)$$

Here $E^0$ is the electromotive force (EMF) at standard pressure, $R$ is the ideal gas constant, $T$ is the temperature, $n$ is the number of electrons transferred in the reaction, $F$ is Faraday constant, and $a_{\text{red}}$ and $a_{\text{ox}}$ are the chemical activities of the reactants.

In ideal systems, chemical activity is linked to concentration. This in turn is related to partial pressures $P$ in gaseous systems. In a hydrogen–oxygen fuel cell, the equation (7.1) can be expressed in a simplified form as:

$$E = E^0 + \frac{RT}{nF} \ln \left( \frac{P_{\text{H}_2} P_{\text{O}_2}^{1/2}}{P_{\text{H}_2O}} \right) \qquad (7.3)$$

The electrochemical reaction is reversible. By applying a potential and running the cell in reverse it is thus possible to split water to generate and separate pure hydrogen and oxygen at opposite electrodes. Similarly, by fixing the oxygen partial pressure at the cathode end, the open-circuit voltage of the cell will be proportional to the hydrogen partial pressure and the device will function as a sensor [1]. Thus, in different applications the requirements of the electrodes may change but the function of the electrolyte remains the same.

In real-life applications, for various reasons, the actual cell voltage is lower than the Nernst voltage given in equation 7.3. The energy losses usually arise from various processes occurring at electrodes and interfaces, and also importantly from the resistance of the electrolyte. The latter effect is related directly to the ability of the electrolyte to transport large quantities of protons rapidly and efficiently, which in turn is linked to the basic chemical properties of the electrolyte material. A proton-conducting solid electrolyte with high conductivity would be advantageous for the construction of a fuel cell with negligible corrosion problems.

## 7.2 Solid proton conductor

A solid proton conductor is a solid material, crystalline or amorphous, which allows the passage of electrical current through the bulk of the material exclusively by the movement of protons, $H^+$ ion [2]. Solid materials which conduct electricity through the migration of polyatomic protonic species also belong to the family of solid proton conductors.

The history of proton-conducting materials dates back to 1806 when C J T von Grotthuss discovered protonic species in an aqueous solution [3]. In 1875, electrical transport was observed in ice [4], which subsequently stimulated extensive studies on proton conduction in ice.

The two most important areas for the usage of solid proton conductors are in energy conversion using a fuel cell and an electrolyzer, where water is reconverted into hydrogen and oxygen.

The following properties of a proton conductor are desirable for application as the electrolytes in a fuel cell [2]:

1. A conductivity higher or equal to 0.1 $(\Omega\text{-cm})^{-1}$ at the working temperature to minimize internal resistive losses.
2. The conduction should be selective to protons.
3. The electrolyte needs to be an insulator electronically to prevent the internal short circuit of the cell.
4. The electrolyte must be stable with respect to thermal and electrochemical decomposition, and stable in the partial water vapor pressure range in the working cell.
5. Easy and economic fabrication.
6. The proton conductor needs to have suitable mechanical and thermal properties, e.g., thermal expansion coefficient, that match the properties of electrodes and other construction elements in the fuel cell.

An electrolyzer in principle is a fuel cell operating in the reverse direction. Water is fed to one of the electrodes, and an externally applied potential difference then decomposes water into $H_2$ and O gas. The material requirements for electrolytes are similar to those for fuel cell applications. A proton conductor can also be used as the ion-specific membrane in sensing devices: for pH measurements in liquids, and partial pressure measurement of $H_2$ in gasses [2].

## 7.3 Materials structure and proton conductivity

Three basic requirements, in general, need to be fulfilled in solid material for exhibiting ionic conductivity [1]:

1. The solid material should have a relatively high concentration of mobile ions.
2. These ions should be able to move relatively easily from their equilibrium position.
3. A path of long-range should exist through which the ions can migrate.

Related to these are several further questions regarding the nature of the mobile ions, their concentration, and the role played by the host structure in the generation of ions and mobility.

The migration of an ion or molecule through a solid involves movement from its equilibrium position in the lattice of the solid. Thus, the mechanistic aspects of diffusion in terms of defects and defect concentration are important in this context. A defect is an atomic level deviation in a material from its ideal crystal structure. The common defects are Frenkel and Schottky point defects and there are other kinds of defects too.

Figure 7.2 shows schematically Frenkel and Schottky defects for a hypothetical ionic material. An atom is displaced from its equilibrium position to an interstitial site giving rise to a Frenkel defect. On the other hand, Schottky defects are a form of charge-neutral non-stoichiometry that leaves lattice sites vacant. In a solid lattice, an

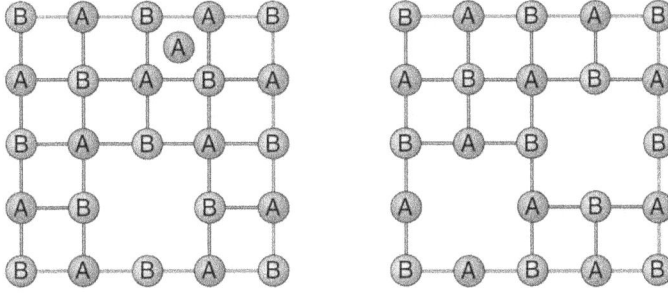

**Figure 7.2.** Frenkel (left) and Schottky (right) defects for the hypothetical solid ionic material AB. Reprinted from reference [1], Copyright (2009), with permission from Elsevier.

ion needs a place to go in order to move. In this context, the central concepts of vacant and interstitial positions are extremely important in the understanding of ionic conductors. The act of ion migration generates a vacant site as the ion moves in the lattice. It thus follows that the mobility and concentration of defects in a solid material are linked to its ion conduction properties [1]. This relationship is expressed as [1]:

$$\sigma = CBe \tag{7.4}$$

Here $\sigma$ is the specific conductivity, $C$ is the concentration of the mobile species, $B$ is the mobility, and $e$ is the charge of the mobile ions. We recall here that mobility $B$ is related diffusivity $D$ via Einstein relation $D = Bk_B T$, and $C$ and $D$ are thermally activity processes [1]:

$$C = C_0 \exp\left(-\frac{E_f}{k_B T}\right) \tag{7.5}$$

$$D = f D_0 \exp\left(-\frac{E_m}{k_B T}\right) \tag{7.6}$$

Here $E_f$ is the enthalpy of formation for a defect pair, $E_m$ is the enthalpy of defect migration, $C_0$ and $D_0$ are the carrier concentration and diffusion coefficient preexponential factors, respectively, $k_B$ is the Boltzmann constant, and $f$ is the correlation factor for the diffusion coefficient.

By combining equations (7.5) and (7.6), the following relation can be obtained [1]:

$$\sigma = f\left(\frac{C_0 D_0 e^2}{k_B T}\right) \exp\left(-\frac{E_a}{k_B T}\right) = \sigma_0 \exp\left(-\frac{E_a}{k_B T}\right) \tag{7.7}$$

Here $E_a$ is the activation energy for the diffusion process. The term $\sigma_0 = f\left(\frac{C_0 D_0 e^2}{k_B T}\right)$ is often called the prefactor.

The difference between an ionic conductor and an insulator lies in the concentration and mobility of defects. This is, however, a continuous scale, which makes the definitions of conductor and insulator quite arbitrary. As a rough guide, materials with a specific ionic conductivity in the range $10^{-10}$ - $10^{-5}$ Siemens (S) $cm^{-1}$ and activation energy in the region 0.6–1.2 eV are classified as ionic conductors, and materials showing specific conductivity $>10^{-4}$ S $cm^{-1}$ and activation energies below 0.6 eV described as fast ionic or superionic conductors [1].

In proton-conducting systems, the exact nature of the mobile ions plays a crucial role. Because of the small radius of the free proton, bare $H^+$ ions are not found in solids under equilibrium conditions. A $H^+$ ion is always covalently bonded to some electronegative atom/ion in the structure. This may be lattice oxygen in solid oxide systems. But in many materials the protons exist as part of a larger molecular species such as water or ammonia, which in some cases may be the mobile charge carriers [1]. It is thus common to classify proton-conducting materials according to actual protonated species—$H^+$, $OH^+$, $H_3O^+$, $NH_4^+$, and so on [1]. There are several other ways to classify the proton conductors, such as the synthesis method, the temperature of operation, and acid-base chemistry, but in the context of the present book we will focus on the mechanistic aspects of proton conduction in terms of the content and state of protons in the host material.

## 7.4 Different classes of solid proton conductors

A simple way of classifying proton-conducting solids is to divide them into (i) the materials that natively contain protons and (ii) the materials that do not and must thus get protonic species from an external source. In terms of defects, these can be described as intrinsic and extrinsic systems, respectively [1]. The intrinsic systems can be further subdivided into those in which the protons are present in a disordered liquid-like region and those in which the protons occupy discrete or semi-discrete fixed crystallographic positions.

Several examples from the different categories are presented together in an Arrhenius-type plot in figure 7.3. The chosen classification system overlaps strongly with that of operating temperature. This indicates that the operating temperature and chemistry are closely linked. Disordered liquid or hydrous-type materials are suitable in the temperature range of 300–373 K, above which they begin to dehydrate. Anhydrous materials perform in the temperature regime 423–573 K range. The extrinsic systems do not depend on hydrogen bonding to maintain their structure, and they are suitable for operation in the temperature range above 773 K. This is an open area of very active research activities. In the subsections that follow we will give short introductions of various classes of materials with a focus on the mechanism conductivity. The readers are referred to numerous review articles [5–9] on the subject for more detailed information.

### 7.4.1 Disordered type hydrous systems

A well-known example of disordered-type hydrous systems is the proton exchange membrane or polymer electrolyte membrane (PEM) [6]. A PEM consists of solid-state inert polymers, mostly with acidic functionalities. In a fuel cell using PEMs, only the protons are allowed to pass through from the anode to the cathode, and not the electrons. PEMs are electronically nonconductive. Generally, sulfonated

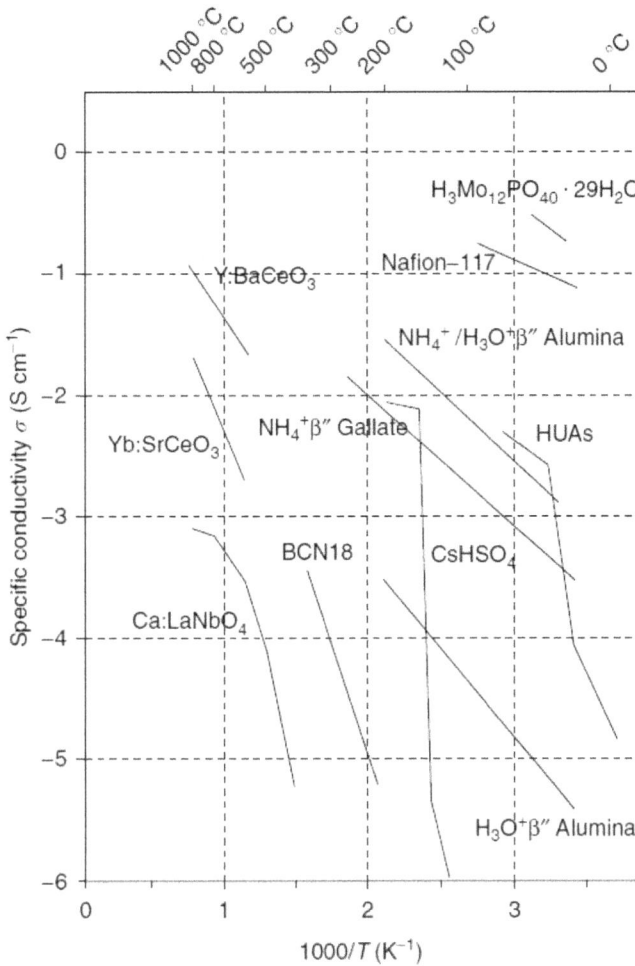

**Figure 7.3.** Arrhenius-type plot showing specific conductivity ($\sigma$) as a function of reciprocal temperature for some representative examples of proton-conducting materials. Reprinted from reference [1], Copyright (2009), with permission from Elsevier.

crosslinked polystyrenes or perfluorinated sulfonated ionomers, or sulfonated phenolic resins are used as typical PEMs. Perfluorinated sulfonated polymer for commercial application was developed by Du Pont in 1968, which is now commercially known as Nafion. This state-of-art proton-conducting material shows good chemical and physical stability, and displays high proton conductivity of $10^{-2}$-$10^{-1}$ S cm$^{-1}$ at a moderate temperature in a fully hydrated state, it also has a very high lifetime. However, its widespread commercial application is limited because of several reasons [6]: (i) conductivity quickly decreases when the temperature goes beyond 100 °C because of the removal of the adsorbed water molecule from the membrane; (ii) degradation at a higher operating temperature; (iii) it is cost-intensive; and (iv) it has an amorphous nature.

Several inorganic systems were discovered through 1970s–1980s, displaying high conductivity ($4 \times 10^{-3}$ S cm$^{-1}$) even at room temperature [1, 10]. They of course require high degrees of hydration to provide a conductive medium for proton transport. Classic examples of this class are the hydrated heteropolyacids $H_3PM_{12}O_{40} \cdot n$ $H_2O$ (where M = W, Mo, Si), hydrated hydrogen uranyl phosphate $HUO_2PO_4 \cdot 4$ $H_2O$ (HUP), hydrogen uranyl arsenate $HUO_2AsO_4 \cdot 4H_2O$ (HUA), and the hydrated and doped $\beta/\beta'$- alumina materials.

None of these inorganic hydrous materials seem to be able to offer the combination of performance and stability that the Nafion polymer membrane exhibits in the range of 300–350 K [1]. As in the polymer system, these inorganic materials tend to dehydrate with increasing temperature, sometimes even irreversibly. The best of the systems at low temperatures is the heteropolyacids, which have been reported to show performance on the order of Nafion at ambient temperature. The HUP, HUA, and $\beta$-alumina systems are relatively inferior at room temperature. However, the $\beta$-alumina materials offer a significant advantage over the other systems in that they retain their structural water, and thus their conductivity, at considerably high temperatures, up to approximately 500 K for the $\beta'$-structured $(NH_4)_{1.67-y}(H_3O)_y Mg_{0.67}$ $Al_{10.33}$ $O_{17}$.

In structural terms, the heteropolyacids can be visualized as globular arrays of $(PM_{12}O_{40})^{3-}$ ions that are connected via protonated water layers (see figure 7.4). The HUP/HUA and the $\beta$-aluminas are better described in terms of inorganic layers that are separated by liquid-like layers of protonated water (see figure 7.4). The latter examples are particularly interesting in that the hydronium ($H_3O^+$) incorporated into the structure can be replaced by other cations, such as $NH_4^+$ [1]. In all these cases and in the majority of heavily hydrated proton-conducting systems, the conduction of protons takes place predominantly by what is termed the vehicle mechanism [1]. In this mechanism, instead of discrete protons, the entire cation assembly ($H_3O^+$, $NH_4^+$ etc.) is mobile. In an operating electrolyte, the uncharged molecules (water and ammonia) pick up protons at the anode, transport them to the

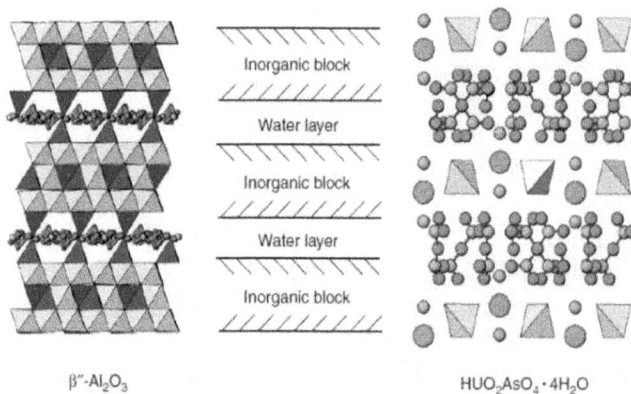

$\beta''$-Al$_2$O$_3$                    HUO$_2$AsO$_4 \cdot$4H$_2$O

**Figure 7.4.** The structures of hydrated $\beta'$-alumina and hydrogen uranyl arsenate. These are visualized as alternating layers of inorganic and liquid-like layers, with proton transport occurring in the latter. Reprinted from reference [1], Copyright (2009), with permission from Elsevier.

cathode, and then return. This can be imagined as a two-way road traffic, with protonated assemblies running in one direction (anode to cathode) and uncharged species returning (cathode to anode) [1]. The dynamics of molecular transport through this liquid-like layer thus control the proton transport in hydrous solid proton conductors.

It may be noted here that the anionic/inorganic structure in all these materials provides more than just a physical structure for the conducting network. The anionic framework in these materials plays an important role in the protonation of the liquid-like conducting layer [1]. In combination with hydrogen bonding within the layers, it influences the mobility of the protonic species. HUP/HUA undergoes several structural transitions in the range of 150–300 K, which relate directly to increasing mobility of the protonic species within the water-containing layer [1]. It is only in the highest temperature phases above $\sim$270 K that the layers can be properly described as liquid-like and the materials become superionic conductors.

### 7.4.2 Anhydrous hydrogen-containing systems

Most prominent among the anhydrous hydrogen-containing systems is the group of materials known as solid acids or acid salts. These are compounds based around oxyanions such as phosphate ($PO_4^{2-}$), sulfate($SO_4^{2-}$), or selenate ($SeO_4^{2-}$), which are linked together via hydrogen bonds [1]. In comparison to the hydrous materials described above, the hydrogen content in these systems is lower. They also have much different proton environments, often with one proton per anion. A notable member of this class is mono- and dihydrogen sulfates and phosphates $MHSO_4$ and $MH_2PO_4$ (where M = group I metal), particularly $CsHSO_4$ and $CsH_2PO_4$ [1]. They show an interesting 'super protonic' phase transition at $\sim$414 K for $CsHSO_4$, and $\sim$503 K for $CsH_2PO_4$. Above those respective temperatures, their specific proton conductivity is enhanced by several thousand times. However, they are soluble in water and highly ductile in the high-temperature range. These properties make their use in real applications somewhat challenging. Nevertheless, fuel cells operating with both $CsHSO_4$ and $CsH_2PO_4$ electrolytes have been successfully demonstrated [1].

In anhydrous materials, proton transport takes place by a hopping mechanism known as the Grotthuss mechanism, which was originally introduced to explain the transport of protons in water. The term Grotthuss-like is now widely used to describe the conduction mechanisms in systems where a formal transfer of a proton occurs between two otherwise static anions [1]. This involves a two-step process, where a transfer and a rotation with the transfer step are usually mediated by hydrogen bonding between the proton and the accepting and donating anions. Within this framework, the dramatic increase in proton mobility in the $CsHSO_4$ and $CsH_2SO_4$ materials is explained as the super protonic phase transition [1]. The left-hand panel of figure 7.5 shows the low-temperature structure of $CsHSO_4$ along with the proton conduction route (Cs is shown in blue, S in pink, O in gray, H in red). In the temperature region below the super protonic phase transition, the structure consists of chains of static $SO_4^{2-}$ anions connected via hydrogen bonds (shown in red

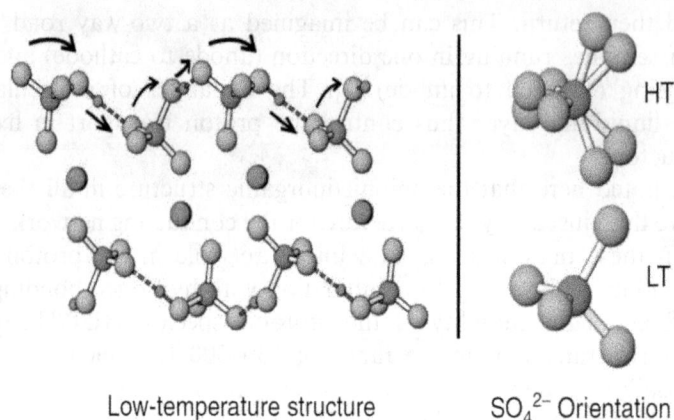

**Low-temperature structure**          **$SO_4^{2-}$ Orientation**

**Figure 7.5.** The low-temperature structure of $CsHSO_4$ (left-hand panel), and low-temperature and high-temperature orientations of $SO_4^{2-}$ anions in $CsHSO_4$ (right-hand panel). Reprinted from reference [1], Copyright (2009), with permission from Elsevier.

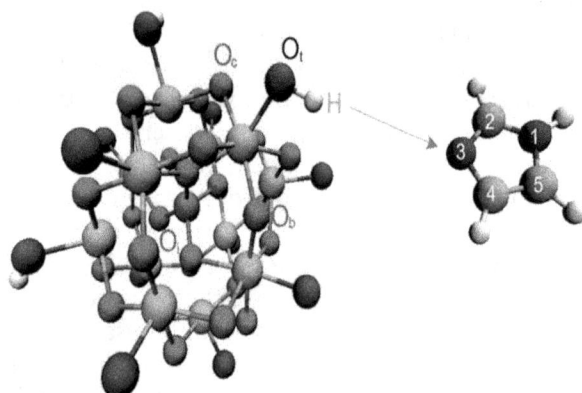

**Figure 7.6.** Atomic structure and atom labeling of (left-hand) the Keggin unit of the 12-tungstophosphoric acid $H_3WPA$ and (right-hand) imidazole in its neutral form. Reproduced from [12]. CC BY 4.0.

dashes in figure 7.5). In the high-temperature phase, the structure allows additional freedom of local movement for $SO_4^{2-}$ anions with different orientations becoming structurally equivalent (see the right-hand panel of figure 7.5)). The ions can then rotate more freely, and thus enables proton transfer via a Grotthuss-like mechanism.

The strongest among the solid acids is 12-tungstophosphoric acid, which forms in the so-called Keggin structure [11, 12]. In this structure one central phosphorous (P) atom is surrounded by four oxygen (O) atoms, that are connected to 12 corners shared $WO_3$ units, resulting in the molecular formula $H_3 [P(W_3O_{10})_4] \cdot nH_2O$ (see figure 7.6). This acid shows a very large proton conductivity of up to $0.18$ S cm$^{-1}$ at room temperature. While the precise location of the acidic protons ($H^+$) on the surface of the Keggin anion in the anhydrous state has been a subject of debate, it seems that terminal ($O_t$) and bridging ($O_b$) oxygens are the preferred sites [12]. In the hydrated state, protons are located in the water phase as hydronium ($H_3O^+$) or Zundel-like ($H_5O_2^+$) ions.

A novel solid-state proton conductor has recently been synthesized by acid-base chemistry via proton transfer from 12-tungstophosphoric acid to imidazole [12]. The resulting material, named $Imid_3WP$, is a solid salt hydrate, which at room temperature includes four water molecules per structural unit. This is an attempt to tune the properties of a heteropolyacid-based solid-state proton conductor using a mixture of water and imidazole, thus interpolating between water-based and ionic liquid-based proton conductors of high thermal and electrochemical stability. The proton conductivity of $Imid_3WP_4H_2O$ measured at truly anhydrous conditions was $0.8 10^{-6}$ S cm$^{-1}$ at 322 K [12]. This is higher than the conductivity reported for any other related salt hydrate, despite the lower hydration.

## 7.5 Proton-conducting oxides

This class of materials does not intrinsically contain protons as part of their structure or stoichiometry. These are typically high-temperature ceramic oxide materials, where the protons reside as extrinsic dopants, being incorporated either during the synthesis process or existing in equilibrium with hydrogen or water in the ambient atmosphere. A much higher temperature of operation in the 873–1073 K range is achievable with these materials than the water-containing and solid acid systems discussed in earlier sections. They are very suitable for applications, such as high-temperature fuel cells, steam electrolyzers, and as sensors in industrial processes such as metal smelting.

Most of these oxide systems are based on modifications of the cubic perovskite structure shown in figure 7.7. The well-known barium calcium niobate (BCN) and barium zirconate ($BaZrO_3$) systems adopt the ideal cubic structure shown, whereas the doped alkaline-earth cerates ($ACeO_3$, where A = Sr, Ba) take distorted, or tilted, modifications of this structure [1]. Zirconate materials have low thermal

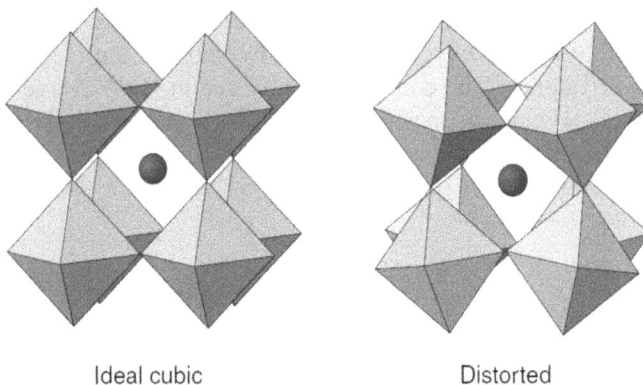

Ideal cubic          Distorted

**Figure 7.7.** The cubic and distorted forms of perovskite. Large cations (e.g., Ca, Sr, Ba) reside on the 12-coordinated site marked in blue. Smaller cations (e.g., Zr, Ce) adopt octahedral coordination as shown in gray. Reprinted from reference [1], Copyright (2009), with permission from Elsevier.

conductivity, low dielectric loss, and very low thermal expansion coefficient, which makes them more favorable for electrochemical devices than other proton-conducting oxide materials [7]. Several non-perovskite structured fast protonic conducting oxides have also emerged. A prominent member of this later class is the accepter-doped $LaNbO_4$ systems. These non-perovskite materials differ significantly in structure from the perovskites. The non-perovskites are based around tetrahedral structural units, rather than the octahedral units from which the perovskite structure is constructed [1]. In addition, their performance from ionic conductivity points of view is not as good as the alkaline-earth cerates and zirconates. However, the non-perovskites have better stability in $CO_2$ and water-containing atmospheres.

Protons are not native to the lattice of all these perovskite and non-perovskite oxides. The incorporation of protons as positively charged hydroxide defects is necessary for proton conduction in these materials. This can occur primarily via the hydration reaction, which in Kröger–Vink notation is expressed as [1, 8]:

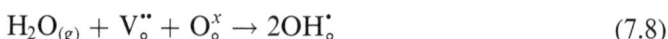

$$H_2O_{(g)} + V_o^{\cdot\cdot} + O_o^x \rightarrow 2OH_o^{\cdot} \qquad (7.8)$$

Here $V_o^{\cdot\cdot}$, $O_o^x$ and $OH_o^{\cdot}$ represent an oxygen vacancy, a lattice oxygen, and a hydroxide defect at a lattice oxygen site, respectively, in Kröger–Vink notation. This defect reaction expressed in relation (7.8) reveals that an observable concentration of oxygen vacancies is necessary for finite proton concentration. Thus, cations with a lower oxidation state than the host cation, i.e., accepter dopants, are essential for introducing extrinsic oxygen vacancies, and in turn obtain significant proton conductivity. A single oxygen vacancy generates two protonic defects, which may then be free to move as charge carriers. The concentration and mobility of incorporated protons depend on several factors, including crystal structure, the characteristics of the host material, and the natures and concentrations of the dopants [8]. A key issue is the concentration of oxygen vacancies because the intrinsic oxygen vacancy concentration arising from the Frenkel and Schottky mechanisms discussed earlier in section 7.3 is too low to generate significant water uptake, and in turn generate fast ionic conductivity [1]. However, these structures, especially perovskite, can tolerate quite high values of oxygen deficiency. It is thus possible to generate vacant oxygen positions by substituting a component cation with one of lower valence. The compensation of the missing charge takes place by the generation of oxygen vacancies. An illustration of this phenomenon can be presented with the example of doping of $Yb^{3+}$ for $Zr^{4+}$ according to the solid solution $BaZr_{1-x}Yb_xO_{3-x/2}$ [1]. In Kröger–Vink notation, this can be expressed as [1]:

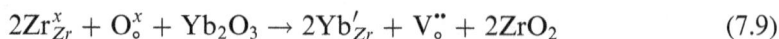

$$2Zr_{Zr}^x + O_o^x + Yb_2O_3 \rightarrow 2Yb_{Zr}' + V_o^{\cdot\cdot} + 2ZrO_2 \qquad (7.9)$$

The protonic conduction in these systems takes place via a Grotthuss-like process with protons migrating between lattice oxygens. The mechanism is illustrated in figure 7.8. The transport is facilitated by a lattice vibration, i.e., phonon, which brings the lattice oxygens sufficiently close for a hydrogen bond to form and transfer to take place. After this transfer, the proton (shown in red in figure 7.7) can rapidly reorientate

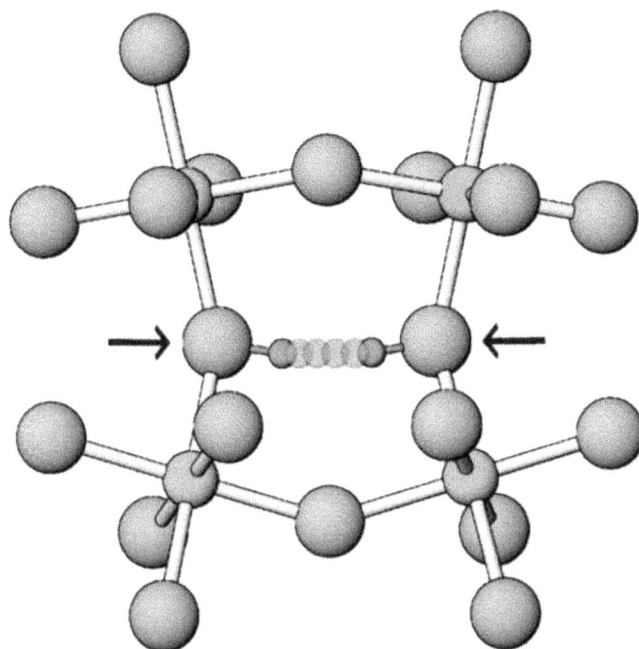

**Figure 7.8.** The perovskite structure of CaZrO₃ (Zr shown in orange, O in gray, Ca is not shown) with the Grotthuss-like transfer mechanism. Reprinted from reference [1], Copyright (2009), with permission from Elsevier.

around the oxygen and make the next hop [1]. The rate-determining step of the transfer of protons between lattice oxygens is thus mediated via coordinated thermal motions of the lattice, with the reorientation step occurring much more easily.

It may be noted that in such oxygen-deficient solids the oxygen vacancies are also mobile defects, and there exists a competition between proton conductivity and oxide ion conductivity. However, the dissolution of water into the structure is an exothermic process. So, proton diffusion is favored at low to intermediate temperatures (typically 673–1073 K). The dissolution of water into the lattice becomes entropically unfavored at higher temperatures, and the materials move over to being predominantly oxide ion conduction [1].

There are alternative structure-type solid oxide systems (e.g., hexagonal perovskite derivatives, brownmillerite, scheelite, etc.), which can efficiently incorporate and enable the transport of protonic defects [9]. Recent reports indicate that such materials can have high ionic conductivity, which is comparable to conventional perovskite conductors.

## 7.6 Hybrid organic networks

A variety of crystalline materials such as metal–organic frameworks (MOFs), coordination polymers, covalent-organic frameworks (COFs), hydrogen-bonded organic frameworks, polyoxometalates (POMs), metalohydrogen-bonded organic frameworks, crystalline organic cages, and their crystalline composites in large numbers have recently come under intense investigation. They can have proton conductivity up to the level of $10^{-2}$ S cm$^{-1}$ (see reference [6] and references therein). We will briefly discuss two of these classes, namely MOFs and CCFs.

### 7.6.1 Metal–organic framework based ionic conductor

We have discussed in chapter 6 (section 6.1.1) that metal–organic frameworks (MOFs) have been recognized as excellent platforms for designing various functional materials via host-guest chemistry or reaction transformation, and they have potential in the fields of hydrogen gas adsorption and energy storage. The application of MOFs as a solid-state conductors has also recently emerged as a subject of new research interest. They have attracted considerable attention because of their insulating character and various intrinsic features (see reference [13] and references within). Many possibilities in chemical compositions, pore structures, and surface functionalities give MOFs with high designability [13]. This in turn gives rise to opportunities for tuning the physicochemical properties of MOFs and exploring for large ionic conductivity.

The merits of MOFs as solid ionic conductors are presented in figure 7.9. First, the very large surface area of MOFs can lead to sufficient contact with other components, while the controllable surface polarity enables the modulation of acid-base interactions in the hybrid systems. This would provide possibilities for the optimization of overall electrochemical properties, such as ion mobility and ion transference number [13]. Second, the periodical crystalline structure with rich porosity offers homogeneously dispersed sites and well-defined pathways for ion diffusion. Thus, such materials would have great advantages over conventional nonporous solids. Third, the microporous feature and ordered channels can aid in

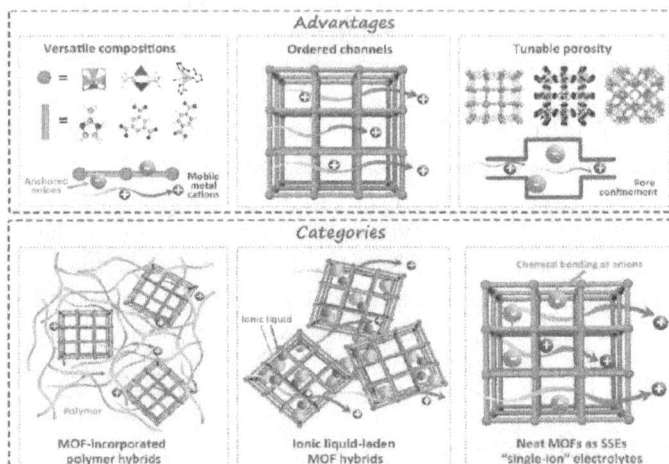

**Figure 7.9.** Schematic presentations of the advantageous features of MOFs as solid ion conductors (top) and the categories of MOF-based solid ion conductors (bottom). Reproduced from [13] with permission from the Royal Society of Chemistry.

controlling the uniformity of cation plating upon the charging/discharging process [13]. This will promote the formation of the stabilized interfacial layer and improve cycling stability. Finally, MOFs with tunable pore size, pore shape, and pore polarity can behave as ion sieves that preferentially promote cation transfer. In addition, the unwanted impurities and by-products in the system could be trapped due to the high surface energy and strong adsorption ability of MOFs. In this way, the side reactions can be avoided. Overall, the modular nature of MOFs in terms of compositions, structures, and internal environments is a powerful combination that provides feasibility for the design of battery-specific solid-state electrolytes. As shown in figure 7.9, the following types of MOF systems are possible: (ii) MOF-incorporated polymer hybrids, (ii) ionic liquid-laden MOF hybrids, and (iii) neat MOFs as solid-state electrolytes. The readers are referred to reference [13] for a detailed discussion on the functioning of such MOF-based systems.

### 7.6.2 Covalent-organic frameworks based ionic conductors

COFs are a newer promising class of solid-state proton-conducting material. We have earlier discussed in chapter 6 (section 6.1.2) that COFs are an emerging class of crystalline porous organic polymers and functional nanostructures, combining several interesting properties, such as high crystallinity, low density, tunable porosity, large surface area, highly ordered, and unique molecular architecture. With such superior properties, COFs have the potential to become versatile proton conductors, both under hydrous and anhydrous conditions.

The film/membrane highly depends upon the compatibility of the substrate with other organic materials during the process of synthesis. To this end, the pure organic nature of COF assists in reducing the agglomeration or precipitation during the synthesis of films, and also provides better compatibility with the polymer matrices. This is a great advantage of COFs compared to MOFs and other crystalline mediums for the preparation of proton-conducting membranes or film. Moreover, the presence of strong covalent bonds in the COF framework improves the mechanical stability of such membranes/films, and gives COFs further advantages over MOFs and other crystalline materials.

The widely followed approaches for developing proton-conducting COFs are either by a modulating backbone of COFs through the installation of several proton-donating species such as highly abundant $-SO_3H$, $-OH$ groups or by incorporation of several protonic species extrinsically (organic acids (phytic acid, $p$-toluene sulfonic acid (PTSA), etc) or mineral acids ($H_3PO_4$, $H_2SO_4$) or POMs ($PW_{12}$) or several N-heterocycles (triazole, tetrazole, imidazole, histamine, pyrrole, etc) inside the well-defined pore channels [6]. The incorporation of external proton sources inside the COF cavities can significantly induce proton conductivity. However, the installation of such proton sources is not only difficult to control but also prone to leaching problems. On the other hand, while intrinsic functionalities can ensure more stable and efficient conductivities with values up to $10^{-2}$ S cm$^{-1}$, the fabrication of such frameworks is quite difficult to achieve. More pragmatic approaches for the maximization of proton-conducting performance may be to combine both the strategies in one COF system or undertake the composite formation approach [6]. The high concentration of several proton carriers inside the pores will not only ensure the formation of infinite H-bonding networks but will also provide low-energy barrier proton-conducting pathways for efficient proton hopping throughout the system. The strategies envisaged in this direction are as follows [6]:

1. Proton conduction is achieved through the intrinsic property.
2. Proton conduction is achieved through the extrinsic property.
3. Proton conduction is achieved through combined properties of both intrinsic and extrinsic both.

We will briefly dwell upon each of the points here, the readers are referred to ref.[6] (and references therein) for a more detailed discussion.

### 7.6.2.1 Proton conduction achieved through intrinsic property

The covalent preinstallation of proton-donating functionalities (e.g., phenolic hydroxyl, imidazole, phosphate, carboxylate, and sulfonate moieties) into the COF backbone during direct synthesis can lead to intrinsic proton sources [6]. This will enable efficient and stable proton-conducting performance, and is thus classified as 'proton conduction achieved through intrinsic property'. These functional groups remain uncoordinated and flanked inside the well-defined pore channels. This is essential to increase the concentration and mobility of the proton carriers. The presence of such functionalities will not only enhance the acidity of the pore channels but will also promote extensive continuous H-bond formation throughout the framework. This will further enhance the efficient proton conduction properties.

### 7.6.2.2 Proton conduction achieved through extrinsic property

The preinstallation of proton-donating functionalities into the COF backbone during direct synthesis is often quite difficult and demands postsynthetic treatments of the pristine framework. In this regard, extrinsic incorporation of some guest protonic species (e.g., triazole, tetrazole, imidazole, histamine, pyrrole, $H_3PO_4$, phytic acid, POMs, etc) onto the pore channels of COFs may significantly induce the proton conduction property and make the extrinsic approach very effective [6]. Moreover, in a few cases, proton-conducting polymers such as Nafion, sulfonated polyether ether ketone(SPEEK), silk nanofibril (SNF), etc (having several anchored protonic groups) could be integrated with some extrinsic proton source-loaded COFs. This further helps to form extensive H-bonding throughout the framework. In turn, rapid and smooth proton transportation is expected, which will further trigger efficient proton conductivity [6].

There are presently two approaches to incorporate extrinsic properties [6]:

1. Extrinsic incorporation of several acids, POMs, or N-heterocycles. Several organic acids (e.g., phytic acid, PTSA, etc), mineral acids (e.g., $H_3PO_4$, $H_2SO_4$), POMs (e.g., $PW_{12}$), or several N-heterocycles (e.g., triazole, tetrazole, imidazole, histamine, pyrrole, etc) have been introduced extrinsically to achieve ultrahigh proton conductivity.

2. Integration with proton-conducting polymers. Here, several proton-conducting polymers (e.g., Nafion, SPEEK, SNF, etc) have been fabricated with the protonic species-loaded COFs. Another possible approach is to install protonic species (e.g., -SO₃H) onto the backbone via postsynthetic modifications.

### 7.6.2.3 Proton conduction achieved through combined intrinsic and extrinsic properties

The proton conductivity of a synthesized material depends on the number of proton carriers presents throughout the framework. Thus, finding a way to increase the proton carrier concentration is the simplest approach to develop a variety of proton-conducting materials with high proton conductivity. To this end, the combined effect of both intrinsic and extrinsic approaches in one COF system may be a very effective way to maximize the proton conductivity.

With a stepwise synthesis strategy, it has been possible to develop a series of highly porous, crystalline, and robust COFs (NKCOF-1 to 4) [14]. A high density of azo groups accompanied by phenolic hydroxyl groups was integrated into the COF structures. This allows azo groups to serve as proton accepters and load sites for added acids, while phenolic hydroxyl groups serve as proton donors facilitating proton conduction. The synthesized COFs exhibit high hydrophilicity and excellent stability in strong acid or base, and in boiling water. By adding $H_3PO_4$, the COFs ($H_3PO_4$@COFs) not only realize an ultrahigh proton conductivity of $1.13 \times 10^{-1}$ S cm$^{-1}$ but also maintained high proton conductivity across a wide relative humidity (40%–100%) and temperature range (20 °C–80 °C) [14].

Certain sections of text in this chapter have been reproduced with permission from [1]. Copyright © 2009 Elsevier B.V. All rights reserved.

## References

[1] Tolchard J 2009 *Electrolytes | Solid: Proton, Encyclopedia of Electrochemical Power Sources* Ed. J. Garche (Amsterdam: Elsevier) 188

[2] Poulsen F W 1980 *An Introduction to Proton Conduction in Solids* Risøo National Laboratory. Risøo-M No. 2244

[3] von Grotthuss C J T 1806 *Ann. Chim.* **58** 54
Codorniu-Hernández E and Kusali P G 2013 *Proc. Nat. Acad. Sci.* **110** 13697

[4] Ayrton W E and Perry J 1875 *Proc. Phys. Soc., London* **2** 171

[5] Kruer K D 2003 *Annu. Rev. Mater. Res.* **33** 333

[6] Sahoo R *et al* 2021 *Adv. Energy Mater.* 2102300

[7] Khalid Hossain M *et al* 2022 *Nanomaterials* **12** 3581

[8] Duan C, Huang J, Sullivan N and O'Hayre R 2020 *Appl. Phys. Rev.* **7** 011314

[9] Fop S 2021 *J. Mater. Chem.* A **9** 18836

[10] Shilton M G and Howe A T 1977 *Mater. Res. Bull.* **12** 701

[11] Nakamura O *et al* 1979 *Chem. Lett.* **8** 17

[12] Martinelli A *et al* 2021 *J. Am. Chem. Soc.* **143** 13895

[13] Zhao R *et al* 2020 *Energy Env. Sci.* **13** 2386

[14] Yang Y *et al* 2020 *Angew. Chem., Int. Ed.* **59** 3678

**IOP** Publishing

Hydrogen
Physics and technology
**Sindhunil Barman Roy**

# Chapter 8

# Superconductivity in hydrogen-based systems

In 1911, Kamerlingh Onnes at Leiden, Holland discovered that mercury (Hg) when cooled with liquid helium below 4.2 K abruptly starts carrying a current with no resistance [1]. This phenomenon was termed superconductivity, and the temperature at which the resistance suddenly dropped is now termed the superconducting transition or critical temperature, $T_C$. Subsequently, the Leiden team discovered that lead and tin were also superconductors, with transition temperatures near 6 K and 4 K, respectively [1]. In 1933, Meissner and Ochsenfeld found that super-conductors expelled a magnetic field when the externally applied magnetic field is below a certain critical strength [2]. The Meissner effect is an essential hallmark of superconductivity. In addition to high temperatures, high magnetic fields also destroy the superconducting state. Over the last hundred years, superconductivity has been found in many classes of materials, including several pure elements, metallic alloys, copper oxide materials or cuprates, iron pnictides, and even in certain covalent systems such as fullerenes and organic compounds.

Fritz London noted that the Meissner effect implied that electrons in super-conductors behave in a collective manner [3]. It is well known that bosons with their integer spin values can behave in this fashion. On the other hand, electrons are fermions that have half-integer spins. Hence, electrons in general are not expected to behave collectively. In 1956, Leon Cooper resolved this apparent contradiction. He demonstrated that the presence of even an arbitrarily small attractive interaction between the electrons in a solid causes the electrons to form pairs [4]. These 'Cooper pairs' behave as effective bosons. Unlike real-space molecules, Cooper pairs consist of electrons in time-reversed momentum states. Consequently, Cooper pairs have zero center-of-mass momentum. A pair of identical fermions is antisymmetric with respect to the exchange of one fermion with another [5]. As a result, the spin and spatial components of the Cooper pair wave function need to have opposite exchange symmetries. Thus, these Cooper pair states are either spin singlets with an even-parity spatial component or spin triplets with odd parity.

doi:10.1088/978-0-7503-5172-0ch8

© IOP Publishing Ltd 2024

In 1957, building on the idea of Cooper pairs, John Bardeen, Leon Cooper, and Robert Schrieffer proposed the first satisfactory microscopic theory of superconductivity [6]. In Bardeen–Cooper–Schrieffer (BCS) theory, it was shown that many Cooper pairs can form in the Fermi sea in solid metal. Despite the electrons repelling each other because of the Coulomb force, at low energies there can be an effective attraction resulting from the electron-ion interaction. This is visualized by considering that in a metal the mobile electrons are detached from the atoms that form the crystalline lattice. These atoms in the metal then become positively charged ions. A mobile electron can then attract the surrounding ions because of their opposite charge. A positive ionic distortion is left in its wake when such an electron moves. This, in turn, attracts a second electron, leading to a net attraction between these two electrons.

A real-space structure also follows directly from the BCS theory [7]. This gives rise to a spherically symmetrical quasi-atomic wave function, with an identical onion-like layered structure for each of the electrons constituting the Cooper pair, with charge layers ~0.1nm and radius ~100 nm for a classic BCS superconductor. This charge modulation induces a corresponding charge modulation in the background ionic lattice. This in turn induces an attractive interaction between these two opposite charge modulations, which produces the binding energy of the Cooper pair [7]. Figure 8.1 schematically shows this attraction between two free electrons mediated by the crystal lattice This mechanism is possible because the ion dynamics is slow relative to the dynamics of electrons. This is a consequence of the fact that the ions are much heavier than the electrons. The interaction at shorter times becomes repulsive because of the Coulomb repulsion between the electrons. this retardation is responsible for limiting superconducting transition temperature ($T_C$).

## 8.1 Bardeen–Cooper–Schrieffer theory of superconductivity

In BCS theory the resulting wave function for the Cooper pairs turns out to be peaked at zero separation of the electrons, i.e., an $s$-wave state. A key aspect of the classic BCS theory is that it applies to the perturbative limit of weak attractive interactions. The theory perfectly describes conventional superconductors for which the attraction between the electrons is ~10 000 times less than the Fermi energy, $E_F$. The results of BCS theory predicted, among other things, the formation of a minimum excitation energy, or energy gap, in a metallic conductor below a critical temperature $T_C$. Various physical properties of conventional superconductors can be understood as consequences of this energy gap. Starting from the model Hamiltonian, BCS theory deduced the major equilibrium predictions that are given in the following subsections. (For a detailed discussion on the BCS theory and the BCS model Hamiltonian, the readers are referred to the textbook by Annett [8].)

### 8.1.1 Superconducting transition temperature

In the absence of any magnetic field, the onset of superconductivity takes place at a critical temperature [5]:

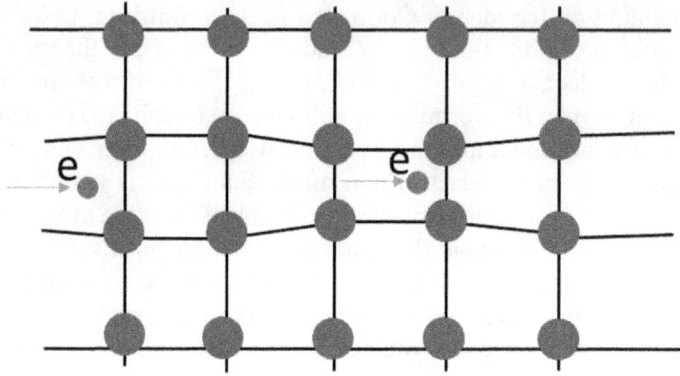

**Figure 8.1.** Real-space schematic diagram of the attraction between two free electrons caused by the lattice.

$$k_B T_C = 1.13\hbar\omega \exp\left(-\frac{1}{N_0 V_0}\right) \tag{8.1}$$

Here $N_0$ is the density of electronic states of a single spin population in the normal metal, and $\omega$ and $V_0$ are the parameters of model BCS Hamiltonian. The effective electron coupling $V_0$ cannot be determined precisely enough to permit an accurate determination of the critical temperature from equation (8.1) because of the exponential dependence. It may be noted that the same exponential dependence in equation (8.1) accounts for the experimentally observed very low superconducting transition temperatures in various metals, even after assuming $\hbar\omega$ of the order $k_B\theta_D$ (where $\theta_D$ is Debye temperature). Thus, the strong dependence of $N_0 V_0$ in equation (8.1) can lead to the critical temperatures in the experimentally observed range, with $N_0 V_0$ ranging from 0.1 to 0.5, i.e., $V_0 n$ in the range 0.1 to 0.5 [5]. The quantity $N_0$ that is widely used in the superconductivity literature is simply $g(\epsilon_F)/2$, where $g(\epsilon_f)$ is the density of the electronic levels in a metal at the Fermi energy $\epsilon_F$. The quantity $N_0$ is of the order of $n/\epsilon_F$, where $n$ is electron density in a metal [5]. It may be noted that the BCS theory predicts a superconducting transition no matter how weak the electron coupling $V_0$. However, as seen from equation (8.1), the superconducting transition temperature will be unobservably low.

A widely-used measure of electron density $r_S$ is defined as the radius of a sphere whose volume is equal to the volume per conduction electron [5]:

$$r_S = \left(\frac{3}{4\pi n}\right)^{1/3} \tag{8.2}$$

The alkali metals with $3.2 \leqslant r_s \leqslant 5.7$ do not show superconductivity, at least within the range of low temperatures currently available. The cancellation between electron–phonon and electron–electron derived terms is almost complete, and in some cases may be overcomplete. This combined with their low Debye temperatures $(\theta_D)$ leads to near-vanishing transition temperatures. In terms of the parameter $\lambda^2 = 1/\pi a_0 k_F = 0$, it is possible to express $N_0 V$ for a metal of valence $Z$ as

$$N_0 V = \frac{1}{2}\lambda^2 \left[ \frac{1}{\alpha^2}\left( 1 + \frac{\lambda^2}{(1/4Z)^{2/3}}\ln\frac{1+\lambda^2}{\lambda^2(1/4Z)^{2/3}} - \frac{\lambda^2}{(1+\lambda^2)(1/4Z)^{2/3}} \right) - \ln\frac{1+\lambda^2}{\lambda^2} \right] \quad (8.3)$$

Here the actual sound velocity has been written as $v_s = \alpha(Zm/3M)^{1/2}v_F$ for metals with ionic mass $M$, electronic mass $m$ and Fermi velocity $v_F$ [5]. The role of the dimensionless quantity $\alpha$ is to correct the standard Bohm–Staver estimate of sound velocity in metals [10]. It may also be noted that when an electron moves in a metal, the other electrons must move out of the way. This is because the Pauli exclusion principle forbids two electrons with the same spin at the same point (this is called exchange interaction), and they also need to minimize the repulsive Coulomb energy. Thus, the electron in a metal is not a bare electron but rather an excitation in the solid or a quasiparticle consisting of a moving electron together with a surrounding exchange-correlation hole [8]. In a metal the effective Coulomb force between quasiparticles is substantially reduced due to screening. In the simplest model of screening in the metal, namely the Thomas–Fermi model, the effective Coulomb interaction takes the form [8]:

$$V_{RF}(\mathbf{r} - \mathbf{r}') = \frac{e^2}{4\pi\epsilon_0 |\mathbf{r} - \mathbf{r}'|}\exp(-|\mathbf{r} - \mathbf{r}'|/r_{TF}) \quad (8.4)$$

Here $r_{TF}$ is the Thomas–Fermi screening length. The screening reduces the effective Coulomb repulsion, and it vanished for $|\mathbf{r} - \mathbf{r}'| > r_{TF}$. The effective Coulomb repulsion is thus substantially weaker than the original $1/r$ potential.

In equation (8.3), the term arising from electron–phonon coupling includes normal and Umklapp processes [5], and both have been calculated in the rigid-ion approximation with Thomas–Fermi screened point-ion potentials [5]. For detailed information on the normal and Umklapp processes in the lattice vibration of a solid, the readers are referred to the solid-state physics textbook by Ashcroft and Mermim [5]. (The importance of Umklapp or interband scattering effects in normal-state transport properties is well known [9].) This rigid-ion approximation becomes exact when applied to metallic hydrogen, whose ions are point charges [5]. Cancellation of the screened electron–electron and electron–phonon derived terms in $N_0 V$ for the monovalent ($Z = 1$) metals is largely a result of their $r_S$ range, in combination with the fact that the observed $\alpha$ values are close to 1 [5].

### 8.1.2 Superconducting energy gap

A formula for zero-temperature superconducting energy gap is expressed as [5]:

$$\Delta(0) = 2\hbar\omega\exp\left(-\frac{1}{N_0 V_0}\right) = 2k_B\theta_D\exp\left(-\frac{1}{N_0 V_0}\right) \quad (8.5)$$

The results expressed in equations (8.1) and (8.5) together give a fundamental formula independent of any phenomenological parameters [5]:

$$\frac{\Delta(0)}{k_B T_c} = 1.76 \quad (8.6)$$

This result agrees well with the experimental results obtained in a large number of superconductors to within about 10%. Notable exceptions are lead and mercury, where the discrepancy is close to 30%. In these superconductors, the results of various other experiments also deviate from the predictions of weak-coupling BCS theory. Those experimental results can be explained better using more elaborate analysis within the framework of a strong-coupling theory [5].

The elementary weak-coupling BCS theory also predicts that near the critical temperature (i.e., $T \approx T_C$) and in the zero-magnetic field, the energy gap vanishes according to the universal law [5]:

$$\frac{\Delta(T)}{\Delta(0)} = 1.74\left(1 - \frac{T}{T_C}\right)^{1/2} \tag{8.7}$$

## 8.2 Strong coupling superconductivity

The expression for $T_C$ (equation (8.1)) obtained from BCS model Hamiltonian does not take the damping effect in the electron–phonon interaction into account, which arises because of the displacements of the heavier ions associated with the phonon modes being slower than the motion of the light electrons. In the early-1960s, Eliashberg introduced a set of equations based on a Green's function formalism, which included this damping effect and could be solved numerically [11]. The Eliashberg equations determine a renormalization function $Z$, and superconducting gap $\Delta$, for which the highest temperature $T$ that yields a nontrivial solution is the critical temperature $T_C$ [12]:

$$Z(\mathbf{k}, iw_n) = 1 + \frac{\pi T}{w_n} \sum_{\mathbf{k}',n'} \frac{\delta(\epsilon_k)\omega_{n'}}{N_0\sqrt{\omega_n^2 + \Delta^2(\mathbf{k}, i\omega_n)}} \times \lambda(\mathbf{k}, \mathbf{k}', n - n') \tag{8.8}$$

$$Z(\mathbf{k}, iw_n)\Delta(\mathbf{k}, iw_n) = \pi T \sum_{\mathbf{k}',n'} \frac{\delta(\epsilon_k)\Delta(\mathbf{k}, iw_{n'})}{N_0\sqrt{\omega_n^2 + \Delta^2(\mathbf{k}, i\omega_n)}}$$
$$\times \left[ \lambda(\mathbf{k}, \mathbf{k}', n - n') - N(0)V(\mathbf{k} - \mathbf{k}') \right] \tag{8.9}$$

Here $\omega_n$ are the fermionic Matsubara frequencies, with a cutoff $\omega_{cut}$ that is typically 10 times the highest phonon frequency of the system, and $\epsilon_k$ are the electronic energy eigenvalues [12]. The anisotropic electron–phonon coupling strength $\lambda(\mathbf{k}, \mathbf{k}', n - n')$, is expressed as:[12]:

$$\lambda(\mathbf{k}, \mathbf{k}', n - n') = \int_0^\infty \frac{2w}{(\omega_n - \omega_{n'})^2 + \omega^2}\alpha^2 F(\mathbf{k}, \mathbf{k}', \omega) \tag{8.10}$$

where the quantity $\alpha^2 F$ is the Eliashberg spectral function [12]:

$$\alpha^2 F(\mathbf{k}, \mathbf{k}', \omega) = N_0 \sum_j |g_{kk'_j}|^2 \delta(\omega - \omega_{k-k'_j}) \tag{8.11}$$

Here $gkk'_j$ stands for the electron–phonon matrix element. These equations are valid for systems whose Fermi surface requires an anisotropic treatment of the electron–phonon interactions.

The Fermi surface anisotropy is fairly weak for most bulk metal systems. In this case, the Eliashberg equations can be simplified to their isotropic formulation by averaging over the momentum vector $\mathbf{k}$. This leads to a set of equations in place of equations (8.8) and (8.9) [12]:

$$Z(iw_n) = 1 + \frac{\pi T}{w_n} \sum_{n'} \frac{\omega_{n'}}{\sqrt{\omega_n^2 + \Delta^2(i\omega_n)}} \times \lambda(n - n') \tag{8.12}$$

$$Z(iw_n)\Delta(iw_n) = \pi T \sum_{n'} \frac{\Delta(iw_{n'})}{\sqrt{\omega_n^2 + \Delta^2(i\omega_n)}} \times [\lambda(n - n') - \mu^*] \tag{8.13}$$

Here $\mu^*$ is a semiempirical parameter describing the residual Coulomb repulsion between electrons in a Cooper pair. In that case, the Eliashberg spectral function is expressed as [12]:

$$\alpha^2 F(w) = \frac{1}{2\pi N_0} \sum_{qj} \frac{\gamma_{qj}}{\omega_{qj}} \delta(\hbar\omega - \hbar\omega_{qj}) \tag{8.14}$$

Here $\gamma_{qj}$ is phonon linewidth, and $\omega_{qj}$ is frequency for a phonon $j$ with wave vector $q$.

McMillan solved Eliashberg equations numerically for 22 systems to determine a simple expression, which could be used to estimate superconducting transition temperatures $T_C$ [13]. The phonon density of state (DOS) of niobium (Nb) was used to model the shape of the Eliashberg spectral function. This yielded the simplified formula based on the Debye frequency $\theta_D$, and the renormalized Coulomb repulsion parameter $\mu^*$ [13]:

$$T_C = \frac{\theta_D}{1.45} \exp\left[ -\frac{1.04(1 + \lambda)}{\lambda - \mu^*(1 + 0.62\lambda)} \right] \tag{8.15}$$

The electron–phonon coupling constant $\lambda$ is derived from the isotropic Eliashberg spectral function as [13]:

$$\lambda = 2 \int_0^\infty \frac{\alpha^2 F(\omega)}{\omega} d\omega \tag{8.16}$$

The McMillan equation often fails to describe the superconducting behavior accurately. Subsequently, Allen and Dynes introduced a modified version of the McMillan equation to estimate superconducting transition temperatures $T_C$ [14]:

$$T_C = \frac{\omega_{\ln}}{1.2} \exp\left[ -\frac{1.04(1 + \lambda)}{\lambda - \mu^*(1 + 0.62\lambda)} \right] \tag{8.17}$$

Here $\omega_{\ln}$ stands for logarithmic average phonon frequency and is expressed as [12]:

$$\omega_{\ln} = \exp\left[\frac{2}{\lambda}\int_0^\infty \frac{\alpha^2 F(\omega)\ln(\omega)}{\omega}d\omega\right] \tag{8.18}$$

The Allen–Dynes equation (equation (8.17)) was based on 217 numerical solutions to the Eliashberg equations for $0 < \mu^0 < 0.2$, $0.3 < \lambda < 10$ and $\alpha^2 F(\omega)$ shapes of Pb and Hg obtained from the known tunneling experiments [12]. An Einstein model was also used, which included no Coulomb repulsion ($\mu^* = 0$) and whose shape of $\alpha^2 F(\omega)$ maximized $T_C$ for given values of $\lambda$ and $\omega_{\ln}$. Allen–Dynes equation considerably reduced the dependence of $T_C$ on the shape of the spectral function, and it is quite accurate for predicting $T_C$ for materials with $\lambda < 1.5$

In a computational study, the Eliashberg spectral function can be calculated for a material using density functional theory (DFT) codes, including the Quantum Espresso, ABINIT, and EPW packages from which values for $\lambda$ can be obtained, and then $T_C$ can be estimated via the Allen–Dynes equation (8.17). However, $T_C$ estimated with equation (8.17) is too low in strongly coupled systems, where $\lambda > 1.5$. In these cases, the usual approach is either a numerical solution to the Eliashberg equations to generate a more appropriate estimate of $T_C$, or the application of correction factors $f_1$ for strong coupling and $f_2$ for shape dependence. The approximate forms for these factors are expressed as [12]:

$$f_1 = \left[1 + \left(\frac{\lambda}{2.46(1 + 3.8\,\mu^*)}\right)^{3/2}\right]^{1/3} \tag{8.19}$$

$$f_2 = 1 + \frac{(\omega_2/\omega_{\ln} - 1)\lambda^2}{\lambda^2 + [1.82(1 + 6.3\,\mu^8)(\omega_2/\omega_{\ln})]^2} \tag{8.20}$$

Here the logarithmic average of the phonon frequency $\omega_{\ln}$ is calculated as in equation (8.18) and $\omega_2$ is expressed as [12]:

$$\omega_2 = \left[\frac{2}{\lambda}\int_0^\infty \omega\alpha^2 F(\omega)d\omega\right]^{1/2} \tag{8.21}$$

Another way to calculate $T_C$ is via the DFT for superconductors (SCDFT) [15]. In this approach, a density functional framework is created based on the electron density of the concerned system, N-body nuclear interactions, and the super-conducting order parameter. This does not rely on empirical parameters such as the effective Coulombic repulsion term $\mu^*$.

## 8.3 Various superconductors

Until the early-1970s the highest known superconducting transition temperature $T_C$ of 18 K was found in A15 intermetallic compound $Nb_3Sn$. This was surpassed in 1975 by another compound, $Nb_3Ge$, in the same A15 family with $T_C$ up to 22 K in sputtered thin films. Figure 8.2 shows a chronological evolution of superconducting

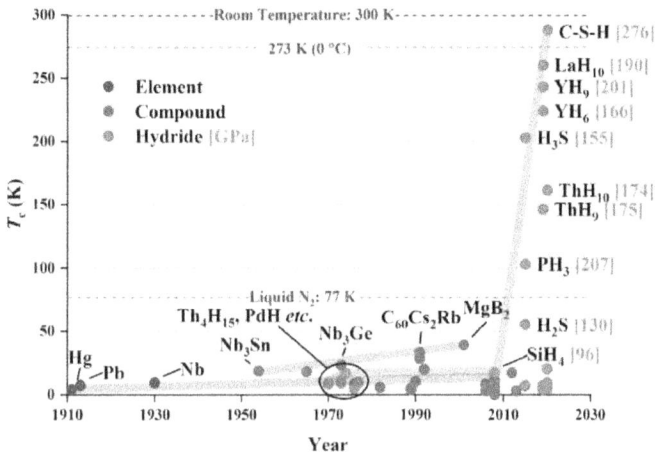

**Figure 8.2.** Chronological evolution of superconducting critical temperature ($T_C$) for various conventional superconductors. Reproduced from [26]. CC BY 4.0.

critical temperature (Tc) for various conventional superconductors. If the possibility of superconductivity in the compound $MgB_2$ had been explored earlier (the other physical properties were investigated), it would have taken the lead with a $T_c$ of 39 K. But this measurement was not done until 2001, whereas the other physical properties were investigated even in the 1950s. However, far from achieving superconductivity at room temperature, these $T_C$s were not even near the boiling temperature of nitrogen (77 K). In the late-1980s, a completely different family of materials (namely the superconducting cuprates) was discovered. A member of the family YBaCuO with $T_C$ of 93 K was the first to cross the boiling point of liquid nitrogen for superconducting transition [16, 17]. Unlike the A15 compounds $Nb_3Sn$, $Nb_3Ge$, and the $MgB_2$ and also numerous other low-temperature superconductors, these superconducting materials belong to a class of unconventional superconductors [18] whose superconducting mechanism are yet to be understood completely and cannot be explained within the realm of the standard Bardeen–Cooper–Schrieffer (BCS) theory. Apart from high $T_C$ cuprate superconductors, other members of the unconventional superconducting class are heavy-fermion superconductors, organic superconductors, and iron-based superconductors [18].

## 8.4 Possible superconducting state in metallic hydrogen

In 1968, Ashcroft proposed theoretically that hydrogen would be a high-temperature superconductor when metalized under pressure [19]. He argued that the light mass of hydrogen—with its concomitant high phonon frequencies, a large concentration of electronic states near the Fermi level, and strong electron–phonon coupling arising from the lack of screening by core electrons—would lead to a superconducting phase existing up to very high temperatures. The anticipated metallic phase of hydrogen would have a zero-pressure density corresponding to an $r_s$ value of about 1.6 and a longitudinal sound velocity $v_s = 1.6 \times 10^6 \mathrm{cm} \ \mathrm{s}^{-1}$ [5]. The velocity of sound in metallic hydrogen is substantial. This in turn would yield a value of $\alpha = 0.45$. For the sake of comparison, values of $\alpha$ for Na, K, Al, and Pb are, respectively, 0.82, 1.00, 0.51, and 0.53 [5].

Coupled with a light ionic mass of hydrogen, the corresponding Debye temperature is $\theta_D = 3.5 \times 10^3$ K. Substitution of the relevant quantities in equation (8.3) gives a reasonably lower limit on $N_0 V = 0.25$. It may be noted that if metallic hydrogen exists in a close-packed hyper crosslinked polymer (HCP) or face-centered cubic (FCC) structure, the screened point-ion potential is large enough to cause a significant zone contact with the Fermi surface. This would increase the Umklapp contribution already underestimated in equation (8.3). In any case, the anticipated value of $T_C$ is pretty enhanced [19]. Ashcroft [5] also remarked that an estimate of the possible upper bound on $T_C$ would hinge on a more detailed estimate of the Umklapp contributions to $N_0 V$. In summary, one can say that with a high Debye temperature that is ensured from the low mass of the proton in metallic hydrogen and assuming a reasonable value for $N(0)V$, from the BCS expression for $T_C$ (equation (8.1)) the superconducting transition temperature in metallic hydrogen can be expected to be very high.

The originally predicted pressure for hydrogen metallization by Wigner and Huntington [20] was indeed quite an underestimate. They treated the compressibility of hydrogen using its zero-pressure value throughout to obtain the pressure value, without accounting for decreasing compressibility with applied pressure. Despite the continuing increase in the predicted pressure of hydrogen metallization [21, 22], even achieving the metallization of hydrogen experimentally in the laboratory has remained quite controversial (see chapter 6). Thus, elemental hydrogenic superconductivity has remained out of reach until now.

## 8.5 Superconductivity in hydrides

In 1970, Gilman [23] proposed that a form of hydrogen in the metallic state might be achieved by changing the dielectric constant of the medium in which it exists. The equilibrium separation distance and the compressibility scale are in proportion to the dielectric constant. This should substantially reduce the pressure needed to make the metallic state of hydrogen stable. It was suggested that the compound $LiH_2F$ is nearly stable under ambient conditions and it would be metal if moderate pressure were applied to it. This compound might have very high hydrogen density, high electrical conductivity, high Debye temperatures in the hydrogen sublattice, and perhaps a high superconducting transition temperature [23].

After more than 30 years, in 2004 Ashcroft returned to this subject and explicitly suggested that compounds with a high hydrogen content might be, in effect, chemically precompressed metallic hydrogen [24]. Ashcroft considered the covalent metallic hydrides with large hydrogen content. Within these hydride phases, the hydrogen sublattice can assume many forms: atomic/hydridic hydrogen, molecular $H_2$, and $H_3^-$ (and more exotic) units, weakly covalent clathrate cages, and more [12]. One ubiquitous group of phases includes those with a mixture of atomic H and molecular $H_2$. The presence of distinct hydrogenic motifs is associated with varying degrees of electron transfer to and from the molecular and atomic units, which strongly influences the electronic structure of the resulting phases.

In group 4a hydrides (see figure 6.8), for simple structures there are eight electrons per unit cell. In such combinations, in a chemical sense hydrogen has already undergone a form of precompression [24]. On subjecting it to further external pressure, it is expected to enter a metallic phase. The electrons from both the hydrogen and the group 4a element may participate in common overlapping bands [24]. Two important criteria are likely to influence the superconductivity in hydrides: (i) a large hydrogen-derived electron density of states at the Fermi level, and (ii) large modifications of the electronic structure in response to the vibration of hydrogen atoms [25, 26]. The hydrogen-poor metal hydrides, in general, do not fit either of these criteria. Ashcroft proposed the possibility of finding good superconductors among hydrides, including those with a large amount $H_2$ molecules. However, once there is a stabilization of $H_2$ molecules in the lattice, hydrides naturally fail to satisfy criterion (i) because hydrogen electrons in a major portion will occupy low-lying energy states far below the Fermi level arising from the formation of intramolecular H–H sigma bonds [26].

The central physical feature of these metal hydride systems lies in the scales of energies associated with their lattice dynamics. As in the case of metallic hydrogen, they are exceedingly high for the less massive protons. But there are also lower frequency branches for the more massive ions [24]. They have much higher charges and lead to appreciably stronger electron couplings. They might also conform to strong-coupling superconductors as sole constituents with phonons as exchange bosons. The overlapping bands of the compressed hydrides will be wide, the density of states generally high, and the electron-ion interactions from protons and group IVa ions quite significant [24]. Within the framework, these systems may be described as wideband insulators or semiconductors. The subsequent attainment of a band-overlap state of contiguous bands would lead to a compensated metal with a Fermi surface in several zones [5]. Due to the wideband character of the metallic phases, Coulomb pseudopotential $\mu^*$ entering into approximations for the superconducting transition temperature is also expected to be favorable for the onset of a superconducting phase. Thus, the arguments supporting metallic hydrogen as a possible candidate for high-temperature superconductivity [24] would be applicable for these hydrides, and also possibly with an additional physical boost from the lower mode frequencies and the further possibility of tuning them to values that are optimal in pursuit of higher $T_C$s [5]. In terms of required pressures, the onset of the metallic phases may well occur at values considerably lower than those required to drive elemental hydrogen metallic.

Ashcroft [24] set some important dynamical energy scales to proceed further. First, the proton plasmon energy for a metallic state of hydrogen is $\hbar\omega_{p,p} = (m_e/m_n)^{1/2}2\sqrt{3}/r_s^{3/2}$ (in Rydbergs). Here $m_n$ is the mass of the nucleon and $m_e$ is the mass of an electron. Second, for a group 4a hydride the equivalent energy is $(m_e/m_n)^{1/2}2\sqrt{3}[8/(A = 4)]^{1/2}/r_s^{3/2}$ (in Rydbergs) for an acoustic plasmon, where $Am_n$ is the mass of the nucleus involved. This energy translates into a somewhat lower scale when screening is incorporated. Third, in the dense hydride, there is also a characteristic optical plasmon energy for the proton subsystem. If the ions are regarded as massive, and if in a long wavelength limit their charge is taken to cancel half that of the standard background, then this energy is just $(m_e/m_n)^{1/2}\sqrt{6}/r_s^{3/2}$ (in Rydbergs) [24]. This is lower than that for protons alone at an equivalent density, but still remains very substantial.

We have seen earlier in section 8.2 that the essential dimensionless quantities entering approximate solutions for $T_C$ of the linearized Eliashberg equations are $\lambda$, which embodies the attractive effects of the electron–phonon coupling, and $\mu^*$, which accounts for the generally repulsive effects of direct Coulomb interactions between electrons. Ashcroft [19] introduced scales for $\lambda$ and $\mu$ for a one-component system using the following elementary arguments. If $V_c$ is the Coulomb interaction, then within the BCS framework a measure of repulsive effects is $\langle N_0 V_c(\mathbf{k}' - \mathbf{k})\rangle$, the average being taken over a spherical Fermi surface of diameter $2k_F$. Here $N_0$ is the density of states per unit volume for a given spin, evaluated at the Fermi energy. If screening is treated in the Thomas–Fermi approximation, and wave vectors normalized to $2k_F$, then:

$$N_0 V_c(\mathbf{x} - \mathbf{x}') = (\gamma^2/2)[(\mathbf{x}' - \mathbf{x})^2 + \gamma^2] \tag{8.22}$$

Here $\mathbf{x} = \mathbf{k}/2k_F$, $\gamma = (k_0/2k_F)^2$ and $k_0$ is the Thomas–Fermi wave vector. The average of equation (8.22) over a sphere of unit diameter then gives $(\gamma^2/2)\ln(1 + 1/\gamma^2)$. Now, Coulomb couplings propagate far faster than ion dynamical equivalents [24]. With the inclusion of this retardation effect, the estimate for $\mu^*$ is expressed as [24]:

$$1/\mu^* = [(\gamma^2/2)\ln(1 + 1/\gamma^2)]^{-1} + \ln(\hbar\omega_{pe}/\hbar w_p) \tag{8.23}$$

Here $\hbar\omega_{pe}$ is the electron plasmon energy and $\hbar\omega_p$ is the corresponding quantity for the ions. Equation (8.23) then yields the well-known value $\mu^* \sim 0.1$.

Equation (8.22) needs to be overcome by the attractive contributions, $\langle N_0 V_{ph}\rangle$, originating with screened electron–phonon coupling $g_q(\mathbf{k}', \mathbf{k})$ associated with the scattering of an electron from $\mathbf{k}$ to $\mathbf{k}'$ by a phonon of wave vector $\mathbf{q}$ to reach a superconducting state. In this case, the average is required of $-N_0 2|g_q(\mathbf{k}', \mathbf{k})|^2/\hbar\omega(\mathbf{k}' - \mathbf{k})$, where for a Debye spectrum and longitudinal modes [24]:

$$g_q(\mathbf{k}', \mathbf{k}) = \frac{4\pi e^2 f(\mathbf{k}', \mathbf{k})}{[(\mathbf{k}' - \mathbf{k})^2 + k_0^2]} i(\mathbf{k}' - \mathbf{k})\left(\frac{\hbar}{2cq} \frac{N/V}{Am_n}\right)^{1/2} \tag{8.24}$$

In equation (8.24), $c = \alpha v_{E_F} \frac{3m_e}{(ZAm_m)^{1/2}}$ stands for the speed of sound. The role of the dimensionless $\alpha$ is to correct the standard Bohm–Staver estimate [10]. The quantity $f(\mathbf{k}', \mathbf{k})$ includes a pseudopotential correction to the familiar point-ion result. Then, for normal intraband ($\mathbf{k}' - \mathbf{k} = \mathbf{q}$) processes, the contribution to $\langle N_0 V_{ph}\rangle$ follows from an average on a sphere of unit diameter but with the restriction $q = |\mathbf{k}' - \mathbf{k}| < k_D$, is expressed as [24]:

$$-N_0 V_c(\mathbf{x} - \mathbf{x'}) = \left[ \frac{f^2(\mathbf{x}, \mathbf{x'})}{\alpha^2} \frac{\gamma^2}{\gamma^2 + (\mathbf{x} - \mathbf{x'})^2} \right] \quad (8.25)$$

The quantity $\alpha$ is $<1$, and when densities are high $f$ reflects the appreciable short-range repulsive region of the pseudopotential. A comparison of equations (8.22) and (8.25) reveals that depending on the system the phonon mechanism will have its important role, and the inclusion of Umklapp terms $((\mathbf{k'} - \mathbf{k}) = \mathbf{q} + \mathbf{K}$, with $\mathbf{K}$ a reciprocal lattice vector) can increase the contribution from equation (8.25) considerably [24].

From the definition of $N_0$ and for $NZ$ electrons in a volume $V$, it also follows that $|g_\mathbf{q}(\mathbf{k'}, \mathbf{k})|^2 \sim (V/Zn)(\epsilon_F \hbar \omega_D)$. Therefore, $g$ is proportional to the geometric mean of the electron and phonon energy scales [24]. Depending on the pressure, the electron energy scale can be 20eV and above, while the phonon energy scale can reach 0.2 eV. Thus, the coupling is large and, depending on the system, the phonon term $\lambda = \langle N_0 V_{\mathrm{ph}} \rangle$ can reach the strong-coupling limit $\sim 1$.

In search for approximate solutions of the Eliashberg equations, the phonon-exchange term (8.24) is written more generally by utilizing the phonon spectral function (see equation (8.11)) in terms, and in turn the phonon-exchange parameter $\lambda$ (see equation (8.16)). The important prefactor in MacMillan equation (8.15) is a characteristic average of the phonon energies originally taken by McMillan to be close to the Debye energy. This will be large for the metallic hydrides [24]. If the metallic group 4a hydrides are in the strong-coupling class, then given their substantial phonon energies the values of the superconducting transition temperature are expected to be significant [24].

The estimated values of $T_C$ from McMillan equation (equation (8.15)) are known to be quite sensitive to the determination of $\lambda$, and hence to $\alpha^2 F$. Bergmann and Rainer [27] have investigated this matter and showed that actual influence of the contributing phonon frequencies on $T_C$ was through the (positive definite) functional derivative $\delta T_C = \delta \alpha^2 F(\omega)$. The frequencies most effective in bringing about changes in $T_C$ were to be found in the vicinity of $2k_B T_C/\hbar$. With the $T_C$s of pure metallic hydrogen being estimated to be more than 100 K, this is an important observation. This is likely to have a considerable bearing on the role of the lower modes of the hydrides. A tuning within this domain may well be brought about by extending the binary hydride $MH_4$ to a ternary, $M^{(1)}_{1-x}M^{(2)}_x H_4$ with considerable flexibility in the choice of the heavy ion masses [24]. The higher domain of frequencies associated with the protons will continue to lead to a very significant Debye temperature $\theta_D$.

These works led to a spurt of activities looking for superconductivity in hydride systems. The three developments that were central to this were [28]: (i) the reliable prediction of the stable structures of the hydrides under pressure, (ii) the accurate computation of their superconducting properties, and (iii) their experimental realization in diamond anvil cells.

## 8.6 Structure and superconductivity of hydrides from first principles

In the early-2000s, with the introduction of the methods for general first-principles structure prediction, the computational discovery of materials with previously unknown structures became popular [28]. Strategies for exploring low-lying configurations of the DFT energy landscapes generated by a state-of-the-art plane wave and pseudopotential codes [29, 30] were adopted for evolutionary and random structure searching [31]. The repeated stochastic generation of structures, followed by careful DFT-based relaxations to the nearby local minima of the Born–Oppenheimer potential, is the starting point for successful first-principles approaches to structure prediction [31]. This is known as ab initio random structure searching if no other steps are taken. It benefits from parallelism and broad exploratory searches. A particular emphasis is placed on the generation of sensible initial structures, in which chemical ideas such as coordination, distances, units, and symmetry are included [31]. For greater utilization of this hard-won information, evolutionary [32] and swarm approaches subsequently took over by utilizing what has already been learned about the energy landscape, while trading some simplicity, parallelism, and exploratory power. The combined application of random and swarm-based searches has been particularly powerful in the study of the hydrides [35, 42]. In combination with general-purpose plane wave DFT codes, databases of reliable potentials covering the entire periodic table, and the arrival of commodity multicore CPUs, first-principles structure prediction has now become widespread and almost routine [31]. Using modern structure prediction methods, it can be easier, faster, and more reliable to rediscover the structures and compute their ground-state energies from the first principles. This approach is even useful for those structure prototypes that might be available in a database.

There have been many striking applications of first-principles structure prediction, in particular to high-pressure phase transitions [25]. In the absence of experimentally derived information, structure prediction has provided the most reliable microscopic models of dense hydrogen itself [28].

The discussions in the previous sections make it clear that the superconducting hydrides are expected to be conventional superconductors with the electron–phonon interaction driving the coupling mechanism, which leads to the condensation of the Cooper pairs. This means that it is possible to perform first-principles calculations of their superconducting critical temperatures using established theoretical and computational approaches [28]. Exploiting the dramatic increase in available computational power, such first-principles calculations have been central to the characterization and understanding of the properties of superconducting hydrides, and also to the prediction of new high-$T_C$ hydride compounds.

Within the framework of DFT, three basic ingredients are required to calculate the $T_C$ once the crystal structure for a given material is known [28]: (i) the Kohn–Sham energies $\epsilon_i$ and wave functions $|\phi_i\rangle$, where $i$ labels a given electronic state; (ii) the phonon frequencies $\omega_\mu$, with a mode index $\mu$; and (iii) the electron–phonon matrix elements $g_{ij}^\mu$, expressed as [36]:

$$g_{ij}^\mu = \left\langle \phi_i \left| \frac{\partial V_{KS}}{\partial u^\mu} \right| \phi_j \right\rangle \tag{8.26}$$

In the above equation (8.26), $u^\mu$ is the atomic displacement according to the normal mode $\mu$ and $V_{KS}$ is the Kohn–Sham effective potential. Phonon frequencies can be calculated within the harmonic approximation, truncating the Born–Oppenheimer energy surface at second order [28]. Linear response theory or finite difference approaches are used to calculate the harmonic force constants. The phonon matrix elements are calculated from linear response or finite difference methods [28].

With the use of the Kohn–Sham energies, phonon frequencies, and electron–phonon matrix elements, the Eliashberg function $\alpha^2F(\omega)$ can be directly evaluated as a phonon density of states weighted by the electron–phonon interaction at the Fermi energy $\epsilon_F$ as [24]:

$$\alpha^2F(\omega) = \sum_{ij\mu} \left| g_{ij}^{\mu} \right|^2 \delta(\omega - \omega_\mu)\delta(\epsilon_i - \epsilon_F)\delta(\epsilon_j - \epsilon_F) \qquad (8.27)$$

The electron–phonon coupling, which measures the strength of the attractive interaction between the electrons and the phonons, is then computed using equation (8.16). Then, the semiempirical Allen–Dynes equation (equation (8.17)) is typically used to predict the critical temperature, with the average logarithmic frequency being $\omega_{ln}$ being computed using equation (8.18). Coulomb pseudopotential $\mu^*$, which accounts for the repulsive electron–electron interaction, is usually taken as a parameter around 0.1, though it can also be explicitly calculated [28].

This approach has been successful in accurately computing the $T_C$ of several compounds, but it suffers from certain limitations that may be important for the superconducting hydrides. First, as we have already discussed, the phenomenological equation tends to systematically underestimate $T_C$ for strong-coupling superconductors with $\lambda > 1$. These difficulties can be surmounted with the direct solution of the many-body Eliashberg equations for the superconducting gap or by adopting a DFT for superconductors, which is an extension of DFT accounting for the superconducting state [28]. A second important limitation is the breakdown of the harmonic approximation used for calculating the phonon frequencies. The electron–phonon coupling constant is a function of the phonon frequencies: $\lambda \sim \sum_\mu 1/\omega_\mu^2$ [28]. In addition, $\lambda$ can be changed substantially if anharmonic effects significantly renormalize the phonon frequencies. The superconducting transition temperature $T_C$ can also change as a result. Substantial anharmonic corrections to $T_C$ are expected in many superconducting hydrides and some candidate phases of hydrogen because of the low mass of hydrogen and its large quantum fluctuations around the equilibrium position (see [28] and references therein).

## 8.7 Developments on the experimental front

The first hydride superconductor $Th_4H_{15}$ with $T_C = 8$ K at ambient pressure was discovered in 1970 [37]. Importantly, that work initiated the search for superconductors among hydrogen-rich compounds as spurred by the proposal of high-temperature superconductivity in a metallic form of hydrogen. Subsequent studies of Pd–H and Pd–Cu–H systems revealed superconductivity below 10 K. The low superconducting transition temperatures were attributed to the negligible contribution from hydrogen to the electronic density of states at the Fermi level of these materials [26].

In 2006, Ashcroft and collaborators first suggested that compressed silane ($SiH_4$) might lead to superconductivity at a pressure lower than that required for pure $H_2$ [38]. Their prediction was backed up by first-principles computations of the expected properties of silane. In the same year, Picard and Needs [39] used a random searching strategy in conjunction with first-principles electronic structure computations to predict the stable high-pressure phases of silane. Their calculations indicated that a first-order phase transition occurred from an insulating/semimetallic phase to a good metal at a pressure achievable within a diamond anvil cell. These theoretical calculations anticipated that the metallic phase of silane might exhibit high-temperature superconductivity in the temperature range 40 K–260 K [38–40]. Experiments confirmed these structural predictions within the next two years. In 2008, Erements *et al* [40] reported the transformation of insulating molecular silane to metal at 50 GPa, which then became a superconductor at a relatively low transition temperature of $T_C = 17$ K at pressures between 96 and 120 GPa. The metallic phase has a hexagonal close-packed structure with a high density of atomic hydrogen, creating a three-dimensional conducting network [40]. Methane ($CH_4$) and germane ($GeH_4$) were also predicted to be high-temperature superconductors, but that is yet to be confirmed experimentally [28]. The next obvious candidates for exploration were the hydrogen storage materials—$LiBH_4$, $NaBH_4$, $NH_3BH_3$, $Si(CH_3)_4$—because of their high hydrogen content, but they did not become metallic, even at very high pressures. One of the hydrides, aluminum hydride ($AlH_3$), was found to be metal experimentally, but it did not become a superconductor at 20 K as predicted by the calculations. Despite these disappointing results, some groups persevered and made some remarkable predictions, most notably that $CaH_6$ would become a superconductor with a $T_C$ of 235 K at 150 GPa [41]. In addition, metallization for $GeH_4$ and superconductivity of $SnH_4$ with Tc close to 80 K at 120 GPa have also been predicted [40].

### 8.7.1 Discovery of superconductivity in hydrogen sulfide

Hydrogen sulfide ($H_2S$) is a prototype molecular system and a sister molecule of water ($H_2O$) [42]. Li *et al* [42] performed an extensive structural study on solid $H_2S$ at pressure ranges of 10–200 GPa through an unbiased structure prediction method. Apart from finding two candidate structures for nonmetallic phases, this study was also able to establish stable metallic structures. Ab initio calculations [43] predicted metallization pressure to be around 111 GPa, which was approximately one-third of the currently suggested metallization pressure of bulk molecular hydrogen. Application of the Allen–Dynes equation (see equation (8.17)) for the metallic structure yielded the possible superconducting state with high $T_C$ values of 191 K to 204 K at 200 GPa, which is among the highest values reported for $H_2$-rich compounds and hydride so far [43].

In 2015, Drozdov *et al* [44] experimentally observed superconductivity in a hydrogen compound around 203 K. This transition temperature was above that predicted for $H_2S$. It was suspected that $H_2S$ disproportionated with temperature, and was likely transformed into $H_3S$ plus sulfur. Almost at the same time, independent calculations also suggested that $H_3S$ may be a high-$T_c$ superconducting compound [43]. $H_3S$ is a good metal with strong covalent bonding between hydrogen and sulfur atoms existing in this compound. This agrees with the general assumption that a conventional superconductor with high $T_C$ should have strong covalent bonding together with high-frequency modes in the phonon spectrum. This assumption is valid, at least, for the superconductor $MgB_2$ with a transition temperature of 39 K [45]. This observed superconductivity was characterized by zero resistance, observation of Meissner effect, and a shift of $T_C$ to lower temperatures with an applied magnetic field. Furthermore, a strong isotope effect was observed through the replacement of $H_2S$ with $D_2S$, which pointed to conventional superconductivity. The superconducting state of $H_3S$ has been further characterized by optical and magnetic measurements (see reference [28] and references therein). There is evidence for the presence of a large electron–phonon mediated superconducting gap in reflectivity measurements, which is in agreement with the reflectivity calculated with the anharmonic $\alpha^2 F(\omega)$. Further magnetic measurements in applied fields up to 65 T at 155 GPa revealed a critical magnetic field consistent with strong-coupling superconductivity with electron–phonon coupling parameter $\lambda \sim 2$. This value is in agreement with first-principles calculations [46].

### 8.7.2 Lanthanide hydrides and beyond

In 2019, two independent groups [47, 48] reported evidence for superconducting transitions as high as above 250 K in a lanthanum hydride at around 150–200 GPa. The compound was synthesized by directly annealing La and $H_2$ gas in a diamond anvil cell or using $BH_3NH_3$ as the hydrogen source [28]. The latter process dramatically simplifies the experiment because only solid samples are used, although the synthesis is less well controlled. Phases with different structures [49] and stoichiometry were synthesized at nearly the same pressure-temperature conditions, which was a severe experimental problem. The final product depends on the kinetics of the transformations. Based on the volume per formula unit, a stoichiometry of around $LaH_{10}$ was estimated. The most probable candidate for such an extraordinary value of superconducting transition temperature is a hydrogen clathrate structure with $LaH_{10}$ stoichiometry. Earlier, first-principles calculations based on DFT had made such predictions and suggested a new family of superconducting hydrides that possess a clathrate-like structure in which the host atom (calcium, yttrium, lanthanum) is at the center of a cage formed by hydrogen atoms. Figure 8.3 presents schematically the crystal structure of $LaH_{10}$. A pure superconducting metallic hydrogen lattice exists for this structure, to which the host La atoms donate electrons [49]. Superconductivity in this newly found lanthanum hydride compound was evidenced by the observation of zero resistance, an isotope effect, and a decrease in critical temperature under an external magnetic field, which suggested an upper critical magnetic field of about 136 tesla at zero temperature [47].

In early-2023, superconductivity was reported on a nitrogen-doped lutetium hydride with a maximum $T_C$ of 294 K at 10 kbar [50]. This is an example of a superconductor at room temperature and near-ambient pressures. The compound

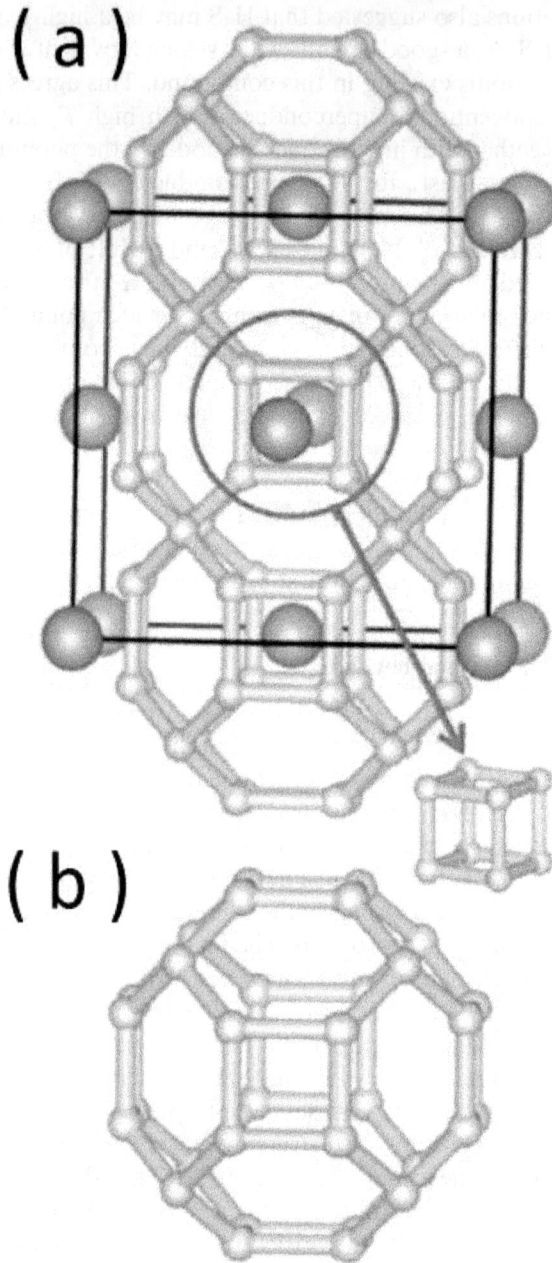

**Figure 8.3.** Schematic presentation of (a) Crystal structure of LaH$_{10}$; (b) hydrogen polyhedron in these structures. Reproduced from [49] with permission of PNAS.

was synthesized under high-pressure high-temperature conditions, and then after full recoverability its material and superconducting properties were examined along compression pathways. The experiments included the study of temperature-

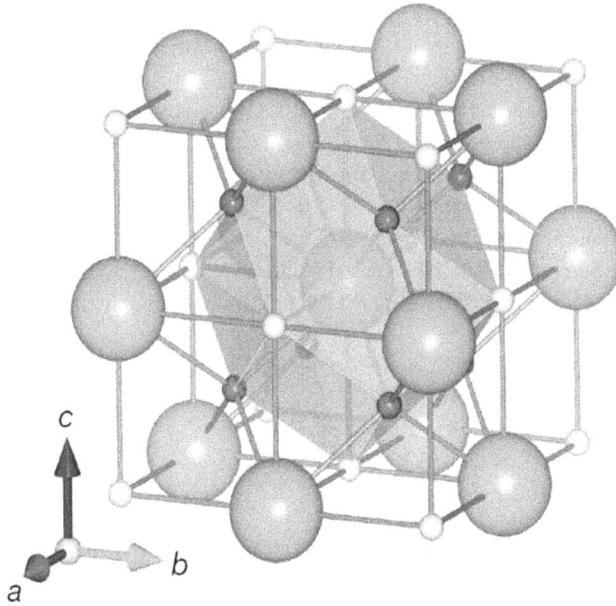

**Figure 8.4.** Schematic presentation of the proposed crystal structure of the superconducting LuH$_{3-\delta}$N$_\epsilon$ phase. Reprinted from [50], Copyright (2023), with permission from Nature Publishing.

dependent resistance with and without an applied magnetic field, magnetization (M) versus magnetic field (H) curve, AC, and DC magnetic susceptibility, as well as heat-capacity measurements [50]. X-ray diffraction experiments, along with energy dispersive x-ray and theoretical simulations provide some insight into the stoichiometry of the synthesized material [50]. The schematic crystal structure of the suggested main phase LuH$_{3-\delta}$N$_\epsilon$ responsible for superconductivity is shown in figure 8.4. The hydrogens in octahedral interstitial sites are shown in white and those in tetrahedral interstitial sites are shown in pink. The lutetium atoms are in green. The coordination polyhedron is shown around the central Lu atom. The cell is shifted by (0.5, 0.5, 0.5) fractional coordinates from the standard setting to better represent the coordination polyhedron. In contrast to the other high-temperature superconducting metal hydrides, where superconductivity shows up at several megabar pressure conditions, this newer discovery of an ambient temperature superconducting material opens up a new direction for superconductivity research. This work was subsequently subjected to severe scientific criticisms on the reliability of data presented in this manuscript, and was ultimately retracted in November 2023 [51].

# References

[1] van Delft D and Kes P 2010 *Phys. Today* **63** 38
[2] Meissner W and Ochsenfeld R 1933 *Naturwissenschaften* **21** 787
[3] London F 1950 *Superfluids* **vol 1** (New York: Wiley)

[4] Cooper L N and Cooper L N 1956 *Phys. Rev.* **104** 1189

[5] Ashcroft N W and Mermin N D 1976 *Solid State Physics* (Philadelphia, PA: Saunders College Publishing)

[6] Bardeen J, Cooper L N and Schrieffer J R 1957 *Phys. Rev.* **106** 162

[7] Kadin A M 2007 *J. Superconduct. Novel Magnet.* **20** 285

[8] Annett J F 2004 *Superconductivity, Superfluids and Condensates (Oxford Master Series)* (Oxford: Oxford University Press)

[9] Ziman J M 1960 *Electrons and Phonons* (New York: Oxford University Press)

[10] Bohm D and Starver T 1950 *Phys. Rev.* **84** 836

[11] Eliashberg G M 1960 *Sov. Phys.-JETP* **11** 966; http://www.jetp.ras.ru/cgi-bin/e/index/r/38/3/p966?a=list

[12] Hilleke K P and Zurek E 2022 *J. Appl. Phys.* **131** 070901

[13] McMillan W L 1968 *Phys. Rev.* **167** 331

[14] Allen P B and Dynes R C 1975 *Phys. Rev.* **B 12** 905

[15] Oliveira L N, Gross E K U and Kohn W 1988 *Phys. Rev. Lett.* **60** 2430

[16] Bednorz J G and Müller K A 1986 *Z. Phys. B Condens. Matter* **64** 189

[17] Wu M K *et al* 1987 *Phys. Rev. Lett.* **58** 908

[18] Norman M R 2011 *Science* **196** 332

[19] Ashcroft N W 1968 *Phys. Rev. Lett.* **21** 1748

[20] Wigner E and Huntington H B 1935 *J. Chem. Phys.* **3** 764

[21] Eremets M I and Troyan I A 2011 *Nat. Mater.* **10** 927

[22] Dalladay-Simpson P, Howie R T and Gregoryanz E 2016 *Nature* **529** 63

[23] Gilman J J 1971 *Phys. Rev. Lett.* **26** 546

[24] Ashcroft N W 2004 *Phys. Rev. Lett.* **92** 187002

[25] Zhang L *et al* 2017 *Nat. Rev. Mater.* **2** 17005

[26] Lv J *et al* 2020 *Matter Radiat. Extremes* **5** 068101

[27] Bergmann G and Rainer D 1973 *Z. Phys.* **263** 59

[28] Pickard C J, Errea I and Eremets M I 2020 *Annu. Rev. Condens. Matter Phys.* **11** 57

[29] Kresse G and Furthmüller J 1996 *Phys. Rev.* **B 54** 11169

[30] Clark S J *et al* 2005 *Z. Krist. Cryst. Mater.* **220** 567

[31] Pickard C J and Needs R J 2006 *Phys. Rev. Lett.* **97** 045504

[32] Oganov A R and Glass C W 2006 *J. Chem. Phys.* **124** 244704

[33] Wang Y, Lv J, Zhu L and Ma Y 2012 *Comput. Phys. Commun.* **183** 2063

[34] Li Y *et al* 2016 *Phys. Rev.* **B 93** 020103

[35] Peng Y *et al* 2017 *Phys. Rev. Lett.* **119** 107001

[36] Giustino F 2017 *Rev. Mod. Phys.* **89** 015003

[37] Satterthwaite B and Toepke I L 1970 *Phys. Rev. Lett.* **25** 741

[38] Feng J *et al* 2006 *Phys. Rev. Lett.* **96** 017006

[39] Pickard C J and Needs R J 2006 *Phys. Rev. Lett.* **97** 045504

[40] Eremets M I 2008 *Science* **319** 1506

[41] Wang H *et al* 2012 *PNAS* **109** 6463

[42] Li Y *et al* 2014 *J. Appl. Phys.* **140** 174712

[43] Duan D *et al* 2014 *Sci. Rep.* **4** 6968

[44] Drozdov A P *et al* 2015 *Nature* **525** 73

[45] An J M and Pickett W E 2001 *Phys. Rev. Lett.* **86** 4366

[46] Errea I 2016 *Nature* **532** 81

[47] Drozdov A P *et al* 2019 *Nature* **569** 528

[48] Somayazulu M *et al* 2019 *Phys. Rev. Lett.* **122** 027001

[49] Liu H *et al* 2017 *PNAS* **114** 6990

[50] Dasenbrock-Gammon N *et al* 2023 *Nature* **615** 244

[51] Dasenbrock-Gammon N *et al* 2023 *Nature* **624** 460

**IOP** Publishing

## Hydrogen
Physics and technology
**Sindhunil Barman Roy**

# Chapter 9

# Hydrogen fusion

A nucleus is, in general, made of protons and neutrons. An ordinary hydrogen atom has a single proton in its nucleus. All other elements have nuclei that contain neutrons as well as protons. Neutrons and protons are collectively known as nucleons. A proton has a positive charge of equal magnitude to that of an electron, and its mass of about 1840 times the mass of an electron [1]. The neutron is uncharged and its mass is slightly greater than that of the proton. The masses of a proton and a neutron are $m_p = 1.672\,623\,1 \times 10^{-27}$ kg and $m_p = 1.674\,9286 \times 10^{-27}$ kg, respectively [1, 2].

It is customary in nuclear and high-energy physics to represent mass in energy units according to the conversion formula $E = mc^2$ [1]. In this convention, the mass of a proton is 938.272 31 MeV $c^{-2}$ and the mass of a neutron is 939.565 63 MeVc$^{-2}$. The energy corresponding to the mass of a particle when it is at rest is called its rest energy [1].

Protons and neutrons are fermions and obey the Pauli exclusion principle, namely no two protons and electrons can reside in the same quantum state. However, one proton and one neutron can exist together in the same quantum state. The number of protons in a nucleus is denoted by $Z$, the number of neutrons by $N$, and the total number of nucleons by the mass number of the nucleus $A$. The number of protons $Z$ is also called the atomic number. The nuclei having the same number of protons but a different number of neutrons are called isotopes; some examples are hydrogen (with one proton), deuterium (one proton and one neutron), and tritium (one proton and two neutrons).

## 9.1 Properties of the nucleus

### 9.1.1 Nuclear radius

The Rutherford scattering experiment provided the first estimates of nuclear sizes [2]. In the Rutherford experiment, an incident alpha particle is deflected by a target nucleus in a manner consistent with Coulomb's law, provided that the distance between them exceeds about $10^{-14}$ m. Coulomb's law is not obeyed for smaller

doi:10.1088/978-0-7503-5172-0ch9
© IOP Publishing Ltd 2024

separations because the nucleus no longer appears as a point charge to the alpha particle. Subsequently, a variety of experiments, including electron and neutron scattering, have been performed to determine nuclear dimensions. An electron interacts with a nucleus only through electric forces, while a neutron interacts only through specifically nuclear forces. Electron scattering provides information on the distribution of charge in a nucleus and neutron scattering provides information on the distribution of nuclear matter [2]. In either case, the de Broglie wavelength of the particle needs to be smaller than the radius of the nucleus under study. It is found that the volume of a nucleus is directly proportional to the number of nucleons it contains, which is its mass number. This suggests that the density of nucleons is very nearly the same in the interiors of all nuclei [2].

The nucleus is such a small particle that a description of it may be more appropriate in terms of the wave function. A rough estimate of nuclear size may be made by locating the region where the wave function has an appreciable magnitude [1]. Experimental studies reveal that the average radius $R$ of a nucleus may be written as [2]:

$$R = R_0 A^{1/3} \tag{9.1}$$

Here $A$ is the mass number, and the value of $R_0 \approx 1.2 \times 10^{-15}$ m $\approx 1.2$ femtometer (fm).

The volume of a nucleus is

$$V = \frac{4}{3}\pi R^3 = \frac{4}{3}\pi R_0^3 A \tag{9.2}$$

The masses of proton and neutron are roughly equal, say $m$. The mass M of the nucleus is roughly proportional to the mass number $A$. Thus, one may write:

$$M = mA \tag{9.3}$$

It is evident from equations (9.2) and (9.3) that mass per unit volume or density of a nucleus is independent of the mass number $A$.

### 9.1.2 Nuclear spin

Protons and neutrons are fermions with spin quantum numbers of $s = \frac{1}{2}$. This leads to a spin angular momenta $S$ of magnitude [2]:

$$S = \sqrt{s(s+1)}\,\hbar = \sqrt{\frac{1}{2}\left(1 + \frac{1}{2}\right)} = \frac{\sqrt{3}}{2} \tag{9.4}$$

Associated with $S$ is spin quantum number $m_s = \pm\frac{1}{2}$. Magnetic moments associated with the spins of protons and neutrons are expressed in nuclear magnetons ($\mu_N$) [2]:

$$\mu_N = \frac{e\hbar}{2m_p} = 5.051 \times 10^{-27} \text{ J T}^{-1} = 3.152 \times 10^{-8} \text{ eV T}^{-1} \tag{9.5}$$

Here $m_p$ is the mass of the proton. The nuclear magneton $\mu_N$ is smaller than the Bohr magneton $\mu_B$ by the ratio of the proton mass to the electron mass, which is 1836 [2]. The spin magnetic moments of the proton and neutron have components in any direction. These are $\mu_{pz} = \pm 2.793\ \mu_N$ for proton, and $\mu_{nz} = \mp 1.913\ \mu_N$. There are two possibilities for the signs of $\mu_{pz}$ and $\mu_{nz}$ depending on whether $m_s$ is $-\frac{1}{2}$ or $+\frac{1}{2}$. The $\pm$ sign is used for $\mu_{pz}$ because $\mu_{pz}$ is in the same direction as the spin $S$, whereas $\mp$ is used for $\mu_{nz}$ because $\mu_{nz}$ is opposite to $S$ [2]. The hydrogen nucleus consists of a single proton, and its total angular momentum is given by equation (9.4).

### 9.1.3 Stability of nucleus

Not all combinations of neutrons and protons form stable nuclei [2]. In general, light nuclei ($A < 20$) contain approximately equal numbers of neutrons and protons. The proportion of neutrons becomes progressively greater in heavier nuclei. The tendency for $N$ to equal $Z$ follows from the existence of nuclear energy levels [2]. Nucleons have spins of $\frac{1}{2}$, and they obey the Pauli exclusion principle. Thus, each nuclear energy level can accommodate two neutrons of opposite spins and two protons of opposite spins. In nuclei, energy levels are filled in sequence, just like the energy levels in atoms, to achieve configurations of minimum energy and therefore maximum stability [2].

It is also to be noted that since protons are positively charged, they repel each other electrically. With more than 10 protons or so, this repulsion becomes so large in nuclei that an excess of neutrons, which produce only attractive forces, is required for stability [2].

## 9.2 Nuclear forces

A new kind of force, called the nuclear force, comes into being when the separation of nucleons comes in the range of the order of a femtometer. In this range, the nuclear force is much stronger than electromagnetic and gravitational forces. Nuclear forces are attractive and they keep the nucleons bound in a nucleus. The protons are charged particles and they exert repulsive Coulomb forces on each other. The neutrons are chargeless neutral particles and they do not have any electric interactions. The nuclear forces are operational between proton and proton, neutron and neutron, and also between proton and neutron. The overall effect of this attractive nuclear force is much stronger than that of the repulsive Coulomb forces between the protons [1]. Thus, the nucleus remains bound.

The nuclear force is not represented by a simple formula, unlike gravitational or electromagnetic force. The nature of the nuclear force is yet to be understood completely and scientists are still working on the details [1]. Some of the qualitative features of nuclear forces are enumerated below [1]:

1. Nuclear forces are short-ranged and effective only up to a distance of the order of femtometer or less. The force between two nucleons decreases rapidly with the separation distance and becomes negligible at the distance of

10 fm. The distance to which the nuclear force is effective is called the nuclear range.

2. Within the nuclear range, nuclear forces on average are $\approx$ 50–60 times stronger than electromagnetic forces.
3. Nuclear forces are independent of charge. The nuclear force between two protons is the same as that between two neutrons or between a proton and a neutron.
4. The force between a pair of nucleons is not solely dependent on the distance between the nucleons. The nuclear force also depends on the direction of the spins of the nucleons. The force is stronger (weaker) if the spins of the nucleons are parallel (antiparallel).

## 9.3 Binding energy

The nucleons are bound together in a nucleus and energy must be introduced to the nucleus for separating the nucleons over large distances. The amount of energy needed for this is called the binding energy of the nucleus. On the other hand, this much energy is released if the initially well-separated nucleons are brought to form the nucleus.

The rest mass energy of a nucleus is smaller than the rest mass energy of its constituent nucleons in the free state [1]. The difference between these two energies is the binding energy. Now, the rest mass energy of a free proton is $m_p c^2$ and that of a free neutron is $m_n c^2$. If the nucleus has a mass $M$, its rest mass energy is $Mc^2$. If the nucleus contains $Z$ protons and $N$ neutrons, the rest mass energy of its nucleons in the free state is $Zm_p c^2 + Nm_n c^2$. Then, the binding energy of the nucleus is expressed as:

$$B = (Zm_p + Nm_n - M)c^2 \qquad (9.6)$$

A useful quantity in nuclear physics is the binding energy per nucleon, which is the binding energy $B$ divided by the mass number. The binding energy of a deuteron containing a proton and a neutron is 2.22 MeV, so the binding energy per nucleon is 1.11 MeV. In the nuclei with increasing mass number, the binding energy per nucleon increases on average and reaches a maximum of around 8.7 MeV for $A \approx$ 50–80 [1]. The binding energy per nucleon then increases slowly with the increase in $A$.

## 9.4 Nuclear fusion

When two light nuclei come close together within the range of attractive nuclear force ($\approx$1 fm), they may combine to form a bigger nucleus. However, bringing together light nuclei within the separation of a femtometer is a rather difficult task. Any bulk material is made up of atoms having a radius of a few angstroms (1 Å $= 10^5$ fm). When the atoms are pushed together, their electrons cause a repulsive force between them. If the electrons are stripped off the atoms, then even the positive charges of their nuclei will cause them to strongly repel each other. In this situation, the possible way out is to heat a gas to an extremely high temperature so that electrons are completely stripped off the atoms, and the nuclei in the gas move at large random speeds. When all the electrons are detached from the atoms, one gets a plasma. In a hot plasma, two nuclei moving toward each other may come close enough to fuse into one nucleus.

Let us now consider two deuterons moving toward each other at equal speeds in a deuteron gas. As they move, the Coulomb repulsion will slow them down. The loss in kinetic energy will be gained in the Coulomb potential energy [1]. The kinetic energy is zero at the closest separation and the potential energy is $\frac{e^2}{4\pi\varepsilon_0 r}$. Let the initial kinetic energy of each deuteron be $K$ and the closest separation is 2 fm, then:

$$2K = \frac{e^2}{4\pi\varepsilon_0(2\,fm)}$$
$$= \frac{(1.6 \times 10^{-19}C)^2 \times (9 \times 10^9 Nm^2 C^{-2})}{2 \times 10^{-15}m}$$

(9.7)

Or,

$$K = 5.7 \times 10^{-14} J \tag{9.8}$$

Assuming the temperature of the deuteron gas to be $T$, the average kinetic energy of random motion of each deuteron nucleus will be $3/2\, k_B T$. Thus, the temperature needed for the deuterons to have an average kinetic energy of $5.7 \times 10^{-14}$ J will be:

$$T = \frac{5.7 \times 10^{-14} J}{1.5 \times 11.38 \times 10^{-23} \text{ J K}}$$
$$= 2.8 \times 10^9 \text{ K}$$

(9.9)

It is well known that nuclear fusion is the main source of energy in the Sun, which it radiates to the Universe, including the Earth. But the temperature inside the Sun is estimated to be about $1.5 \times 10^7$ K, which is well below the temperature estimated for nuclear fusion to take place (see equation (9.9)). There are two main reasons why nuclear fusion takes place in the Sun. One reason is that although the average kinetic energy of the particle is $3/2k_B T$, there are particles in the Sun having energy much higher than that. Second, and more importantly, even if the kinetic energy of two interacting nuclei is less than that needed to bring those within the nuclear range, there is a finite chance of fusion through the process of *quantum barrier penetration* [1].

### 9.4.1 Fusion in the Sun

Summarizing the discussion above, one can say that nuclear fusion is a process in which one or more light nuclei fuse to generate a relatively heavier nucleus. This leads to some mass deficiency, which is released as energy. The quantity of energy released follows Einstein's formula: $E = mc^2$, in which $E$ is the energy in joules, $m$ is the mass difference in kilograms, and $c$ is the speed of light. In the Sun, fusion takes place dominantly by proton-proton cycle as follows [1]:

$$^1\text{H} + {}^1\text{H} \rightarrow {}^2\text{H} + e^+\nu$$
$$^2\text{H} + {}^1\text{H} \rightarrow {}^3\text{He} + \gamma$$
$$^3\text{He} + {}^3\text{He} \rightarrow {}^3\text{He} + 2{}^1\text{H} \tag{9.10}$$
$$4{}^1\text{H} \rightarrow {}^4\text{He} + 2e^+ + 2\nu + 2\gamma$$

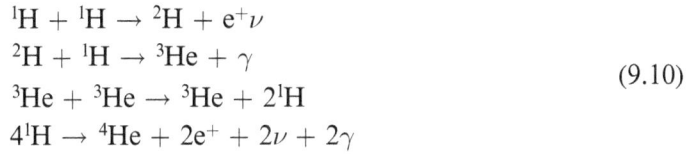

In the above set of reactions, $\nu$ denotes *neutrino* (a chargeless particle with negligible mass) and *gamma* denotes $\gamma$ *ray*. The first two reactions in the above set need to occur twice to produce two $^3\text{He}$ nuclei and initiate the third reaction. In this reaction cycle, effectively four hydrogen nuclei combine to form a helium nucleus. About 26.7 MeV energy is released in this reaction cycle [1]. So, hydrogen is the fuel in the Sun, which burns into helium to release energy. It has been estimated that the Sun has been radiating this energy for the last $4.5 \times 10^9$ years. The Sun will continue to do so till all the hydrogen in it is used up, which is possible after $5 \times 10^9$ years.

### 9.4.2 Fusion in the laboratory

In stellar objects such as the Sun, the fusion fuel remains confined at high temperatures due to gravitational pull. In the laboratory, the problem is to confine the hot plasma in a small volume for an extended period. Meanwhile, producing such a high-temperature environment is itself a challenging task. The main problem is the confinement at such a high temperature. One cannot use solid walls as containers because no solid can sustain the temperature required for a fusion reaction to take place. We will cover these technological problems in part II (chapter 16) of this book.

The easiest thermonuclear reaction that can be handled in the laboratory is the fusion of two deuteron atoms (D–D reaction) or fusion of a deuterium atom with a tritium atom (D–T) reaction [1]:

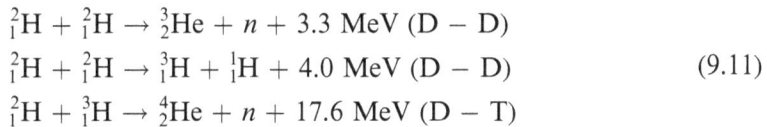

$$^2_1\text{H} + {}^2_1\text{H} \rightarrow {}^3_2\text{He} + n + 3.3 \text{ MeV (D} - \text{D)}$$
$$^2_1\text{H} + {}^2_1\text{H} \rightarrow {}^3_1\text{H} + {}^1_1\text{H} + 4.0 \text{ MeV (D} - \text{D)} \tag{9.11}$$
$$^2_1\text{H} + {}^3_1\text{H} \rightarrow {}^4_2\text{He} + n + 17.6 \text{ MeV (D} - \text{T)}$$

In laboratory fusion, the basic principle is to start with deuterium gas or a mixture of deuterium and tritium, heat the gas to high temperatures, ensure the confinement for a reasonable period, and look for the fusion event.

### 9.4.3 Lawson criterion

Three parameters—namely plasma temperature, density, and confinement time—need to be achieved simultaneously for a sustained fusion reaction to occur in a hot plasma. The product of these is called a fusion product. This product needs to exceed a certain quantity, derived from the so-called Lawson Criterion, named after the British scientist J D Lawson. He showed that in a fusion experiment to get an energy output than the energy input, the fusion reactor should achieve [1]:

$$n\tau > 10^{14} \text{ s cm}^{-3} \tag{9.12}$$

Here $n$ denotes the density of the interacting particle and $\tau$ denotes the confinement time. The quantity $n\tau$ in s cm$^{-3}$ is called the Lawson number. For a fusion reaction to occur, $n\tau \geqslant 10^{16}$ s cm$^{-3}$ for a D–D reaction and $n\tau \geqslant 10^{14}$ s cm$^{-3}$ for a D–T reaction.

## References

[1] Verma H C 2020 *Concepts of Physics* **vol 2** (New Delhi: Bharati Bhawan)
[2] Beiser A 2003 *Concepts of Modern Physics* (New York: McGraw Hill)

# Part II

Hydrogen technology

**IOP** Publishing

Hydrogen
Physics and technology
**Sindhunil Barman Roy**

# Chapter 10

## Applications of hydrogen

It is now well recognized that dihydrogen ($H_2$), commonly known as 'hydrogen', is a potentially carbon-free fuel. For example, it produces only water vapor when consumed in a fuel cell. Hydrogen can also be used as an energy carrier, which can store, move, and deliver energy produced from various other sources. It is an attractive fuel option for transportation and electricity generation applications. The potential applications of hydrogen today include the industrial sector (e.g., metals refining), liquid fuels (e.g., biofuels and synthetic fuels), heat generation, energy storage, and transportation. Hydrogen can be stored, distributed, and used as a feedstock for transportation (trucks, trains, ships, etc.), stationary power processes, or building heat, and also in industrial and manufacturing sectors (e.g., steel manufacturing). This creates an additional revenue stream and stimulates the economy.

Figure 10.1 schematically presents various existing and emerging applications of hydrogen. In this chapter, we will briefly introduce such applications of hydrogen, some of which will then be elaborated further in the subsequent chapters.

## 10.1 Hydrogen for power systems and energy storage

More than half of global final energy consumption and a third of global energy-related carbon dioxide ($CO_2$) emissions originate from heat generation in domestic and industrial activities [1, 2]. In the conventional burning process of natural gas, the majority of the products lead to greenhouse gaseous emissions, which contaminate the environment. The primary source of carbon dioxide emissions is energy consumption, mainly resulting from the burning of fossil fuels. In contrast, the use of hydrogen gas as an alternative fuel to natural gas has proved to be an efficient pathway to reduce greenhouse gaseous emission [2].

If hydrogen is produced using renewable energy sources, it can actively contribute to the decarbonization process in the energy sector. This is due to the reacting nature of hydrogen, whether it is combusted or used in a fuel cell. Currently, typical

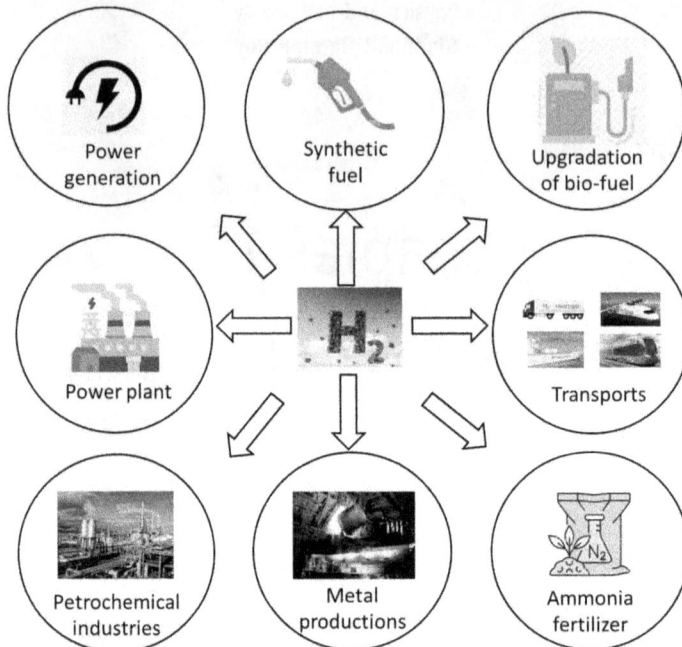

**Figure 10.1.** Schematic presentations of various existing and emerging applications of hydrogen (H$_2$).

nonrenewable sources are used to manufacture hydrogen, with coal, heavy oil/naphtha, and natural gas accounting for 18%, 30%, and 48% of this production, respectively [2]. Roughly one and a half million tonnes of carbon dioxide are released annually as a result of this production process [2]. Furthermore, the expensive natural gas used in the entire power-generating infrastructure needs a sizable amount of area for storage. In this sense, hydrogen seems to be a viable option for fueling gas turbines.

There are many benefits of using hydrogen in the central heating system of a building rather than natural gas. These include comparable operating activity and a higher rate of heat generation with less hazardous emissions [1]. When hydrogen is burned, its combustion velocity is higher than that of natural gas [1]. Thus, hydrogen can be used as a fuel feed stream to power sophisticated burners with specialized technical requirements.

The excess electricity produced by power plants can be converted into hydrogen, which can then be chemically changed into chemicals utilized in many industrial applications or directly used in the natural system [2]. In the aerospace sector, hydrogen can also be utilized alone or in conjunction with oxygen as a propellant. Due to its high energy output, the hydrogen-liquid/oxygen combination is better suited for space applications. Water is released during the hydrogen combustion process. This attribute, along with its superior efficiency as compared to gasoline, makes hydrogen suitable for use as an automotive fuel.

Hydrogen-based energy storage offers a lot of promise for medium- to long-term energy storage. Although the production of renewable energy has increased significantly in recent years, its availability is frequently seasonal and erratic. Therefore, it is necessary to store the generated renewable energy in a way that is resilient to fluctuations in those energy sources. The three most common forms of energy storage are power-to-power, power-to-heat, and power-to-gas [2]. In light of this, hydrogen provides excellent energy storage capabilities, a long storage period, and flexibility. It has the ability to absorb excess energy production and lessen energy volatility. Practically speaking, it can address concerns related to seasonal changes and the economic issues. Hence, hydrogen has the ability to store excess electrical energy when production rates are higher than needed and during periods when the cost of power is lowest, allowing it to be reused in the opposite circumstances [2]. On the other hand, in an environment with a high energy demand, hydrogen can be used to generate electricity.

Hydrogen has a greater capacity for storage than batteries. Unlike batteries, which have a maximum lifespan of hours, hydrogen can last for weeks or months. Hydrogen can also be used to convert renewable resources into energy under various seasonal and climatic conditions. According to estimates, hydrogen has a storage capacity of up to megawatt-hours, or even terawatt-hours, which is quite large when compared to batteries with a typical capacity of kilowatt-hours [2].

Globally, a large number of hydrogen power storage facilities have been built, demonstrating the potential of hydrogen technology. Orsted in Denmark, SoCalGas in the USA, and Underground Sun Storage in Austria are a few examples of power facilities designed to produce and store hydrogen [2]. The energy produced by renewable solar and wind resources is stored below the surface of Earth in the Underground Sun Storage project. Any excess power is stored for use in future emergencies and reprocessed into hydrogen using the electrolysis process. The Orsted facility was built to run electrolyzers by feeding them excess electricity generated by wind farms The SoCalGas project successfully produced methane inside a bioreactor by directly converting the hydrogen generated by the solar electric system.

With the employment of electrolyzers and fuel cells, hydrogen can be utilized as a tool to supply the energy grid with essential services, such as frequency maintenance and voltage strengthening. Additionally, by modifying its power factor to satisfy local voltage support requirements, hydrogen-based equipment can help with voltage support requirements [2]. This can be achieved by using an inverter or rectifier monitoring systems.

A process known as 'power-to-gas' uses electrical energy to create flammable gas [2]. Hydrogen is a combustible gas with a large power density, and thus it can lead to useful power-to-hydrogen technologies. In this process, the hydrogen generated by an electrolyzer can be converted into methane by the methanation process. This is then either pumped to the natural gas grid operating system or stored for subsequent usage.

Hydrogen fuel cells are efficient and environmentally-friendly energy generators. Through the redox processes, fuel cells can be employed as integrated electrochemical devices to transform the chemical energy into its electrical equivalent [2]. Water is the final byproduct of fuel cells powered by hydrogen. We will discuss fuel cells in more detail in chapter 15.

Several power systems can benefit from the use of fuel cells to increase their efficiency and lower their overall production costs, including co-generation systems (i.e., heat + power / cold + power) or tri-generation systems (i.e., cold + heat + power) [2]. Co-generation is the sequential production of two distinct useful forms of energy from a primary single source. For instance, fuel cells can produce enough power to meet electrical demand, while the heat that is emitted can be used for heating purposes. The overall efficiency will therefore be roughly 95% [2]. A co-generation fuel-cell system is made up of different components, including fuel processors, power supplies, heat recovery units, energy (thermal/electrochemical) storage units, control devices, additional apparatus (i.e., pumps), and stacks. In a tri-generation system, a single primary source achieves the required cooling using thermally-driven equipment.

## 10.2 Hydrogen as a transportation fuel

The transport sector accounts for one-fifth of the global primary energy demand, and currently almost entirely relies on petroleum. This transport sector is the second-largest source of $CO_2$ emissions, after electricity and heat generation, and is responsible for about 25% of global emissions. It is also one of the most challenging to decarbonize. This is due to its distributed nature and the advantages of fossil fuels in terms of high energy densities, ease of transportation, and storage.

Several countries have now identified hydrogen as a sustainable future fuel for decarbonizing the transportation sector. Renewable hydrogen can have an enormous impact on sustainable mobility applications, whether by powering fuel-cell electric vehicles (FCEVs) such as cars, trucks, and trains, or as a feedstock for synthetic fuels for ships and airplanes [3].

### 10.2.1 Road

Hydrogen-rich fuels can be converted into electricity in fuel cells through a chemical reaction. Instead of a battery, FCEVs can use fuel cells to power electric motors and operate near-silently without any tailpipe emission. Fuel cells also have a lower material footprint compared to lithium batteries. Internal combustion engine technology converts fuel into kinetic energy at roughly 25% efficiency. In comparison, a fuel cell can mix hydrogen with air to produce electricity at up to 60% efficiency. However, at more than 80%, pure EVs are more efficient than both. Overall, the key advantages are short refueling times, lower added weight for stored energy, and zero tailpipe emissions.

Long-distance and heavy-duty vehicles offer the greatest potential. However, the high costs of ownership and a lack of enabling infrastructure are key challenges that must be addressed through policy support, technological innovation, and financial investment [3]. The challenges of low utilization of refueling stations can be met with captive fleets of taxis, buses, and trucks, and there are ongoing efforts in various countries in this direction. Figure 10.2 shows some such examples of hydrogen buses that are operational in India and England.

(a)          (b)

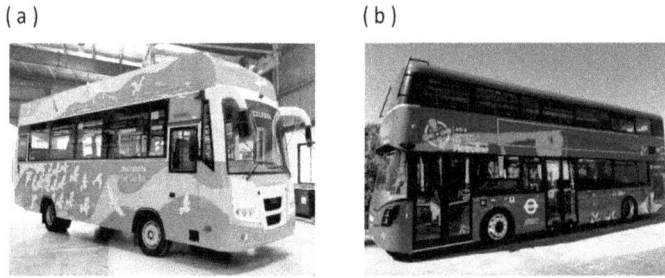

**Figure 10.2.** Hydrogen buses plying the roads of (a) India (image source: GrowNXT Digital; https://www. grownxtdigital.in/general/news/india-hydrogen-fuel-cell-buses-launch//) and (b) England (image source: Business Wire; https://www.businesswire.com/news/home/20210623005632/en/Nel-and-Linesight-Collaborate-on-Hydrogen-Fueled-Bus-Programme-for-Transport-for-London-TfL.).

### 10.2.2 Railway

Railways are one of the most energy-efficient and clean transport modes. Trains carry 9% of global motorized passengers and 7% of freight but account for only 3% of energy demand and 1% of $CO_2$ emissions for the overall transportation sector [3]. Renewable hydrogen-powered trains have much promise in rail freight, especially where long distances and low network utilization do not justify the high costs of track electrification. Hydrogen trains use hydrogen battery engines, where hydrogen and oxygen are used to produce electricity to charge the battery of the train. Tanks for the hydrogen gases and the fuel cells are stationed underneath the trailer cars of the train and give a range of around 1000 km.

Hydrogen trains also hold promise due to flexible bi-mode operations, allowing them to run both on electrified and conventional railway lines. Innovation in compressing and storing hydrogen will, however, be required for economic viability and scalability. Efforts are being made in this direction in various countries. Germany was the first to introduce hydrogen trains into commercial service in 2018 (see figure 10.3), and China was the second country in the world and the first in Asia. India aims to introduce its first hydrogen train by the end of 2024.

### 10.2.3 Shipping

Shipping is one of the most efficient forms of freight transport. At the same time, it accounts for about 3% of overall global and 11% of transportation-related $CO_2$ emissions. Other gaseous emissions such as nitrogen oxide and sulfur oxide are also associated with shipping activities [2]. Moreover, the leakage of heavy fuels in the aquatic environment poses a threat to the ecosystem. Renewable hydrogen can overcome such limitations. Hydrogen can be employed in the shipping sector in two ways: (i) fuel cells or (ii) internal combustion engines. Fuel cells are suitable for ships sailing for long distances and can supply the ancillary energy requirements of larger ships in comparison to the other battery-powered ones. However, the cost involved at present is higher in comparison to fossil fuels, and there is the challenge of cargo volume loss due to fuel storage. In terms of energy content parity, while batteries

**Figure 10.3.** Hydrogen train introduced by European railway manufacturer Alstom (image credit: Alstom).

**Figure 10.4.** Norwegian hydrogen-powered ferry developed in 2012 (image credit: LMG Marin).

require 64 times more volume than marine diesel oil, hydrogen requires 3 times more [3]. The deployment of global refueling networks also needs to be addressed. A sustainable ferry designed by LMG Marin in 2021 (see figure 10.4) and operated by Noled, Norway is the world's first liquid hydrogen-powered vessel.

### 10.2.4 Aviation

Petroleum forms the majority of the fuel used in the aviation sector. Presently, aviation accounts for around 3% of global energy-related $CO_2$ emissions and 12% of emissions in the transport sector. It adds further to the overall contribution to global warming through emissions other than $CO_2$, such as nitrogen oxides and soot. Refrigerated hydrogen fuel can be potentially better than petroleum as aviation fuel because it emits fewer greenhouse gaseous emissions. Furthermore, a hydrogen-fuel-operated aircraft is characterized by minimal maintenance costs, long lifetime of engines, high energy content, and better combustion [3]. Overall, liquid hydrogen has great promise as an energy-efficient aviation fuel for reducing greenhouse gaseous emissions, thus resulting in a significant improvement in air quality. However, there are some constraints in using hydrogen as an aviation fuel, such

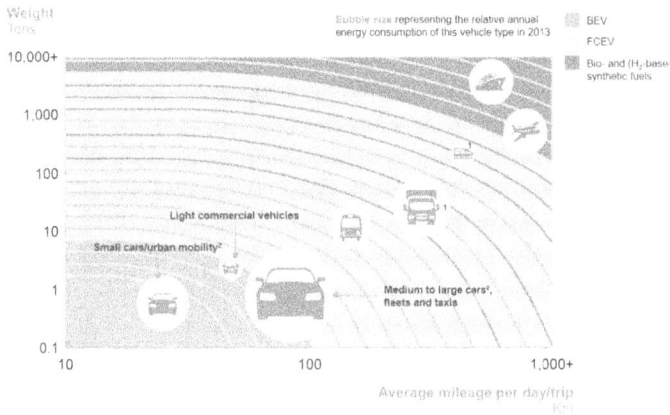

**Figure 10.5.** Hydrogen applications in the mobility sector. Reproduced from [3].

as depressed ignition energy, high flammability, and the possibility of unburned traces forming that result in metal embrittlement [2]. Significant R&D is needed to resolve these issues.

In summary, hydrogen can complement existing efforts to electrify road and rail transportation and provide a scalable option for decarbonizing shipping and aviation [3]. Figure 10.5 summarizes the mobility segments for which battery electric vehicles (BEVs), FCEVs, and vehicles running on bio- or hydrogen-based synthetic fuels are most applicable.

## 10.3 Production of hydrocarbon fuels

Synthesis gas (syngas) is a mixture of carbon monoxide and hydrogen in various ratios. It is a product of different thermochemical conversion processes, e.g., pyrolysis, gasification etc., and can be utilized in two ways: (i) direct fuel or (ii) transformed into transportation fuels via Fischer–Trospch synthesis process and syngas fermentation [2]. The two strategies are categorized as gas-to-liquid transformation strategies that can generate hydrocarbon fuels and alcohols based on the syngas feedstock stream [2].

Reaction temperatures and pressures of 200–350 °C and 1.5–4 MPa, respectively, are required for Fischer–Trospch synthesis. The process takes place in three main stages: (i) syngas production, (ii) syngas treatment, and (iii) transformation into hydrocarbon fuels associated with their upgrading. Other valuable products such as paraffin, naphtha, etc can also be produced during the production process of transportation fuel. The as-produced green fuels based on the Fischer–Trospch process have many advantages over petroleum-based fuels [2]. These include excellent burning characteristics, elevated smoking points, and freedom from heavy contaminants. The physicochemical properties of resultant fuels depend on the reaction conditions, i.e., reactor type, heating rate, residence time, etc.

Syngas fermentation or biorefining process is regarded as the interconnection between the biochemical and thermochemical scenarios [2]. It can give rise to value-added products (i.e., alcohols) from syngas by flexibly employing several groups of microorganisms at different reaction temperatures of 37–40 °C and 55–90 °C for mesophilic (i.e., clostridium autoethanogenum) and thermophilic (i.e., moorella thermoacetica), respectively. The feedstock of syngas can be simply converted into alcohols in two subsequent stages. The first stage is the production of acetyl coenzyme A, which is followed by transformation into ethanol. Other alcohols and chemicals (i.e., acetate, butanol, and formate) can be synthesized by acetogenic bacteria. There are several operational advantages in syngas fermentation over the Fischer–Trospch process [2]:

1. No costly pretreatment step is necessary.
2. Versatility of the process with different biomass compositions.
3. Independence on the hydrogen/carbon monoxide ratio in the feedstock upstream.
4. High selectivity of as-used microorganisms.
5. Moderate ambient working parameters with no necessity for catalysts to be used, with the associated poisoning problems.

However, syngas fermentation process involves various operational challenges: (1) weak solubility of the gas in the liquid state, (2) complicated bioreactor design, (3) existence of impurities, and (4) low yield of production. In summary, integration between different thermochemical, biochemical, and hydrothermal routes can effectively compensate for the shortage of individual techniques and maximize productivity.

## 10.4 Refining of crude oil and petroleum products

Hydrogen is used as an upgrading agent for crude oil products and petroleum distillates in terms of hydrocracking and hydroprocessing processes. Hydrocracking is the technique of treating heavier hydrocarbons with hydrogen in order to simultaneously divide them into lighter derivatives and improve the hydrogen/carbon ratio [2]. During the hydroprocessing process, numerous treatment procedures are used to extract various heteroatoms from petroleum products, including nitrogen, sulfur, oxygen, and heavy metals. These processes encompass hydrodemetallization, hydrodeoxygenation, hydrodesulfurization, and hydrodenitrogenation [2]. Refining is accomplished by catalyzing a reaction between hydrogen and the upstream feedstock, which includes heavy oils and petroleum products. As a result, the aromatic (C–C) bonds get saturated and these impurities are eliminated. Fuel upgrading is a direct result of the process of removing these contaminants from feedstocks.

## 10.5 Production of ammonia

Ammonia is an important ingredient used in industrial fertilizer processes. It is also utilized in a number of different industries, including fuel cells, polymer processing, explosives, refrigerant, medicines, and gas sensors. The synthesis of ammonia involves a catalytic reaction between hydrogen and nitrogen elements through the

Haber process [2]. Furthermore, the ammonia synthesis process predominantly uses steam gas reforming. However, this is not an environmentally-friendly method to obtain the hydrogen needed for the process. As a result, there is growing interest in alternative environmentally-friendly and sustainable methods for producing ammonia, such as photocatalytic nitrogen fixation (also known as artificial photocatalysis) and electrochemical hydrogen manufacture technologies. Hydrogen can be generated from water employing an electrolysis process using renewable wind and solar energy. This reduces harmful greenhouse gaseous emissions.

## 10.6 Metallurgical industries

Hydrogen is utilized in the metallurgical industries as a reducing agent and to create oxy-hydrogen fires, which allow metals to be extracted from their ores. The process of creating oxy-hydrogen flames involves allowing hydrogen and oxygen to react at extremely high temperatures (3000 °C). This is utilized for cutting and welding tasks in nonferrous metals. Hydrogen has a well-known capacity to extract metals from the aqueous solutions containing their salts. The metals could be incorporated into a composite material or powdered for use in metallurgical applications in the future [2].

Hydrogen can interact chemically with the other elements of the periodic table in three different ways [2]: (i) the elements of the Ia and IIa groups form an ionic bond; (ii) the elements of the VIa, VIIa, and VIII groups form an interstitial solid solution; and (iii) the elements of the IIIa, IVa, and Va groups form a metallic bond. Furthermore, the self-trapping of metals and the propensity of hydrogen to capture free electrons can lead to electrostatic shielding. The process of metal–hydrogen interaction is also efficiently facilitated by the small particle size of hydrogen [2].

## References

[1] Dodds P E *et al* 2015 *Int. J. Hydrog. Energy* **40** 2065
[2] Osman A I *et al* 2021 *Environ. Chem. Lett.* **20** 153
[3] De Blasio N, Hua C and Nunez-Jimenez A 2021 *Sustainable Mobility: Renewable Hydrogen in the Transport Sector, Belfer Center for Science and International Affairs* (belfercenter.org/ENRP)

# Chapter 11

# Methods of hydrogen production

Hydrogen is the most abundant element in the Universe, yet it is not easily found in its elemental form in nature. Dihydrogen ($H_2$), commonly known as hydrogen (and henceforth will be termed so), is only scarcely present in the Earth's atmosphere (0.07%) and is quite rare in the Earth's crust (0.14%). However, it may be produced from any primary energy source and used as fuel in a fuel cell or for direct combustion in internal combustion engines, with water being the only by-product.

Hydrogen can be produced by a variety of processes. These production processes can be divided into two major categories, namely conventional and renewable technologies, according to the raw materials used [1]:

1. The processing of fossil fuels is an essential element in the conventional technologies, and includes the methods of hydrocarbon reforming and pyrolysis. The chemical techniques involved in the hydrocarbon reforming process are steam reforming (SR), partial oxidation, and autothermal SR.

2. In renewable methods, hydrogen is produced by using renewable resources, such as water or biomass. The techniques that use biomass as a fuel can be further classified into two broad subcategories: biological and thermochemical processes. In contrast to the key biological processes of direct and indirect biophotolysis, dark fermentation, photo-fermentation, and sequential dark and photo-fermentation, thermochemical technology primarily consists of pyrolysis, gasification, combustion, and liquefaction. The other class of renewable technologies splits water by processes such as electrolysis, thermolysis, and photo-electrolysis to make hydrogen, with water as the only material input.

A schematic illustration of hydrogen creation from various energy sources is shown in figure 11.1. The process of producing hydrogen in each of these technologies requires extracting and separating hydrogen in the form of $H_2$ molecules at the state of purity needed for that particular use. The hydrogen

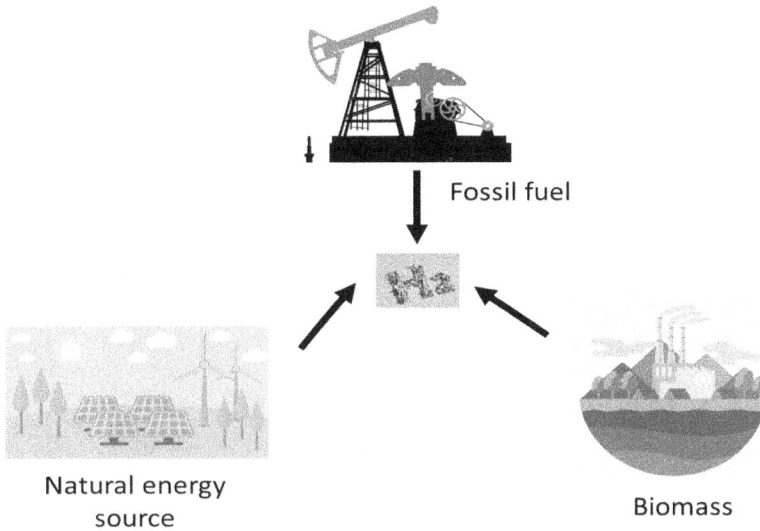

**Figure 11.1.** Schematic representation of hydrogen generation from different sources.

produced by various methods is color-coded according to the amount of $CO_2$ (one of the primary drivers of global warming) released and its influence on the environment. We will talk about this environmental cleaning issue before going into the details of the hydrogen generation technology.

## 11.1 Environmental cleanliness and hydrogen color coding

In the literature, the hydrogen cleanness level is generally described with color coding; mainly gray, blue, and green. This color coding is described below and is based only on the production route, i.e., the origin of the hydrogen [2, 3]:

1. Gray hydrogen is produced using fossil fuels such as natural gas. In this route, 1 tonne of hydrogen production gives rise to 10 tonnes of $CO_2$.
2. Blue hydrogen is produced from fossil fuels like gray hydrogen but $CO_2$ released during the production of hydrogen is stored underground. Capture and storage of $CO_2$ mitigates environmental emissions but increases the overall cost of blue hydrogen production.
3. Green hydrogen is usually produced from 100% renewable sources, such as wind or solar energies, with a lower carbon footprint.
4. Yellow hydrogen is distinguished from green hydrogen in that the former is produced by using electricity supplied from the grid. The total greenhouse gas emissions of yellow hydrogen mainly depend on how the electricity has been generated and on the energy mix.
5. Black or brown hydrogen is produced from black (bituminous) or brown (lignite) coal via coal gasification. The carbon footprint of this production technique is considerably high in comparison to the other methods.
6. Turquoise hydrogen is produced from the thermal decomposition of natural gas, i.e., methane pyrolysis or cracking by spitting methane into hydrogen

and carbon at a temperature range from 600 °C to 1200–1400 °C. In contrast with gray hydrogen, the production process of turquoise hydrogen produces black carbon soot as a by-product instead of carbon oxide emissions in gray hydrogen.

7. Purple, pink, and red color codes indicate the use of nuclear energy to produce hydrogen. Hydrogen can be produced from water by using electricity, heat, or through electrolysis, a thermochemical cycle, or a hybrid cycle. Purple hydrogen is produced from nuclear electricity through electrolysis, the color code red indicates hydrogen generation through thermochemical cycles, while the pink color code indicates generation through hybrid cycles such as high-temperature electrolysis or hybrid thermochemical cycle.

8. Aqua hydrogen is produced from oil sands (natural bitumen) and fields without emitting any $CO_2$. The production process utilizes electricity to supply air separation units, with the consequent injection of oxygen into the oil reservoir. There a water-gas shift reaction takes place at 350 °C. The generated synthesis gas and $CO_2$ are left underground, while hydrogen is extracted using palladium alloy membranes. Aqua hydrogen exploits fossil fuel energy but does not emit $CO_2$.

9. White hydrogen is produced by solar thermochemical water splitting. This technology exploits thermal solar power and water, which are both renewable.

Figure 11.2 presents a summary of the main hydrogen color codings and its association with the hydrogen production route. A few conclusions can be drawn from the analysis of hydrogen color coding with the available data [3]: (i) black, brown, and yellow hydrogen have the highest carbon footprints; (ii) blue and turquoise hydrogen have the potential to achieve negative GHG emissions; (iii) green and nuclear-based hydrogen currently have the lowest carbon footprints;

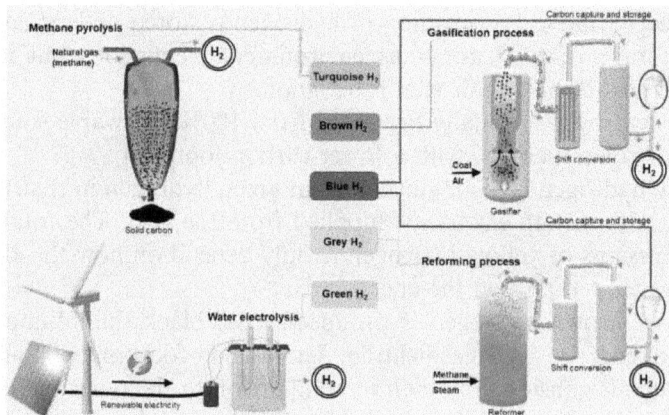

**Figure 11.2.** Hydrogen color coding for various technological processes. Reproduced from [2]. CC BY 4.0.

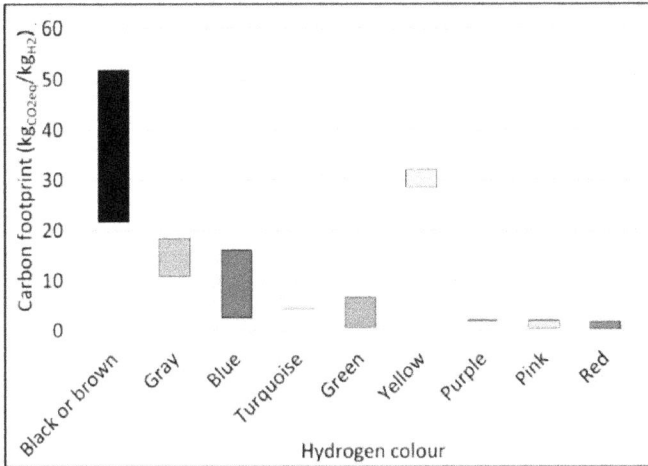

**Figure 11.3.** Carbon footprints of hydrogen with different color codings. Reproduced from [3]. CC BY 4.0.

but (iv) only green, white, and blue (if produced from biomethane) hydrogen can be considered renewable. Figure 11.3 presents carbon footprints of hydrogen with various color codings.

The color coding is, however, not precise for the complete characterization of hydrogen because it assumes that green hydrogen always has lower carbon emissions than blue or gray hydrogen [2]. This is not strictly correct in all cases. For example, blue hydrogen is regarded as less safe than green hydrogen, although it does not release carbon during the entire process, while green hydrogen may do. Moreover, the $CO_2$ capture and storage (CCS) technique used in the blue hydrogen reduces toxic emissions significantly. On the other hand, the manufacture of photovoltaic panels as renewable energy technology has a significant carbon footprint and generates various types of waste, liquid, and gaseous byproducts that are hazardous to the environment [2]. A solar panel has a 30-years lifespan, and after that it must be treated as a particular waste.

Environmental impacts for steam methane reforming with water electrolysis using wind, solar photovoltaic, hydropower, solar thermal, and biomass gasification as energy sources have been compared in a recent study [4]. It was concluded that among all the technologies evaluated, solar photovoltaic electrolysis had the most damaging environmental effects. This is because of the significant acidification potential in the photovoltaic panel production phase and the relatively poor efficiency of photovoltaic systems.

This discussion indicates that the life cycle of the equipment used is crucial, along with accurate measurement of the greenhouse gas emissions in the entire production process. This is necessary to evaluate how green is green hydrogen and how blue is blue hydrogen [2]. An improved model for hydrogen color coding has been proposed recently, which consists of a hydrogen cleanness index followed by the number of depth levels [5]. For example, 80 green-4 means that this hydrogen is produced via renewable resources but is only 80% green, due to emissions related to the process.

The number after the color, which in this case is 4, indicates that greenhouse gas emissions ($CO_{2-e}$) linked with the purification during the production process have been considered [2].

## 11.2 Hydrogen production from fossil fuels

Currently, fossil fuels have a dominant role in the production and supply of hydrogen worldwide. Approximately 48% of hydrogen is created from natural gas, 30% from naphtha and heavy oils, and 18% from coal [1]. The most advanced and widely-used techniques in this field at the moment are pyrolysis and hydrocarbon reforming.

To produce hydrogen from fossil fuels, a membrane reactor is a necessary component [6]. A membrane is a structure that permits mass transmission in the presence of gradients in driving forces, such as concentration, pressure, temperature, electric potential, etc [1]. Depending on the regime used for separation, membranes can be classified as dense, porous, or ion-exchange. The two main classes of membranes are biological and synthetic. Furthermore, the synthetic membranes can be classified as inorganic (ceramic or metallic) and organic (polymer). High hydrogen selectivity, high permeability to function at high flows and constrained surfaces, and strong chemical and structural stability are all necessary for a membrane to be used in hydrogen generation [6]. The common geometries of tubular and planar membranes are shown in figure 11.4. Essential components of a composite membrane are a sufficient porosity support that permits gas passage and a barrier that limits inter-diffusion in the metallic support [1, 6].

### 11.2.1 Hydrocarbon reforming methods

In hydrocarbon reforming methods, the hydrocarbon fuel is converted into hydrogen through some reforming techniques. Steam reforming (SR) is the name given to the endothermic reaction that occurs when steam is added to hydrocarbons as a reactant during the reforming process. Partial oxidation is the term used to describe the exothermic reaction when oxygen is the other reactant. The process that results from

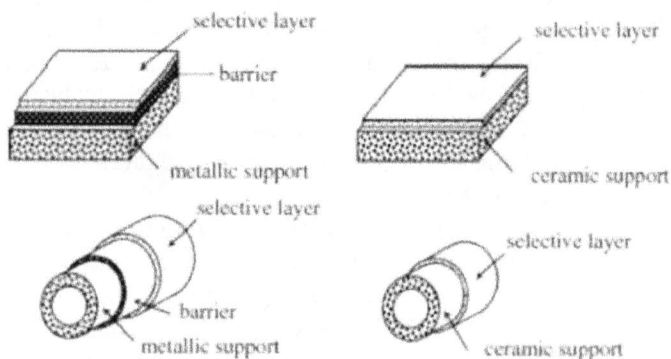

**Figure 11.4.** The geometries of planar and tubular composite membranes. Reprinted from [1], Copyright (2017), with permission from Elsevier.

combining these two processes is called an autothermal reaction. The desulfurization unit, the reforming and cleanup sections, and the ancillary units such as pumps, compressors, expanders, heat exchangers, coolers, combustor, etc are parts of a typical reforming plant.[1].

### 11.2.1.1 SR method

During the SR process, hydrocarbons and steam are catalytically transformed to carbon oxides and hydrogen. The production of reforming or synthesis gas (syngas), water-gas shift (WGS), and methanation or gas purification are the three main stages in the process. A steam reformer can be fed with various mixes of light hydrocarbons, such as ethane, propane, butane, pentane, and light and heavy naphtha, as well as natural gas and other methane-containing gases [1]. Prior to the reforming step, there must be a desulfurization stage if the feedstock contains organic sulfur compounds. This is to avoid poisoning the catalyst used as a reformer, which is often nickel-based. The operating parameters of the reforming process are adjusted at high temperatures, pressures up to 5MPa, and steam-to-carbon ratios of 3.5 in order to obtain the desired purity of the hydrogen product and avoid coking formation on the catalyst surface. The gas mixture enters a heat recovery stage after the reformer and is then fed into a WGS reactor. The mixture either passes through a $CO_2$-removal stage and methanation, or a pressure swing adsorption (PSA), after the CO interacts with the steam to produce additional hydrogen. The end product is $H_2$ with a higher purity of near 100%. The $CO_2$ emissions can be lowered significantly by $CO_2$ capture and storage (CCS). This is the process through which $CO_2$ is captured and injected in geological reservoirs or the ocean [1].

The main chemical reactions that take place in SR are:

$$\text{Reformer: } C_nH_m + nH_2O \rightarrow nCO + \left(n + \frac{1}{2}m\right)H_2 \tag{11.1}$$

$$\text{WGS Reactor: } CO + H_2O \rightarrow CO_2 + H_2 \tag{11.2}$$

$$\text{Methanator: } CO + 3H_2 \rightarrow CH_4 + H_2O \tag{11.3}$$

Equation (11.1) is used to model the steam reactor of methane by substituting $n = 1$ and $m = 4$. With a conversion efficiency of 74%–85%, steam methane reforming (SMR) is the most popular and advanced technique for producing hydrogen on a large scale [1]. By integration of SMR with an optional membrane, either directly inside the reaction environment or applied downstream to reaction units, the technology is further enhanced to produce significant volumes of $H_2$. Combination of the chemical reaction and gas separation in a single unit and usage of palladium-based membrane reactors lead to more benefits.

Figure 11.5 presents a schematic of a multiple membranes reactor for methane SR. The molecular hydrogen produced in the reformer is transported by adsorption and atomic dissociation on one side of the membrane. It is then followed by dissolution and diffusion in the membrane, and finally desorption on the other side [6]. The Pb-based membrane reactors enable methane conversions up to 90%–95% at lower temperatures 450–550 °C as opposed to 850–900 °C of traditional SMR.

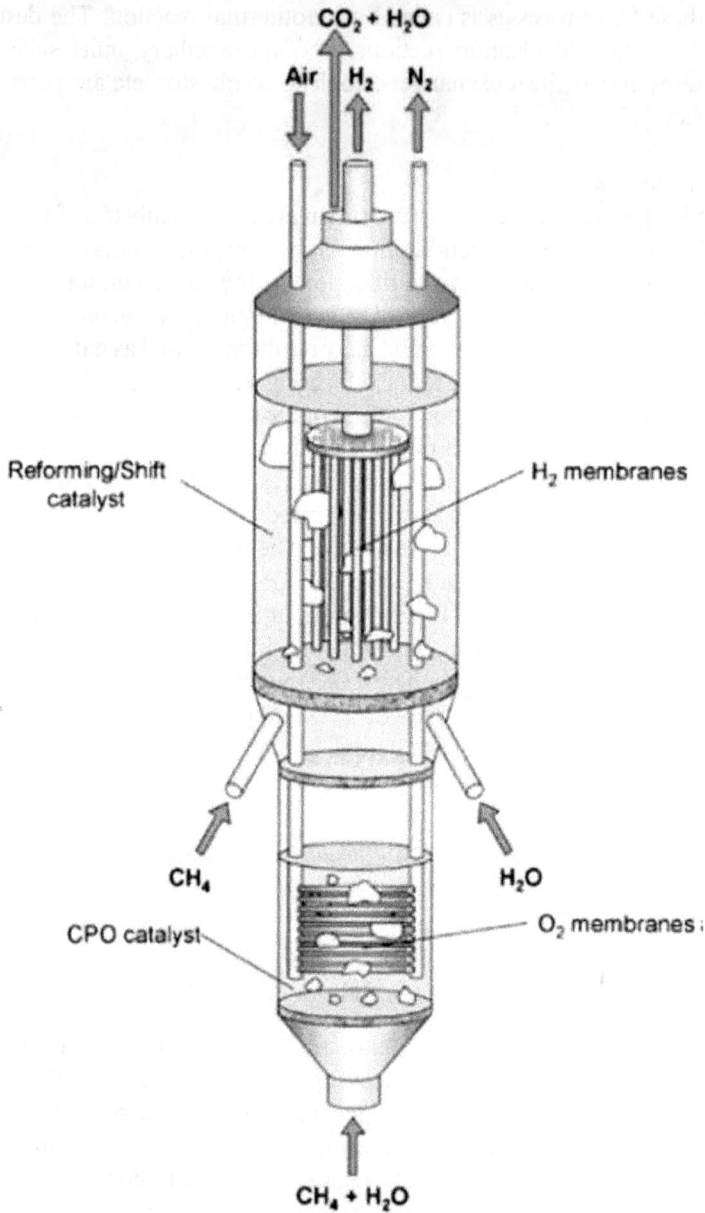

**Figure 11.5.** Multiple membranes reactor for methane SR. Reprinted from [6], Copyright (2011), with permission from Springer.

### 11.2.1.2 Partial oxidation method

Steam, oxygen, and hydrocarbons are transformed into hydrogen and carbon oxides via the partial oxidation (POX) process. Methane to naphtha are typical feedstocks used in the catalytic process, which runs at around 950 °C [1]. Methane, heavy oil, and coal are among the hydrocarbons used in the non-catalytic process, which runs between 1150 °C and 1315 °C [1]. Hydrocarbon feedstock is partially oxidized using pure $O_2$ that is obtained after sulfur removal. Following that, the generated syngas is treated in a manner identical to that of the product gas in the SR process. However, the cost of the oxygen plant along with the additional costs of desulphurization process make it quite capital-intensive. Equations (11.4) and (11.5) provide the catalytic and non-catalytic reforming processes, respectively, whereas equations (11.6) and (11.7) illustrate the chemical reactions of methanation and WGS [1]:

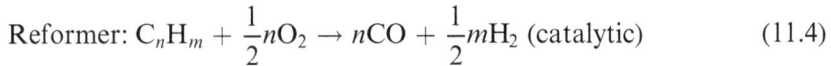

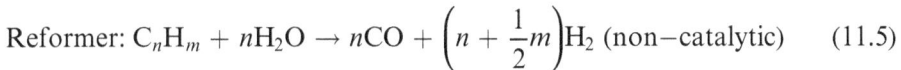

$$\text{Reformer: } C_nH_m + \frac{1}{2}nO_2 \rightarrow nCO + \frac{1}{2}mH_2 \text{ (catalytic)} \tag{11.4}$$

$$\text{Reformer: } C_nH_m + nH_2O \rightarrow nCO + \left(n + \frac{1}{2}m\right)H_2 \text{ (non--catalytic)} \tag{11.5}$$

$$\text{WGS Reactor: } CO + H_2O \rightarrow CO_2 + H_2 \tag{11.6}$$

$$\text{Methanator: } CO + 3H_2 \rightarrow CH_4 + H_2O \tag{11.7}$$

POX technology is the most suitable for producing hydrogen from heavy oil residues and coal. Residual fuel oils can be represented by applying $n = 1$ and $m = 1.3$ to equation (11.5), and at a pressure of 6 MPa, the typical composition of syngas is 46% $H_2$, 46% CO, 6% $CO_2$, 1% $CH_4$ and 1% $N_2$.

Equation (11.5) can be used to represent coal feed by putting $n = 1$ and $m = 0$. The process is known as coal gasification. This is a prominent method through which hydrogen is obtained from coal. The reaction mechanisms of coal gasification have some resemblance to that of the POX of heavy oils distribution. However, there is a substantial economic impact from the additional processing of the comparatively unreacted fuel as a solid and the removal of the large amounts of ash.

### 11.2.1.3 Autothermal reforming method

In the autothermal reforming (ATR) method, the exothermic partial oxidation provides the heat and endothermic SR induces hydrogen production. Steam and oxygen or air are injected into the reformer, enabling the reforming and oxidation reactions to occur simultaneously. The process is represented in equation (11.8):

$$C_nH_m + \frac{1}{2}nH_2O + \frac{1}{4}nO_2 \rightarrow nCO + \left(\frac{1}{2}n + \frac{1}{2}m\right)H_2 \tag{11.8}$$

Figure 11.6 presents a flow diagram of the ATR of methane representing the process explained in equation (11.8) with $n = 1$ and $m = 4$. The thermal efficiency is 60%–75%, while the optimum operating value has been calculated at around 700 °C inlet temperature, for sulfur–carbon ratio S/C = 1.5 and oxygen–carbon ratio $O_2$/C = 0.4. Here, the maximum hydrogen yield is about 2.8 [1].

### 11.2.2 Hydrocarbon pyrolysis

In hydrocarbon (CH) pyrolysis, the only source of hydrogen is the hydrocarbon itself, which undergoes thermal decomposition through the following general reaction:

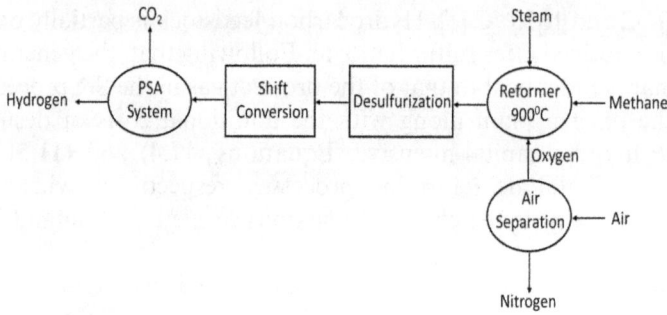

**Figure 11.6.** Flow diagram showing the process of autothermal reforming of methane.

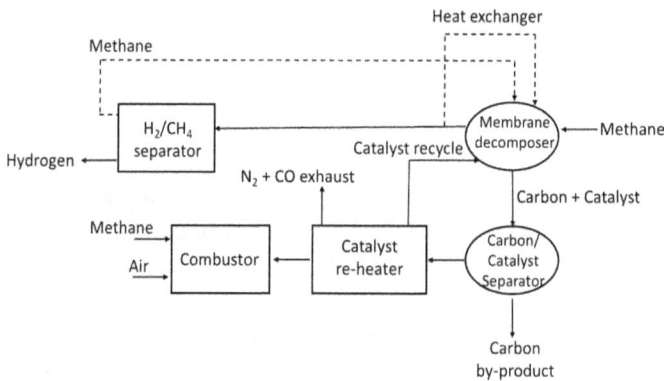

**Figure 11.7.** Flow diagram showing the process of methane pyrolysis.

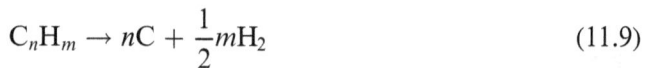

$$C_nH_m \rightarrow nC + \frac{1}{2}mH_2 \qquad (11.9)$$

In light liquid hydrocarbons (the boiling point between 50 °C and 200 °C), thermo-catalytic decomposition is carried out with the production of elemental carbon and hydrogen. On the other hand, in the case of heavy residual fractions with boiling point higher than 350 °C, hydrogen is produced in a two-steps scheme, namely hydrogasification and cracking of methane [1]:

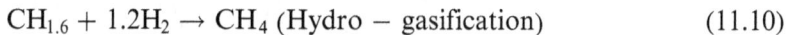

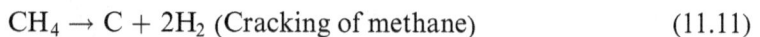

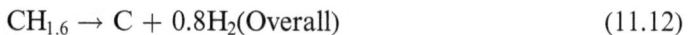

$$CH_{1.6} + 1.2H_2 \rightarrow CH_4 \text{ (Hydro} - \text{gasification)} \qquad (11.10)$$

$$CH_4 \rightarrow C + 2H_2 \text{ (Cracking of methane)} \qquad (11.11)$$

$$CH_{1.6} \rightarrow C + 0.8H_2 \text{(Overall)} \qquad (11.12)$$

A flow diagram for the methane pyrolysis process is shown in figure 11.7. Natural gas ($CH_4$) is directly decarbonized at atmospheric pressure and up to 980 °C in an air- and water-free environment. The energy needed to make one mole of hydrogen (37.6 kJ mol$^{-1}$) is less than that required by the SMR approach (63.3 kJ mol$^{-1}$), and it may be achieved by burning about 15%–20% of the hydrogen generated throughout the process [1]. Furthermore, WGS and $CO_2$ elimination steps are not included in pyrolysis. Additionally, carbon management replaces the energy-intensive stage of CCS, which can be stored on land or underwater for future use, or it can be employed in the chemical and metallurgical sectors [1]. At a given temperature, the decarbonization conversion can be increased by continuously removing hydrogen through membrane separation. Usually, Pd–Ag alloys are employed for this because they lower the operating temperature and minimize the development of coke. Large hydrocarbon pyrolysis plants require comparatively less capital investment than steam conversion or partial oxidation processes. As a result, the cost of producing hydrogen is reduced by 25–30%.

## 11.3 Hydrogen production from renewable sources

Hydrogen can be produced using a variety of methods using renewable resources. We will briefly discuss here some biomass-based technologies and approaches related to water splitting.

### 11.3.1 Biomass process

Biomass is a primary energy source that is renewable. Energy crops and crop residues, timber from forests and forest residues, grass, industrial residues, animal and municipal waste, and a variety of other resources can be used for useful energy production [1]. Plant-derived biomass is classified as organic matter. Through the process of photosynthesis, the energy of sunlight is stored in the chemical bonds of these organic materials. When biomass is used to produce energy, $CO_2$ is emitted. However, this gaseous emission amount is the same as the amount that is absorbed by the living organisms [1].

There are two ways to produce hydrogen from biomass: thermochemical reactions and biological activities. Biological processes require less energy and are more friendly to the environment. Nevertheless, depending on the raw materials utilized, they offer modest rates and yields (mol hydrogen/mol feedstock) of hydrogen. By contrast, thermochemical processes are comparatively quick and provide a larger stoichiometric yield of hydrogen, with gasification being a promising option based on economic and environmental considerations [1].

#### 11.3.1.1 Thermochemical processes

Biomass is converted into hydrogen and hydrogen-rich gases in thermochemical processes. The two primary steps in thermochemical technology are gasification and pyrolysis. Both conversion procedures provide $CH_4$ and CO among other gaseous products. Through the use of SR and the WGS reaction, these gases can be further processed to produce additional hydrogen.

Biomass pyrolysis produces gaseous chemicals, solid charcoal, and liquid oils by heating the biomass to a temperature between 650 and 800 K and a pressure between 0.1 and 0.5 MPa [1], which occurs in the complete absence of oxygen. There are several exceptions, such as when partial combustion is permitted to supply the thermal energy needed for the process. Methane and other hydrocarbon gases that

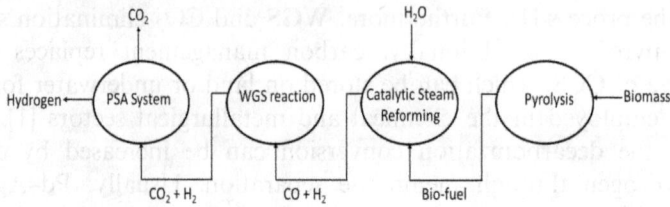

**Figure 11.8.** Flow diagram showing the process of biomass pyrolysis.

are produced can be steam reformed. WGS process are used to produce even more hydrogen. PSA is used to achieve the necessary level of purified hydrogen subsequent to conversion of CO to $CO_2$ and $H_2$. The following equations represent the different steps [1]:

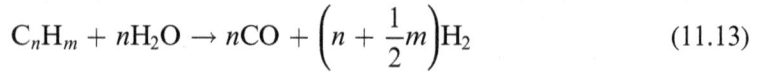

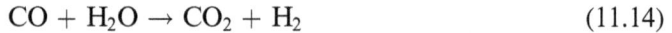

$$C_nH_m + nH_2O \rightarrow nCO + \left(n + \frac{1}{2}m\right)H_2 \tag{11.13}$$

$$CO + H_2O \rightarrow CO_2 + H_2 \tag{11.14}$$

A flow diagram of the biomass gasification process is shown in figure 11.8. The yield of hydrogen production from biomass is determined by the type of feedstock, the catalyst utilized, the temperature, and the duration of residence.

The process of converting biomass into a gaseous fuel, or syngas, by thermo-chemical means in a gasification medium such as air, oxygen, or steam is known as biomass gasification [1]. The temperature and operating pressure range for gasification are 500–1400 °C and 33bar, respectively, depending on the plant scale and the ultimate use of the generated syngas. The three primary reactor types used for biomass gasification are fixed bed, fluidized bed, and indirect gasifiers. They are differentiated based on the flow and velocity of the gasifying agent. Equations (11.15) and (11.16) describe the transformation of biomass into syngas, when it reacts with air or steam, respectively [1]:

$$\text{Biomass} + \text{Air} \rightarrow H_2 + CO_2 + CO + N_2 + CH_4 + \text{other CHs} + \text{tar} + H_2O + \text{char} \tag{11.15}$$

$$\text{Biomass} + \text{Steam} \rightarrow H_2 + CO + CO_2 + CH_4 + \text{other CHs} + \text{tar} + \text{charcoal} \tag{11.16}$$

A flow diagram of the biomass gasification process is shown in figure 11.9. After the biomass is converted into syngas, the gas combination is further processed in the same manner as the product gas of the pyrolysis process. The primary factors influencing the amount of hydrogen produced are the kind of biomass, temperature, steam-to-biomass ratio, particle size, and catalyst type. Compared to rapid pyrolysis, steam gasification yields far more hydrogen. With an overall thermal-to-hydrogen efficiency of up to 52%, this is an effective method of renewable hydrogen production.

### 11.3.2 Biological methods

The majority of biological functions require comparatively less energy because they function at room temperature and atmospheric pressure. They make use of

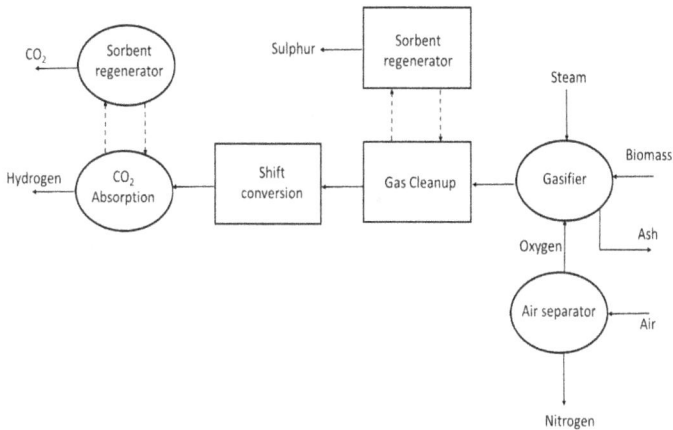

**Figure 11.9.** Flow diagram showing the biomass gasification process.

sustainable energy sources. Because biological processes can use a variety of waste materials as feedstock, they can aid in the recycling of trash. [1].

The major biological processes for producing hydrogen gas include direct and indirect biophotolysis, photo and dark fermentations, and multistage or sequential dark and photo-fermentation. Water is the feedstock for photolysis, a process in which certain bacteria or algae use their hydrogenase or nitrogenase enzyme systems to directly create hydrogen. Using bioprocessing technologies, biomass is fed into fermentative processes that transform carbohydrate-containing materials into organic acids and ultimately hydrogen gas [1].

The same concepts that underpin photosynthesis in plants and algae are modified in biophotolysis to produce hydrogen gas. Green plants facilitate only $CO_2$ reduction because they lack the enzymes necessary to catalyze the creation of hydrogen. On the other hand, under the right circumstances, algae can create hydrogen because they have the enzymes necessary for doing so. Water molecules can be split into hydrogen ions and oxygen via direct and indirect biophotolysis, respectively, by green and blue-green algae [1].

Green algae use direct biophotolysis, or photosynthesis to break down water molecules into hydrogen ions and oxygen. The direct biophotolysis flow diagram is displayed in figure 11.10. The hydrogenase enzyme [1] then transforms the produced hydrogen ions into hydrogen gas. Due to the high sensitivity of this enzyme to oxygen, the oxygen content must be kept below 0.1%. At full sunlight intensities, 90% of the light that is captured by the photosynthetic apparatus (chlorophyll and other pigments) is not used for photosynthesis; instead, it is released as heat or fluorescence. Using mutants originating from microalgae, one can prevent this 'light-saturation effect'. They are said to have a lower pigment content with less chlorophyll and a good tolerance to oxygen, both of which contribute to a larger production of hydrogen. The conversion of water to hydrogen by green algae may be represented by the following general reaction [1]:

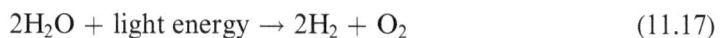

$$2H_2O + \text{light energy} \rightarrow 2H_2 + O_2 \qquad (11.17)$$

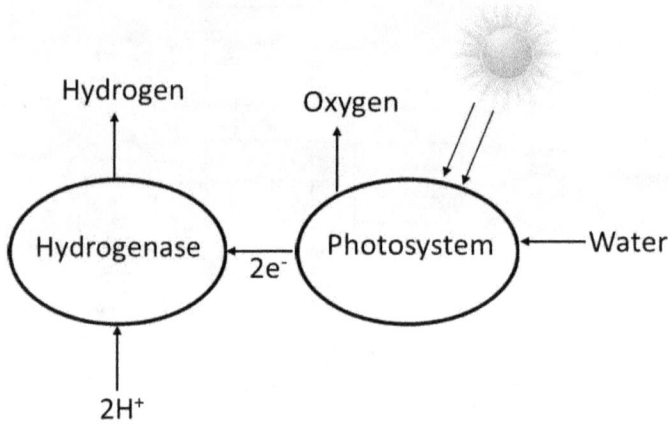

**Figure 11.10.** Flow diagram showing the direct biophotolysis process.

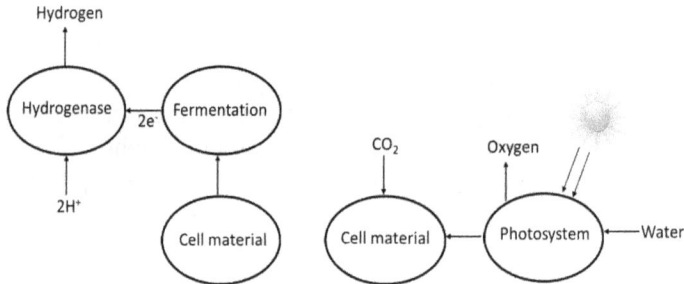

**Figure 11.11.** Flow diagram showing the indirect biophotolysis process.

In indirect biophotolysis, the general reaction for hydrogen formation from water by cyanobacteria or blue-green algae is represented by the following reactions [1]:

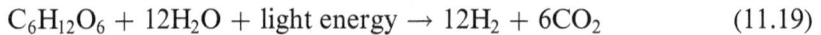

$$12H_2O + 6CO_2 + \text{light energy} \rightarrow C_6H_{12}O_6 + 6O_2 \qquad (11.18)$$

$$C_6H_{12}O_6 + 12H_2O + \text{light energy} \rightarrow 12H_2 + 6CO_2 \qquad (11.19)$$

The flow diagram for the indirect biophotolysis process is shown in figure 11.11. Nitrogenase enzymes and hydrogenase both contribute to the synthesis of hydrogen. The production rate of synthesis is similar to hydrogenase-based production by green algae. Although indirect biophotolysis technologies are still in their infancy, they are thought to be a cost-effective and environmentally-friendly way to use water, which is a renewable resource, and consume $CO_2$, which is one of the air pollutants [1]. The primary shortcomings of the bio-hydrogen generation method include its low $H_2$ production rate, need for a large surface area to gather enough light, and lack of waste utilization.

Fermentations are biochemical processes that take place with or without oxygen. They perform microbial transformations of organic feed materials producing alcohols, acetone, and $H_2$ in minimal amounts, as well as $CO_2$ [1]. These techniques for producing bio-hydrogen make use of waste materials and offer low-cost energy generation along with concurrent waste treatment.

The dark fermentation process mainly involves anaerobic bacteria on carbohydrate-rich substrates. The fermentation occurs in oxygen free (anoxic) dark conditions. The following reactions represent the dark fermentation process [1]:

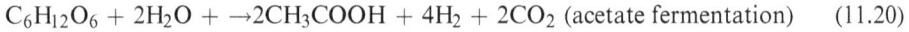

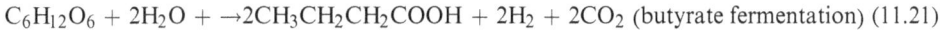

$$C_6H_{12}O_6 + 2H_2O + \rightarrow 2CH_3COOH + 4H_2 + 2CO_2 \text{ (acetate fermentation)} \quad (11.20)$$

$$C_6H_{12}O_6 + 2H_2O + \rightarrow 2CH_3CH_2CH_2COOH + 2H_2 + 2CO_2 \text{ (butyrate fermentation)} \quad (11.21)$$

With glucose serving as the model substrate, acetic and butyric acids make up more than 80% of the total end products. Theoretically, the yields of $H_2$ in acetate and butyrate-type fermentation are 4 and 2 mol per mole of glucose, respectively. The flow diagram for the dark fermentation process is shown in figure 11.12. The ideal source for this procedure is glucose, which is not easily accessible in big quantities and is quite expensive. But one may get glucose from agricultural wastes. Natural starch-containing materials and cellulose, the main component of plant biomass, can also be used as substitutes. The pH of the starting materials has a significant impact on how much $H_2$ is produced during the dark fermentation process; for best results, the pH should be kept between 5 and 6 [1]. Moreover, since the generation of $H_2$ tends to decrease with increasing pressure, the hydrogen must be extracted as it is produced. The advantage here is that a rather straightforward procedure that does not depend on the presence of light sources is used to achieve the dark fermentation. Because of this, the technology does not need a lot of area, and it can produce hydrogen consistently day and night from a wide range of potentially useful substrates, including refuse and waste products.

The second biochemical process is photo-fermentation. This is achieved via solar energy and organic acids in nitrogen-deficient conditions. Some photosynthetic bacteria can convert organic acids such as acetic, lactic, and butyric acids into hydrogen and $CO_2$ because they include nitrogenase. The total conversion from acetic acid, the reactant, to $H_2$ is shown below [1].

$$CH_3COOH + 2H_2O + \text{light energy} \rightarrow 4H_2 + 2CO_2 \quad (11.22)$$

The photo-fermentation process flow diagram is shown in figure 11.13. The production rate and $H_2$ yield are stimulated by an increase in light intensity. On the other hand, light conversion efficiency may also be reduced. A significant issue occurs when using industrial effluents for $H_2$ generation because the color of waste water may reduce light penetration. Furthermore, pretreatment is necessary before using industrial effluents due to the presence of toxic materials such as heavy metals. In general, more hydrogen is produced in a lit environment than in a dark one. The main obstacles preventing the photo-fermentation approach from being

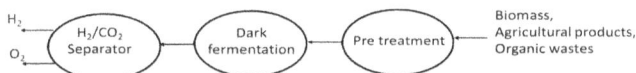

**Figure 11.12.** Flow diagram showing the dark fermentation process.

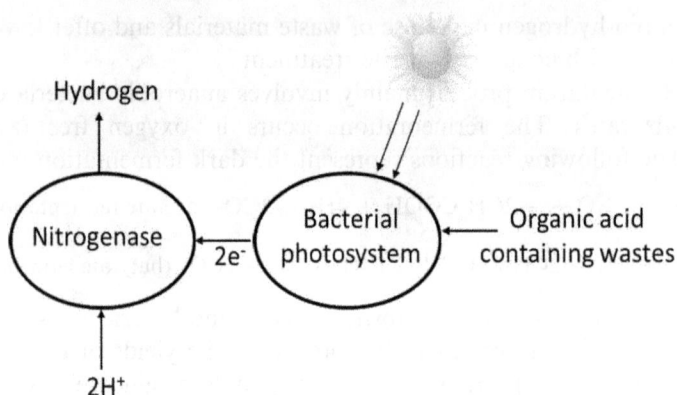

**Figure 11.13.** Flow diagram showing the photo-fermentation process.

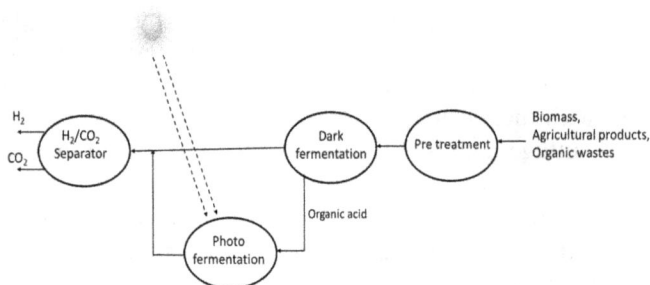

**Figure 11.14.** Flow diagram showing the sequential dark and photo-fermentation process.

widely used are the efficiency of solar energy conversion, the need for complex anaerobic photo-bioreactors that span broad areas, and the limited availability of organic acids.

By adopting a hybrid system, it is possible to attain larger hydrogen production yields with a lower light energy requirement. Both anaerobic or non-photosynthetic and photosynthetic bacteria make up such a hybrid system. A flow diagram of the successive dark and photo-fermentation process is shown in figure 11.14. Anaerobic bacteria that produce hydrogen in the dark can break down a range of carbohydrates. Photosynthetic bacteria may then be able to create more hydrogen from the resultant organic acids. Sequential dark/photo-fermentation is another term for the combination of dark and photo-fermentation. The following relation can be used to depict this two-stage procedure. [1]

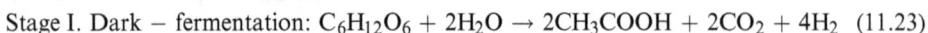

Stage I. Dark − fermentation: $C_6H_{12}O_6 + 2H_2O \rightarrow 2CH_3COOH + 2CO_2 + 4H_2$  (11.23)

Stage II. Photo − fermentation: $2CH_3COOH + 4H_2O \rightarrow 8H_2 + 4CO_2$  (11.24)

The largest documented yield in practice is 7.1 mol $H_2$/mol glucose, although the theoretical limit of hydrogen production is 12 mol of hydrogen produced per mol of glucose [1]. The temperature, which increases $H_2$ yield, and pH level, which needs to be between 4.5 and 6.5 and above 7 for fermentative and photosynthetic and fermentative bacteria, respectively, are the primary factors affecting $H_2$ production.

### 11.3.3 Electrolysis

Electrolysis is an efficient and well-known technique for splitting water. Electricity provides the necessary energy input for the endothermic reaction process. Figure 11.15 shows a flow diagram of the water electrolysis process. An electrolyte-immersed cathode and an anode are the essential components in a conventional electrolysis unit. When an electrical current is applied the water splits, producing hydrogen at the cathode. Oxygen is evolved on the anode side through the following reaction [1]:

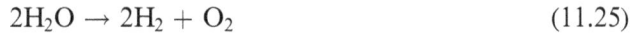

$$2H_2O \rightarrow 2H_2 + O_2 \tag{11.25}$$

The most common electrolysis technologies are alkaline, proton exchange membrane (PEM), and solid oxide electrolysis cells (SOEC). When water is added to a PEM electrolyzer, it splits into protons (hydrogen ions, $H^+$) at the anode, which then moves through the membrane to the cathode and forms $H_2$, and oxygen, which returns back with the water [1]. This PEM process is described by the following relation [1]:

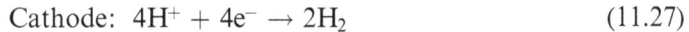

$$\text{Anode: } 2H_2O \rightarrow O_2 + 4H^+ + 4e^- \tag{11.26}$$

$$\text{Cathode: } 4H^+ + 4e^- \rightarrow 2H_2 \tag{11.27}$$

In alkaline and SOEC, water is introduced at the cathode. There the water is split into $H_2$, which is separated from water in an external separation unit, and hydroxide ions ($OH^-$), which in turn travel through the aqueous electrolyte to the anode to form $O_2$. A part of the electrical energy input in SOEC systems is converted into

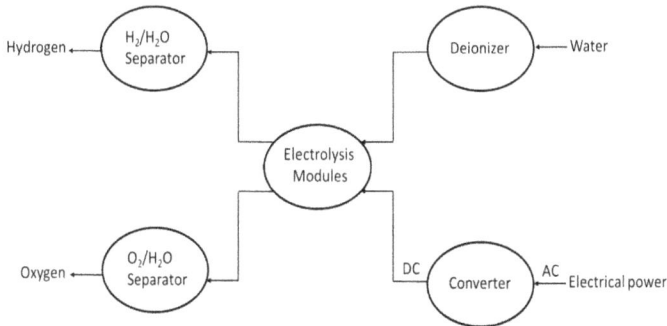

**Figure 11.15.** Flow diagram showing the process of water electrolysis.

thermal energy. As a result, the temperature increases, and consequently the $H_2$ is left in an unreacted steam stream. This process in SOEC is described by the following relation [1]:

Water is added at the cathode in alkaline and SOEC, and there the water is split into $H_2$ and hydroxide ions. Hydrogen is separated from the water in an external separation unit. The hydroxide ions then pass through the aqueous electrolyte to the anode, where they combine to generate $O_2$. In SOEC devices, some electrical energy input is transformed into heat energy. The temperature rises as a result, leaving the $H_2$ in an unreacted stream of steam. This process in SOEC is described by the following relation [1]:

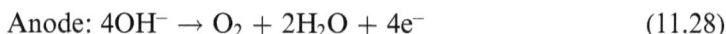

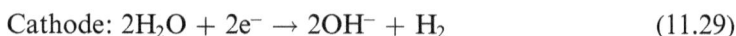

$$\text{Anode: } 4OH^- \rightarrow O_2 + 2H_2O + 4e^- \tag{11.28}$$

$$\text{Cathode: } 2H_2O + 2e^- \rightarrow 2OH^- + H_2 \tag{11.29}$$

Thus, extremely pure hydrogen can be produced from water by electrolysis. However, because electrolyzers consume a lot of electricity, it becomes less appealing in large-scale production. Currently, this technology contributes with a share of about 5% to the total hydrogen generation [1]. On the other hand, the $H_2$ that is produced is the cleanest energy carrier when the electrical energy originates from renewable sources such as solar, wind, and hydropower. This can be used to store excess electricity, and increase plant-load factor and efficiency in small-scale applications [1].

### 11.3.4 Thermolysis

Hydrogen and oxygen are produced when water is heated to a high temperature through a process known as thermochemical water splitting, or thermolysis. Water does not decompose until it reaches a high enough temperature (usually above 2500 °C) to make the Gibbs energy function ($\Delta G$) zero, which makes it energetically favorable to separate hydrogen from the equilibrium mixture. The process of single-stage water decomposition is described by the following relation [1]:

$$2H_2O \rightarrow 2H_2 + O_2; \text{T} > 2500 \text{ °C} \tag{11.30}$$

Single-stage water decomposition requires a significant amount of primary energy, which is difficult to get from sustainable heat sources. Consequently, a number of thermochemical water-splitting cycles have been suggested in an effort to reduce the temperature and boost overall effectiveness. [1].

Thermochemical cycles constitute one of the most promising processes through which heat is converted into chemical energy in the form of hydrogen, which consists of a series of chemical reactions at different temperatures. Two examples of such a thermochemical cycle are shown below [1]:

- Multistage Cu-Cl cycle:

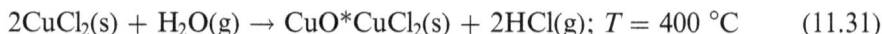

$$2CuCl_2(s) + H_2O(g) \rightarrow CuO*CuCl_2(s) + 2HCl(g); T = 400 \text{ °C} \tag{11.31}$$

$$CuO*CuCl_2(s) \rightarrow 2CuCl(l) + 0.5O_2; \; T = 500 \; °C \qquad (11.32)$$

$$4CuCl(s) + H_2O \rightarrow 2CuCl_2(aq) + 2Cu(s); \; T = 25 - 80 \; °C \qquad (11.33)$$

$$CuCl_2(aq) \rightarrow CuCl_2(g); \; T = 100 \; °C \qquad (11.34)$$

$$2Cu(s) + 2HCl(g) \rightarrow 2CuCl(l) + H_2(g); \; T = 430 - 475 \; °C \qquad (11.35)$$

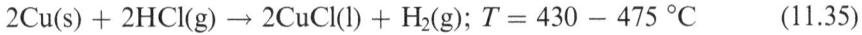

- Novel two−step $SnO_2$/SnOcycle:

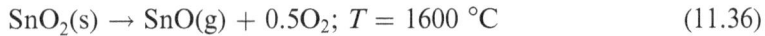

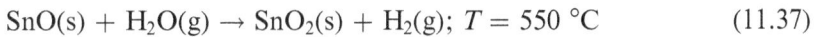

$$SnO_2(s) \rightarrow SnO(g) + 0.5O_2; \; T = 1600 \; °C \qquad (11.36)$$

$$SnO(s) + H_2O(g) \rightarrow SnO_2(s) + H_2(g); \; T = 550 \; °C \qquad (11.37)$$

Solar heat and nuclear energy can be used to generate the necessary high temperature. Here, solar energy is obviously important, and parabolic reflectors such as trough, tower, and dish systems can be used to obtain large-scale concentrations of solar energy [1]. Figure 11.16 shows the flow diagram for a thermochemical water-splitting process powered by the Sun. With no greenhouse gas released into the atmosphere and fulfilling a minimum temperature requirement of 550 °C, the Cu–Cl and Mg–Cl cycles appear to be the most promising low-temperature thermochemical cycles available today. The intensity of solar light determines the energy efficiency and hydrogen production rates. It should be emphasized that factors including toxicity, chemical availability and cost, material separation, and corrosion issues must be taken into account when calculating the cost of producing hydrogen, in addition to the capital expenditure for the required equipment.

### 11.3.5 Photo-electrolysis

The process known as photo-electrolysis, or photolysis, breaks down water into hydrogen and oxygen by absorbing visible light energy with the aid of certain

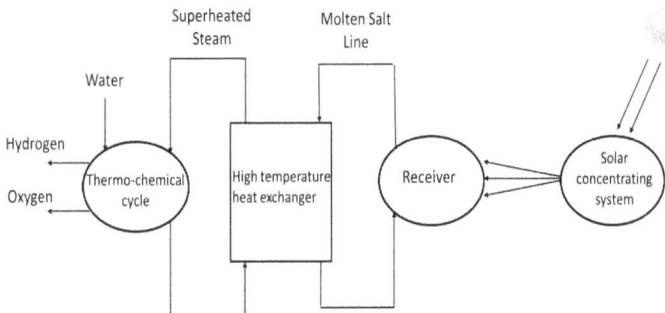

**Figure 11.16.** Flow diagram showing the process of solar-based thermochemical water splitting.

photocatalysts. Certain semiconducting materials have the ability to absorb sunlight, and when this happens the process of splitting water is comparable to electrolysis. An electron-hole pair is created on the semiconducting surface of the anode when a light quantum or photon with energy larger than or equal to the bandgap of the relevant semiconductor is incident on the semiconducting surface. The electron-hole pair is subsequently separated by the electric field that exists between the semiconductor and the electrolyte. The holes stay at the anode, where they split water into oxygen $O_2$, which stays back with water, and hydrogen ions $H^+$, which move through the electrolyte to the cathode. Solar photon-produced electrons travel to the cathode via an external circuit, where they combine with $H^+$ to make hydrogen $H_2$. The process flow diagram for photo-electrolysis is displayed in figure 11.17. The following relations encapsulate the energetic view of hydrogen generation in this process [1]:

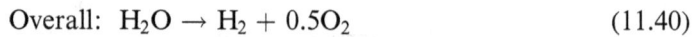

$$\text{Anode: } 2p^+ + H_2O \rightarrow 0.5O_2 + 2H^+ \tag{11.38}$$

$$\text{Cathode: } 2e^- + 2H^+ \rightarrow H_2 \tag{11.39}$$

$$\text{Overall: } H_2O \rightarrow H_2 + 0.5O_2 \tag{11.40}$$

The free energy necessary for the decomposition of water into $H_2$ and $O_2$ is about 1.23 eV. However, in order to separate electrons from the hole without the need for an external bias potential, semiconductors with a high bandgap energy are necessary. As a result, the overall efficiency is significantly reduced. After looking at a number of materials, it was discovered that a combination of SiC and $TiO_2$ work well to create a self-driven system with appropriate band positions and a photo-conversion efficiency of roughly 0.06% [1].

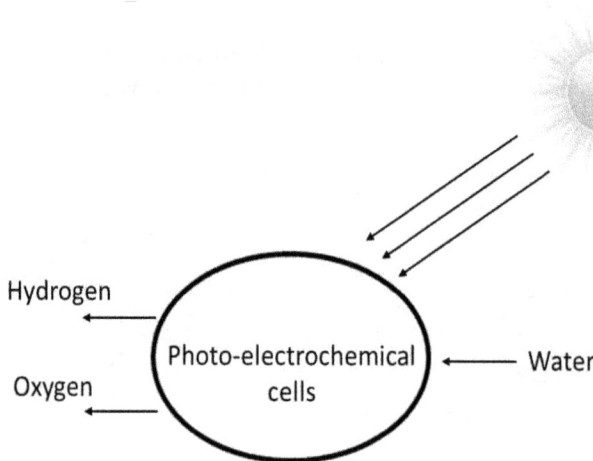

**Figure 11.17.** Flow diagram showing the process of photo-electrolysis.

## 11.4 Comparison of various $H_2$ production processes

A comparison of the main benefits, drawbacks, and efficiency of the various hydrogen production procedures is shown in table 11.1. SMR is currently the most economical method to produce hydrogen. SR has the lowest operating temperature requiring no oxygen, which is followed by ATR and POX. On the other hand, the benefit of hydrocarbon pyrolysis is that it is a reduced-step and emission-free process, requiring treatment for only carbon as a by-product. These approaches offer greater advantages because, as table 11.1 shows, they make use of an established and highly developed process for generating hydrogen using infrastructure that is already in place. In addition, they offer up to 85% more efficient energy conversion than renewable techniques. They are therefore typically the most viable approach. Combining membrane reactors with $H_2$ generation from fossil fuels can result in further gains in terms of efficiency and the maximum temperature of operation. Relatively low temperatures (about $\approx 550$ °C) can be achieved by concentrated solar energy sources [1] or gas turbine exhaust gases used in a combined cycle for the production of power and hydrogen. The main drawbacks are the reliance on fossil fuels and a $CO_2$ by-product is emitted into the atmosphere during the reforming process. Furthermore, these methods are heavily dependent on the price of fossil fuels because they are presently used as both the reactants and the fuel for the process.

**Table 11.1.** Comparison of the different hydrogen production processes. Continued to table 11.2. Reprinted from [1], Copyright (2017), with permission from Elsevier.

| Process | Efficiency (in %) | Major advantages | Major disadvantages |
|---|---|---|---|
| SR | 74–85 | Most developed technology, existing infrastructure. | $CO_2$ by-product, dependence on fossil fuels. |
| POX | 60–75 | Proven technology, existing infrastructure. | $CO_2$ by-product, dependence on fossil fuels. |
| ATR | 60–75 | Proven technology, existing infrastructure. | $CO_2$ by-product, dependence on fossil fuels. |
| CHs pyrolysis | — | Emission-free, reduced-step procedure. | Carbon by-product, dependence on fossil fuels. |
| Biomass pyrolysis | 35–50 | $CO_2$-neutral, abundant and cheap feedstock. | Tar formation, varying $H_2$ content due to seasonal availability and feedstock impurities. |
| Biomass gasification | — | $CO_2$-neutral, abundant and cheap feedstock. | Tar formation, varying $H_2$ content due to seasonal availability and feedstock impurities. |
| Biophotolysis | 10 | $CO_2$ consumed, $O_2$ is the only by-product, operation under mild conditions. | Requires sunlight, low $H_2$ rates and yields, requires large reactor volume, $O_2$ sensitivity, high raw material cost. |

**Table 11.2.** Comparison of the different hydrogen production processes. Continued from table 11.1. Reprinted from [1], Copyright (2017), with permission from Elsevier.

| Process | Efficiency (in %) | Major advantages | Major disadvantages |
|---|---|---|---|
| Dark fermentation | 60–80 | $CO_2$-neutral, simple, can produce $H_2$ without light, contributes to waste recycling, no $O_2$ limitation. | Fatty acids removal, low $H_2$ rates and yields, low conversion efficiency, requires large reactor volume. |
| Photo-fermentation | 0.1 | $CO_2$-neutral, contributes to waste recycling, can use different organic wastes and wastewaters. | Requires sunlight, low $H_2$ rates and yields, low conversion efficiency, requires large reactor volume, $O_2$ sensitivity. |
| Electrolysis | 40–60 | No pollution with RES proven technology, existing infrastructure, abundant feedstock, $O_2$ is the only by-product, contributes to RES integration as an electricity storage option. | Low overall efficiency, high capital costs. |
| Thermolysis | 20–45 | Clean and sustainable, abundant feedstock, $O_2$ is the only by-product. | Elements' toxicity, corrosive problems, high capital costs. |
| Photo-electrolysis | 0.06 | Emission-free, abundant feedstock, $O_2$ is the only by-product. | Requires sunlight, low conversion efficiency, noneffective photocatalytic material. |

Biomass is a sustainable fuel that is abundantly available practically everywhere. With an efficiency of 35%–50%, thermochemical pyrolysis and gasification offer a viable method of producing hydrogen. The fermentative methods also offer the benefit of producing hydrogen and recycling garbage at the same time. In terms of $CO_2$ concentration, these modes are neutral. Because photosynthesis is a part of the process, the biophotolysis is seen as $CO_2$ consumed. In terms of efficiency, dark fermentation may compete with all of the techniques listed in table 11.1. On the other hand, sunlight is necessary for photo-fermentation and biophotolysis. The feedstocks' impurities and their seasonal availability are the main drawbacks of thermochemical processes. Because biological processes yield modest rates, a bigger reactor volume is needed for them. Because of this, biomass could only be used on a large scale through thermochemical pathways such as gasification and pyrolysis, whereas biological processes could be employed on a smaller scale for centralized waste recycling and treatment or local $H_2$ production [1].

Water-splitting pathways are clean and sustainable. They produce only $H_2$ and $O_2$ from one of the most abundant raw materials in the world, namely water. Electrolysis only causes an emission of $CO_2$ when fossil fuels are required to generate the needed electric power for the process. When renewable energy sources such as solar and wind are used, they can produce hydrogen and improve the load factor of a plant. Thus, it also contributes to the increase in the integration of renewable energy sources. As per table 11.1, thermochemical cycles provide a more efficient way of producing hydrogen, with the main limitations being the toxicity of the elements involved.

Pathways that split the water are sustainable and clean. They use water, one of the most plentiful basic elements in the world, to create only $H_2$ and $O_2$. $CO_2$ is only released during the electrolysis process if fossil fuels are used to provide the electricity necessary. Utilizing renewable energy sources, such as solar and wind power, can increase load factor of a plant and produce hydrogen. Therefore, it also plays a part in the growing integration of renewable energy sources. Table 11.1 shows that thermochemical cycles offer a more effective method of producing hydrogen. The primary drawback here is the toxicity of the constituent elements.

Photo-electrolysis is a less efficient process, with an efficiency of only 0.06%. This is due to the lack of effective photocatalytic materials. Comparing all methods based on their primary energy source, solar-based processes have less conversion efficiency. Following the biological methods, which operate under mild conditions, water-splitting methods provide moderate temperatures and conversion efficiencies when nuclear is the preferred energy source for both water electrolysis and thermochemical water-splitting [1]. The less efficient method is photo-electrolysis, which has an efficiency of just 0.06%. The lack of effective photocatalytic materials is the cause of this. When comparing all of these techniques according to the main energy source, solar-based processes have a lower conversion efficiency. In line with the mild operating conditions of biological systems, water-splitting methods provide moderate temperatures and conversion efficiencies when nuclear is the preferred energy source for both water electrolysis and thermochemical water-splitting [1].

## 11.5 High-temperature conversion in nuclear power plants

Nuclear power facilities are able to produce hydrogen [7]. The created hydrogen has a high purity and can be used in hydrogen fuel cells, which have a 99.999% purity requirement. Hydrogen is created using nuclear power through the use of SOECs and/or thermochemical conversion. Relatively new fourth generation high-temperature reactors are well suited for producing hydrogen.

In high-temperature electrolysis, helium is the primary coolant and is used to drive a gas turbine to provide the electricity needed for electrolysis [7]. In addition, some of the heat contained within the primary coolant is transferred using a heat exchanger to the water before electrolysis. This addition of this high-temperature process heat decreases the amount of energy that must be converted from heat (from the nuclear reactor) to electricity (from the generator) and then to chemical energy in the hydrogen molecule. This results in an increase in the overall efficiency of the process. High-temperature reactors with operating temperatures up to 1000 °C can be coupled with certain thermochemical processes to produce hydrogen with an efficiency of around 50% [7].

Several thermochemical processes can be used for hydrogen production in a high-temperature nuclear reactor. These include sulfur–iodine cycle and copper–chlorine cycle.

The copper–chlorine cycle involves the following equations [7]:

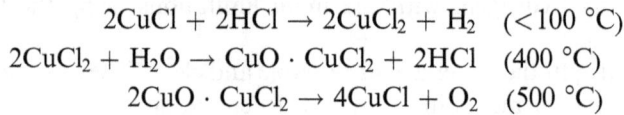

$$2CuCl + 2HCl \rightarrow 2CuCl_2 + H_2 \quad (<100 \ °C)$$
$$2CuCl_2 + H_2O \rightarrow CuO \cdot CuCl_2 + 2HCl \quad (400 \ °C)$$
$$2CuO \cdot CuCl_2 \rightarrow 4CuCl + O_2 \quad (500 \ °C)$$

The copper–chlorine cycle can use low-grade waste heat and requires a relatively low voltage. The temperature requirement is only up to 500 °C. The reactants are available relatively easily, and they react to completion without side reactions.

The sulfur–iodine cycle is a purely thermal process, which can be described with the following equations [7]:

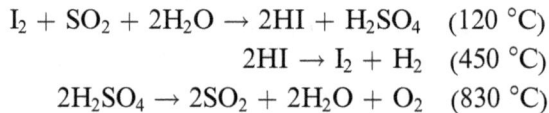

$$I_2 + SO_2 + 2H_2O \rightarrow 2HI + H_2SO_4 \quad (120 \ °C)$$
$$2HI \rightarrow I_2 + H_2 \quad (450 \ °C)$$
$$2H_2SO_4 \rightarrow 2SO_2 + 2H_2O + O_2 \quad (830 \ °C)$$

This sulfur–iodine cycle is suitable for large-scale hydrogen production. It has a high thermal efficiency of up to 50% and is an all-fluid process. Thus, it is possible to expand and to have a continuous operation.

There are some challenges that need to be met in order for nuclear energy to produce hydrogen in large-scale quantities. The main drawback is that nuclear power plants are highly capital-intensive and are not available worldwide. Where it is available, the long-term storage of nuclear waste remains an important issue to be solved if nuclear energy is to become viable source of hydrogen production.

# References

[1] Nikolaidis P and Poullikkas A 2017 *Renew. Sustain. Energy Rev.* **67** 597
[2] Osman A I *et al* 2021 *Environ. Chem. Lett.* **20** 153
[3] Ustolin F, Campari A and Taccani R 2022 *J. Mar. Sci. Eng.* **10** 1222
[4] Al-Qahtani A *et al* 2021 *Appl. Energy* **281** 115958
[5] Han W-B *et al Electrochim. Acta* **386** 138458
[6] de Falco M, Marrelli L and Iaquaniello G (ed) 2011 *Membrane Reactors for Hydrogen Production Processes* (Berlin: Springer)
[7] Sheffield J W, Martin K B and Folkson R 2014 Alternative Fuels and Advanced Vehicle Technologies for Improved Environmental Performance ed R Folkson (Amsterdam: Elsevier) [in]

**IOP** Publishing

Hydrogen
Physics and technology
**Sindhunil Barman Roy**

# Chapter 12

## Methods of hydrogen storage

Hydrogen storage is a very important technological problem for the deployment of hydrogen technology in various applications, including stationary power, portable power, and transportation. Hydrogen has the highest energy per unit mass of any fuel. However, one kilogram of hydrogen gas at ambient temperature and atmospheric pressure occupies a volume of 11 m$^3$. So, its density is low at ambient temperature, which results in a low energy per unit volume. Hydrogen storage with such a low density of 0.09 kg m$^{-3}$ is one of the key issues restricting its widespread use. Storing hydrogen efficiently, economically, and safely is one of the challenges for enabling hydrogen to become an economic source of energy.

Hydrogen can be stored physically as either a gas or a liquid. Storage of hydrogen as a gas typically requires high-pressure (5000–10 000 psi) tanks. Pressures of up to 77 MPa can be achieved by using standard piston-type mechanical compressors, but the work needed for the compression is much higher than 2.21 kWh kg$^{-1}$, providing a gravimetric and volumetric density of 13 wt% and lower than 40 kg m$^{-3}$, respectively [1]. Storage of hydrogen as a liquid requires cryogenic temperatures because the boiling point of hydrogen at one-atmosphere pressure is −252.8 °C or about 20 K. Liquefaction of hydrogen can be achieved through a double-step process of compression and cooling in a heat exchanger. Hydrogen can also be stored by adsorption on the surfaces of solids or by absorption within solids. Figure 12.1 presents an overview of the hydrogen storage technologies that are presently available, namely high-pressure gas compression, hydrogen liquefaction, and hydrogen storage in various kinds of solids.

The most suitable form of hydrogen storage depends on the particular application. Large-scale storage facilities are necessary for maintaining adequate reserves between production facilities and small-scale storage facilities. A typical example of small-scale storage is the application of hydrogen in the transportation sector, which requires storage at a volume that can be accommodated within the vehicle and also at a weight that does not limit the performance of the vehicle. Storage volume also

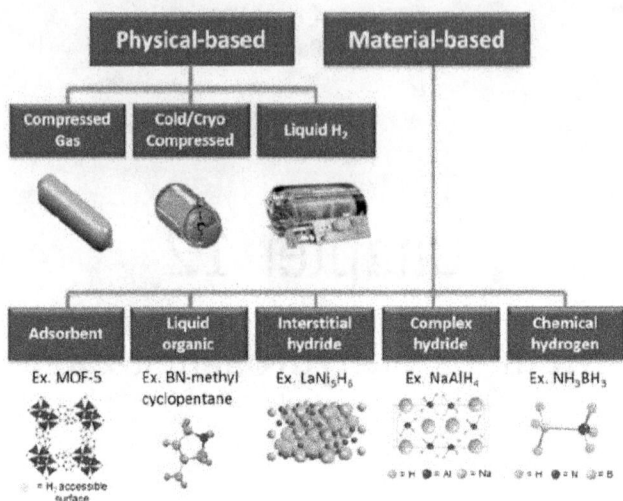

**Figure 12.1.** An overview of hydrogen storage technologies. Reprinted from [2], Copyright (2019), with permission from Springer.

needs to be restricted for building-integrated applications. In the sections that follow, we will discuss various types of both large-scale and small-scale hydrogen storage methods.

## 12.1 Large-scale storage

### 12.1.1 Liquid hydrogen storage

The boiling point of hydrogen at one-atmosphere pressure is 20 K ($<-253$ °C) with a density of close to 71 kg m$^{-3}$ [3]. This density is significantly higher than that of compressed hydrogen at 70 MPa. Liquid hydrogen storage thus requires refrigeration to a temperature of 20 K, and needs to be stored in specially designed insulated cryogenic tanks under pressure with provisions for cooling, heating, and venting. Tank sizes are in the range from 1.5 m$^3$ (100 kg) to 75.0 m$^3$ (5000 kg) [4].

The energy requirements of hydrogen liquefaction are high, and is about 30% of the heating value of hydrogen. The heating value is the amount of heat produced by a complete combustion of fuel and is measured as a unit of energy per unit mass or volume of substance (e.g., kcal kg$^{-1}$, kJ kg$^{-1}$, kWh kg$^{-1}$, J mole$^{-1}$, and Btu m$^{-3}$) [2]. The theoretical work required to liquefy hydrogen gas starting from room temperature is 3.23 kWh kg$^{-1}$, but the estimated technical work is about 15.2 kW h kg$^{-1}$ [2, 5]. This is in comparison to about 6.0 kWh kg$^{-1}$ for compression to 70 MPa. Another drawback is the gasification of liquid hydrogen inside the cryogenic vessel. This is an inevitable loss, even with a perfect insulation technique. A heat source of the gasification is the exothermic reaction of the conversion from ortho-hydrogen to para-hydrogen. This heat of conversion is 519 kJ kg$^{-1}$ at 77 K, and 523 kJ kg$^{-1}$ at temperatures lower than 77 K [5]. This is greater than the latent heat of vaporization (451.9 kJ kg$^{-1}$) of normal hydrogen at the normal boiling point.

The initial capital investment for a new storage facility is high due to the need for liquefaction machinery, as well as storage. The cost of larger plants with higher liquefaction capacities is more to start with, but the cost of hydrogen and the energy needed to liquefy hydrogen decreases per kilogram of hydrogen liquefied [4]. Liquefaction capacities range from 100 kg h$^{-1}$ to 10 000 kg h$^{-1}$ and typical on-site storage capacities are from 115 000 kg to 900 000 kg [4]. This kind of cryogenic storage of hydrogen is economical at large production rates and long-term storage. Liquid hydrogen is likely to remain the main technique of large-scale bulk, stationary hydrogen storage for the foreseeable future [2].

### 12.1.2 Underground storage

Natural underground formations such as aquifers and depleted natural gas and man-made caverns or cavities are appealing low-cost options for stationary hydrogen storage on a large scale [3, 4]. A schematic of three such interesting possibilities is presented in figure 12.2. These are salt dome intrusions, cavities in solid rock formations, and aquifer bends.

Figure 12.2(a) shows a salt cavity, which may be formed in salt deposits by flushing water through the salt. This process has already been used to store compressed natural gas. Salt domes are salt deposits extruding upwards towards the surface, therefore allowing cavities to be formed at modest depths [3]. Man-made salt caverns are created by pumping fresh water into a salt dome and dissolving the salt. The resulting cavern can then be used to store hydrogen. The rock cavities shown in figure 12.2(b) may be either natural or excavated. Their walls need to be properly sealed to ensure airtightness. The creation of excavated rock cavities is considerably more expensive than salt caverns.

Aquifers are an underground layer of water-bearing permeable rock or sand trapped between layers of impermeable rock [3, 4]. They allow underground water to flow along the layer. Hydrogen gas may be stored in aquifers by displacing water

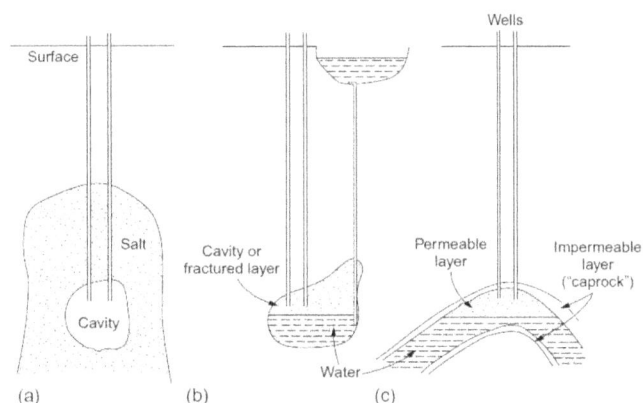

**Figure 12.2.** Types of underground compressed gas storage facilities: (a) storage in salt cavity, (b) rock storage with compensating surface reservoir, and (c) aquifer storage. Reprinted from [3], Copyright (2018), with permission from Elsevier.

(see figure 12.2(c)). This happens when the geometry of the aquifer includes upward bends, allowing a gas pocket to be held in place by water on the sides [3]. The aquifer must have layers above and below with little or no hydrogen permeability. Aquifers suitable for hydrogen storage are found in many parts of the world, except for areas dominated by rock all the way to the surface. In such areas, relatively more expensive rock storage needs to be used.

Choosing the sites for underground gas stores is a fairly involved process. The detailed properties of the cavity cannot be guaranteed based on geological test drillings and modeling alone. These properties will not be fully known until the installation of the facility is complete. It may also turn out that the salt cavern may be unable to sustain an elevated pressure and does not live up to its expectations. For a natural rock cave or a fractured zone created by explosion or hydraulic methods, the stability of the structure is uncertain until a full-scale pressure test is conducted. The decisive measurements of permeability in aquifers can only be made at a finite number of places. But there can be a rapid change in permeability over small distances of displacement [3].

There are two other factors that influence the stability of a cavern: the variation in the temperature and the pressure. The hydrogen gas may be precooled before injecting it into the cavern to keep the cavern's wall temperature nearly constant. Otherwise, the compression may be performed so slowly that the temperature only rises to the level prevailing on the cavern walls. However, this isothermal compression is impractical for most applications because excess power needs to be converted at the rate at which it comes. Therefore, most storage systems employ one or more hydrogen gas cooling steps. To deal with the problem of pressure variations (when the store holds different amounts of energy) the possible solution is to store the hydrogen at constant pressure but variable volume. This is a possibility offered naturally in some aquifers. This may be achieved in an underground rock cavern by connecting the underground reservoir to an open surface reservoir (see figure 12.2(b)). In this case, a variable water column takes care of the variable amounts of hydrogen stored at the constant equilibrium pressure prevailing at the depth of the cavern [3].

There are certain advantages for storing hydrogen in salt caverns [6]: (i) salt surrounding the caverns is highly impermeable and virtually leakproof where the only possibility for gas loss is escape of gas through leaky wells; (ii) salt does not react with hydrogen; (iii) withdrawal or discharge of hydrogen is highly flexible in rate, duration, and volume with lower requirements of gas cushion to avoid rock breakage; and (iv) caverns are a mature, financeable storage technology that has been successfully used to store compressed gases for over seven decades.

An approximately constant working pressure in the aquifer storage system shown in figure 12.2(c) would correspond to the average hydraulic pressure at the depth of the hydrogen-filled part of the aquifer. If the cavern store with compressed gas at pressure $P$ is considered as a cylinder with a piston at positions describing the volumes $V_0$ and $V$, then the stored energy $E$ may be expressed as [3]:

$$E = - \int_{V_0}^{V} P dV \tag{12.1}$$

In the case of the aquifer, energy $E$ is simply equal to the pressure $P$ times the volume of hydrogen gas displacing water in the aquifer. This volume equals the physical volume $V$ times the effective porosity $p$, which is the fractional void volume accessible to incoming hydrogen gas. The energy stored may be written as [3]:

$$E = pVP \tag{12.2}$$

Typical parameter values are $p = 0.2$ and $P$ around $6 \times 10^6$ Nm$^{-2}$ at depths of some 600 m, with useful volumes of $10^9$–$10^{10}$ m$^3$.

The permeability of the aquifer determines the time required for charging and emptying hydrogen gas. This permeability is the proportionality factor between the flow velocity of a fluid or gas through the sediment and the pressure gradient causing the flow [3]. The assumed linear relation is expressed as:

$$\nu = -K(\eta\rho)^{-1}\frac{\partial P}{\partial s} \tag{12.3}$$

Here $\nu$ is the flow velocity, $K$ is the permeability, $\eta$ is the viscosity of the fluid or gas, $\rho$ is its density, $P$ is the pressure, and $s$ is the path length in the downward direction. The permeability $K$ has the dimension of $m^2$ in SI units and has to exceed $10^{11} m^2$ if filling and emptying of the aquifer storage are to take place in a matter of hours rather than days. Sediments such as sandstone are found with permeabilities ranging from $10^{10}$ to $3 \times 10^{12}$ m$^2$ [3].

There are additional losses in practice. For example, the cap-rock bordering the aquifer region may not have negligible permeability, which can lead to possible leakage loss. In the pipelines leading to and from the aquifer, friction may cause a loss of pressure, as may losses in the compressor and turbine [3]. About 15% of losses are typically expected in addition to those of the power machinery.

## 12.2 Small-scale storage

Small-scale storage is needed to distribute and utilize hydrogen in various applications. Small-scale storage deals with relatively small amounts of hydrogen for relatively short storage periods [4]. Vehicles running on hydrogen, whether fuel cell or internal combustion, require onboard storage of hydrogen, as well refilling from storage at a fueling station. Small-scale hydrogen storage is also required for residential and portable power applications, such as home fueling, stationary fuel cells, and emergency backup power units.

### 12.2.1 Compressed hydrogen gas storage

The most common hydrogen storage form today is compressed hydrogen gas. Specially designed tanks capable of withstanding the storage pressures are required to store compressed hydrogen gas. Standard cylindrical tanks withstand pressures of 10–20 MPa, and fuel cell stores for motorcars can handle pressures in the range

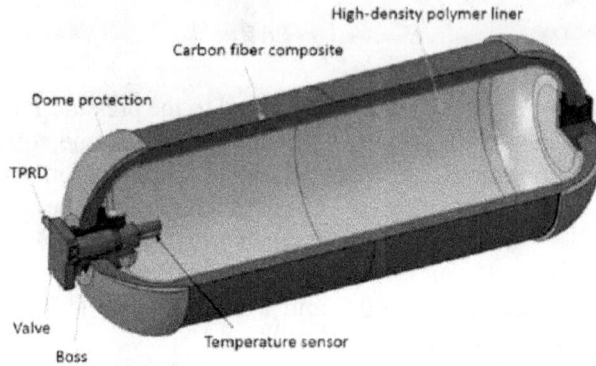

**Figure 12.3.** Schematic presentation of a polymer-lined carbon-fiber tank. Reprinted from [2], Copyright (2019), with permission from Springer.

25–70 MPa. Hydrogen storage tanks are usually made of steel or aluminum-lined steel. Composite carbon-fiber tanks are more suitable for vehicle application, where weight is a matter of concern. A typical design of such tanks involves a carbon-fiber composite shell with an inside polymer liner and outside reinforcement. Figure 12.3 presents the schematic of a polymer-lined carbon-fiber tank. The density of hydrogen when compressed, at 35.0 MPa is about $23 \text{ kg m}^{-3}$ and at 70.0 MPa is about $38 \text{ kg m}^{-3}$. This leads to an energy density of $767 \text{ kWh m}^{-3}$ at 27 °C and 35 MPa [4].

Compression of hydrogen gas takes place at a filling station, where hydrogen is either produced on-site or received from a pipeline. The method of compression dictates the energy requirement. The work required for isothermal compression at temperature $T$ from pressure $P_1$ to $P_2$ can be expressed as [3]:

$$W = AT log(P_2/P_1) \tag{12.4}$$

The above expression is derived from the modified gas law:

$$PV = AT \tag{12.5}$$

Here $A$ is the gas constant $R = 8.314 \text{ J mol}^{-1} \text{ K}^{-1}$ times an empirical, pressure-dependent correction valid particularly for hydrogen and which decreases from 1 at low pressures to around 0.8 at 70 MPa [3]. Adiabatic compression requires more energy input and the preferred solution is a multistage intercooled compressor [3]. This reduces the energy input to about one-half for a single stage.

The density of compressed hydrogen is lower than that of liquid hydrogen. Thus, the volume of the storage tank is an important technological issue. Compression of hydrogen is an energy-intensive process, but it consumes only a third of the energy that liquefaction needs. However, the cost of compressed storage tanks must also be taken into account. The cyclic loading of tanks, which tend to heat up as they are filled with compressed hydrogen, reduces the tank's life [4]. For more information on compressed hydrogen storage vessels and the related safety issues, the readers are referred to the article by Zheng *et al* [7] and the book edited by de Miranda [8].

Power requirements in portable equipment such as cameras, mobile and smart-phones, or laptop computers (which are currently covered by batteries), can be met with a small fuel cell and some 10–20 g of directly or indirectly stored hydrogen. This will prolong operational time by a factor of 5–10 [3].

### 12.2.2 Cryogenic and cryo-compressed hydrogen

Hydrogen vehicles can benefit from the greater energy density of cryogenic hydrogen when compared to compressed hydrogen. Hydrogen stored as a compressed gas occupies a relatively large volume. However, a significant problem in using liquid hydrogen is the boil off from the liquid hydrogen storage in a vehicle that is not in constant use due to the heat release associated with ortho- to para-$H_2$ conversion. The hydrogen tank pressures need to be controlled by venting valves. The insulation properties of the hydrogen tank influence the boil off, which will start after a dormancy period of a few days and then proceed at a level of 3% - 5% per day [3]. The lower values of the boil off rate are obtained by installing heat exchangers between outgoing cold hydrogen and air intake to regulate pressure. The boil off also leads to safety issues for cars parked in a garage. Thus, hydrogen boil off limits the usefulness of liquid vehicle storage except in those cases where the vehicle is used regularly. Furthermore, liquid hydrogen evaporates very easily during delivery, transfer, and refueling.

In a relatively new approach, cryogenic high-pressure vessels combine existing storage technologies to capture the advantages of both cryogenic and high-pressure storage [9]. This technology concept encompasses storing hydrogen in a pressure vessel that can operate at temperatures down to 20 K and high pressures, e.g., 350 atm. This cryogenic high-pressure vessel can house liquid hydrogen, compressed gaseous hydrogen, or fluid hydrogen at elevated supercritical pressures, namely cryo-compressed hydrogen. Stored in the latter form, the supercritical fluid can exceed the density of liquid hydrogen.

Figure 12.4 presents the density and volume versus temperature phase diagram of hydrogen. The dotted lines represent high-pressure isobars. Solid lines represent the theoretical minimum energy or the thermomechanical exergy necessary for com-pression and cooling of hydrogen from ambient conditions (300 K and 1 atm) to any desired temperature and density [9]. As we have discussed earlier, hydrogen is often stored as a compressed gas (red dot in figure 12.4) at ambient temperature (horizontal axis), very high pressure (dotted lines), and relatively low density (vertical axis). As indicated in figure 12.4, hydrogen is much more compact as a cryogenic liquid (represented by a blue dot) but with a higher energetic cost (indicated by solid lines) to compress and/or liquefy hydrogen. Cryogenic pressure vessels have the flexibility to operate across a broad region (shaded in light blue) of the phase diagram (figure 12.4). They can be fueled with gaseous hydrogen at a low energetic cost when energy or fuel cost savings are important, and with liquid hydrogen or cryogenic hydrogen at elevated supercritical pressures when long driving range is desired. This technology also addresses three key vehicle-related issues arising from the highly volatile nature of liquid hydrogen: (i) evaporative

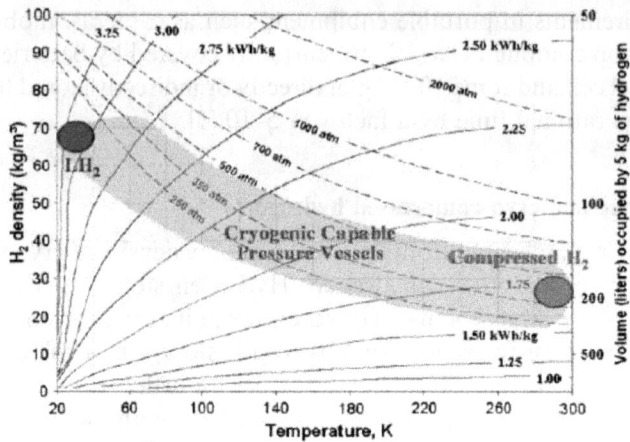

**Figure 12.4.** Density and volume versus temperature phase diagram of hydrogen. Reprinted from [9], Copyright (2010), with permission from Elsevier.

losses after a short period of inactivity, (ii) cumulative evaporative losses for short daily driving distances, and (iii) the risk of being stranded due to fuel evaporation after parking for long duration.

Electricity consumption in the storage cycle can be reduced substantially if the cryogenic-capable pressure vessel is filled with compressed hydrogen. The energy advantage of using ambient-temperature gaseous hydrogen in place of liquid hydrogen for short trips can be seen in figure 12.4, which indicates that compression at ambient temperature is the most efficient method to densify hydrogen. This is because the isothermal work of compression rises nearly logarithmically with density. Compression of hydrogen to 250–1000 atm requires 1.5–2.0 kWh kg$^{-1}$, which is nearly half the theoretical energy needed to liquefy hydrogen by cooling to 20 K (3.25 kWh kg$^{-1}$) [9]. Figure 12.4 applies to so-called 'normal hydrogen' (25% para-hydrogen and 75% ortho-hydrogen), which is the equilibrium phase of hydrogen at 300 K. This neglects the additional necessary energy (0.62 kWh kg$^{-1}$) to convert normal hydrogen to 99.8% para-hydrogen (equilibrium phase at 20 K) in the liquefaction process. In practice, the energy requirements to compress and/or liquefy hydrogen are typically two to three times more than the theoretical minimum [9].

Cryogenic pressure vessels have several potential safety advantages. The most notable is the significantly lower theoretical burst energy of low-temperature hydrogen. There will be a sudden release of mechanical energy in the case of a sudden vessel rupture due to the adiabatic expansion of high-pressure hydrogen gas to atmospheric pressure. Figure 12.5 shows the theoretical maximum mechanical energy released by hydrogen gas at three temperatures, 80 K, 150 K, and 300 K. Hydrogen stored at 70 atm and 300 K will release a maximum mechanical energy of 0.55 kWh kg$^{-1}$ if suddenly expanded to atmospheric pressure [9]. This maximum energy release increases only slightly with much higher hydrogen pressures. Raising the vessel's pressure to 1000 atm causes an increase in maximum mechanical energy release by only 10%, while shrinking the vessel's volume and strengthening the

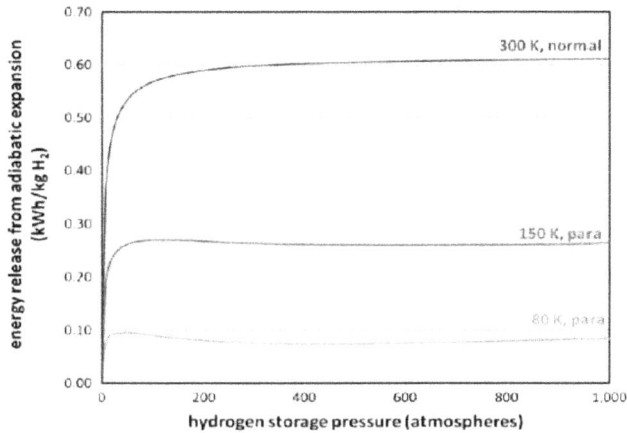

**Figure 12.5.** Maximum mechanical energy (per kilogram of hydrogen) released upon instantaneous expansion of hydrogen gas from a pressure vessel as a function of initial storage pressure at 80 K, 150 K, and 300 K. Reprinted from [9], Copyright (2010), with permission from Elsevier.

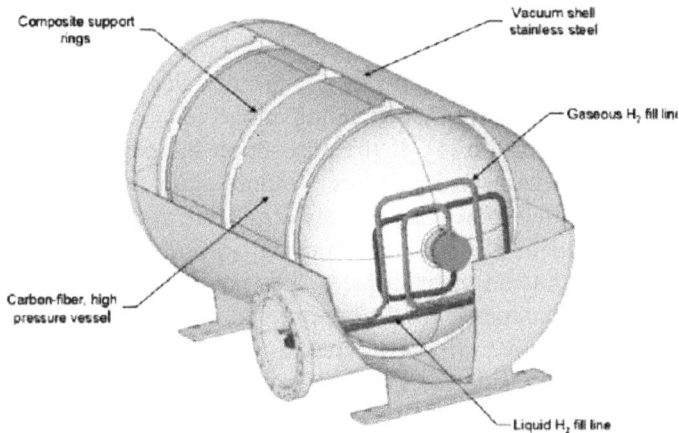

**Figure 12.6.** Schematic diagram of a generation 2 cryogenic pressure vessel for storing hydrogen. Reprinted from [9], Copyright (2010), with permission from Elsevier.

vessel's walls many times over [9]. Together with the high hydrogen storage density of cryogenic temperatures, the low burst energy permits smaller vessels to be used. These can be better packaged onboard to withstand automobile collisions. Figure 12.6 presents a schematic diagram of a second-generation cryogenic pressure vessel design. An inner vessel consisting of an aluminum-lined, carbon fiber-wrapped pressure vessel stores the compressed hydrogen gas. This vessel is surrounded by a vacuum space filled with numerous sheets of highly reflective metalized plastic (which minimizes heat transfer into the vessel), and an outer jacket of stainless steel. This vacuum jacket provides a second layer of protection, eliminating environmental impacts over the lifetime of the pressure vessel.

Vacuum jacketing also provides expansion volume to adjust with shocks from hydrogen release. Cryogenic pressure vessels avoid the fast fill heating and over pressures (up to 25%) that are typical of ambient-temperature vessels. As a result, they operate with higher safety factors, especially because driving the automobile cools the remaining hydrogen fuel and reduces average hydrogen pressures further over typical driving and refueling cycles [9].

## 12.3 Solid-state storage

Solid-state storage is effective at storing large amounts of hydrogen at moderate temperature and pressure. We have discussed in chapter 6 that hydrogen can combine with solids, either physically or chemically, through the mechanisms termed physisorption and chemisorptions, respectively. In physisorption, the hydrogen molecule sticks to the solid surface, whereas in chemisorptions the hydrogen chemically reacts with the solid forming a hydride. We focused on the scientific aspects related to hydrogen in various kinds of solid matrices in chapter 6. In this chapter, we will discuss various classes of materials that are important for solid-state hydrogen storage technology.

### 12.3.1 Physisorption

In physisorption, the adsorbed hydrogen gas can be released reversibly. There are different mechanisms of adsorption depending on the geometry of the adsorbent and the temperature of adsorption [5]. If the adsorption happens on an open surface, then the multilayer mechanism dominates. Volume filling would happen in a pore narrower than 2nm. If a pore size is between 2 nm and 50 nm, then capillary condensation can happen. The mechanism of adsorption in a pore size larger than 50 nm is the same as that on open surfaces. The possibility of condensation of the adsorbed adsorbates is assumed in all these mechanisms. This possibility does not exist at temperatures above critical temperatures, and a different mechanism of adsorption needs to be considered. There is only one mechanism for the adsorption of supercritical gases on any kind of adsorbent, which is the monolayer surface coverage [5]. This mechanism corresponds to the single type of isotherms of supercritical adsorption. In this mechanism, the first molecular layer of adsorbate is fixed on the solid surface due to the interaction between gas and solid. When a layer of adsorbate completely covers the surface, more gas molecules are adsorbed above the first layer due to the interaction between the same species of adsorbate molecules. Thus, the second layer is formed, and subsequently the other layers. The interaction force involved in the formation of the first layer of molecules is thus different from that felt in the second and subsequent layers. This difference is reflected in the heat of adsorption of different layers. The experiment with nitrogen and carbon black showed that the heat of adsorption of the first layer is 11–12 kJ mol$^{-1}$ (0.11–0.12 eV) and it drops to the latent heat of condensation, 5.56 kJ mol$^{-1}$ (0.058 eV), in the subsequent layers [5]. The subsequent layers from the second cannot exist at temperatures above the critical temperature. This is due to a classical law of physics which states that a gas cannot be liquefied at above critical temperatures, no matter how high the pressure applied [5]. Significant physisorption is therefore observed only at relatively low temperatures because the monolayer adsorbate is itself fixed by a weak interaction.

Two basic rules are followed in the physisorption of supercritical gases: (i) the monolayer adsorption mechanism, and (ii) the exponential decrease of adsorption with the increasing temperature [5]. It follows that the adsorption capacity of hydrogen on a material depends on the specific surface area of the material and that higher temperatures will lower the adsorption capacity. The determination of specific surface areas is a major problem in the characterization of porous and finely-dispersed solids. The Brunauer, Emmett, and Teller (BET) theory is commonly used to evaluate the gas adsorption data and estimate a specific surface area, which is expressed in units of area per mass of sample ($m^2/g$). The adsorption isotherm can be used to accurately quantify specific surface area of a solid material. The entire storage capacity of a porous substance is not defined by the adsorption capacity alone. Compression in the void space and adsorption on solid surfaces both contribute to storage capacity. Figure 6.1 shows a schematic illustration of many physically bound hydrogen storage systems.

### 12.3.1.1 Nanoporous carbon materials

Nanoporous carbon materials are promising for hydrogen storage because of their high porosity, low density, and cost efficiency. We have discussed in chapter 6 that nanoporous carbon materials are available in nature in a variety of forms, e.g., activated carbon (AC), carbon nanotubes (CNTs), and carbon nanofibers (CNFs). They have a wide range in their material structures and synthesis techniques, and offer different compositions, pore sizes, surface areas, and functionalities for hydrogen storage [10, 11].

AC is a viable option for hydrogen gas storage because it is inexpensive and widely available. It exhibits a high degree porosity, which results in a sizable surface area. Monolayer adsorption on the surface is usually the first step in the physical adsorption of hydrogen in carbon materials. The physical adsorption capacity of AC is improved by its large surface area, especially at cryogenic temperatures and high pressures [10].

The weak van der Waals force is at the origin of the absorption interaction. Consequently, modulation of AC is required to raise the heat of adsorption between AC and hydrogen molecules in order to enhance the gas intake capability. Theoretically, the hydrogen uptake of AC could achieve 4.0 wt% at 77 K but less than 1.0 wt% at room temperature and 100 bar [10]. Chemical modifications starting from a polypyrrole precursor and subsequent potassium hydroxide (KOH) treatment lead to a high surface area as 3000–3500 $m^2\,g^{-1}$ and hydrogen storage capacity of up to 7.03 wt% at 77 K and 20 bar [10]. Doping with transition metals, such as platinum (Pt), and palladium (Pd) in AC demonstrated 1.10 wt% and 5.50 wt% hydrogen intake capabilities at 298 K with 100 bar and 80 bar, respectively [10]. Some green processes have also been employed to obtain AC starting from plant fibers, coconut shells, or oilseeds as the raw materials. ACs produced by using fruit bunch can have up to 2.14 wt% hydrogen intake at 77 K and 19 bar [10].

Hydrogen storage in CNTs has been the subject of intense investigation since the 1990s. CNTs have diameters in the range of a nanometer. There are single-wall carbon nanotubes (SWCNTs) and multiwall carbon nanotubes (MWCNTs) consisting of nested SWCNTs. Hydrogen can be stored in the microscopic pores or within the tube structures of CNTs. In SWCNTs, all carbon atoms are exposed to the surface. This makes SWCNTs the material with the highest surface-to-bulk atom ratio, and therefore highly surface active [11]. These nanostructure carbon materials also have large amounts of void spaces in the form of pores. Hydrogen can be accommodated in large quantities in such pores. Hydrogen can also be physisorbed in SWCNT bundles on various sites, such as external wall surfaces, grooves, and interstitial channels [11]. Thus, SWCNTs can fulfill the requirement of large energy density in mobile applications.

Interest in the hydrogen storage properties of nanostructure carbon materials was triggered in the late-1990s by the pioneering work by Dillon *et al* reporting a gravimetric storage capacity between 5 and 10 wt% in SWCNTs [12]. The experiments on hydrogen adsorption in CNTs can be roughly classified into two categories, which depend on the method of storage: gas-phase hydrogen storage and electrochemical hydrogen storage [11]. In gas-phase hydrogen storage techniques, a macroscopic sample of nanotubes, typically weighing less than 1 gram, is exposed to pure hydrogen gas under various temperature and equilibrium pressure conditions. The amount of hydrogen adsorbed by nanotubes is then measured gravimetrically. In the second method, hydrogen is electrochemically stored in carbon nanotubes. An electrochemical cell consists of CNTs as the working electrode, platinum (Pt) as the counter electrode, and an appropriate electrolyte. In such a system, hydrogen is stored in the CNT electrode by the reduction of water at a suitable potential [11]. The pristine-carbon nanotubes exhibit better hydrogen storage capacity at cryogenic temperatures than at room temperature. Experimental results showed that the hydrogen storage capacity of SWCNTs and MWCNTs reached a moderate capacity of approximately 1 wt% at ambient temperature and pressure [10].

The hydrogen storage capacity of MWCNTs is significantly improved at room temperature with the application of pressure, e.g., 2.0 wt% at 40 bar, 4.0 wt% at 100 bar, and 6.3 wt% at 148 bar. Metal doping can also enhance the storage capability of CNTs. Lithium- and calcium-doped MWCNTs offer a hydrogen uptake up to 20 wt% and 14 wt%, respectively, at room temperature and pressure of 1 bar. Doping on MWCNTs is also possible with elements such as calcium (Ca), cobalt (Co), iron (Fe), nickel (Ni), and palladium (Pd), and storage capacities under ambient conditions reach 0.3 wt%, 1.05 wt%, 1.5 wt%, 0.75 wt%, and 0.4 wt% hydrogen intake, respectively. Defects such as the pentagon-heptagon pair, the substitution of heteroatoms (B, N, or P), and topological distortion improve the hydrogen adsorption binding energies and storage capability of SWCNTs [10].

Low cost and commercially available CNFs are a potential candidate for solid hydrogen storage. They can be mass produced relatively easily with simple methods of preparation. The hydrogen storage capability of CNFs ranged from 0.7 wt% to 6.54 wt% at room temperature and approximately 100 bar. CNFs synthesized by chemical activation treatment using hydroxide salts, carbonate salts, zinc chloride, and phosphoric acid are of particular interest due to their increased surface area and controllable pore sizes. Doping CNFs with metals such as nickel (Ni) show an enhanced hydrogen uptake of 2.2 wt% at 298 K and 100 bar [10].

In summary, carbon nanoporous materials with their diversified structures and natural abundance are highly attractive for material design and production to match the requirement of hydrogen storage purposes. The hydrogen storage capability of these materials depends highly on the methods of fabrication methods, shapes, impurity contents, oxygen-containing functionalities, and adsorbed (doping) species [10]. The cost-efficient mass production of stable nanostructure carbon materials is very suitable for commercial applications. Table 12.1 summarizes the hydrogen storage properties of some selected nanoporous carbon materials.

### 12.3.1.2 Metal–organic framework

We have studied in chapter 6 that metal–organic frameworks (MOFs) are crystalline porous materials consisting of metal ion clusters and organic ligands. The use of MOFs for hydrogen storage applications was first reported in the early-2000s [10].

**Table 12.1.** Hydrogen storage properties of some selected nanoporous carbon materials. Reproduced from [10]. CC BY 4.0.

| Carbon material | Storage conditions Temperature (K)/pressure (bar) | BET surface area ($m^2\ g^{-1}$) | Hydrogen capability (wt%) |
|---|---|---|---|
| AC (Maxsorb) | 77/30 | 3306 | 5.70 |
| AC (AX-21) | 77/60 | 2745 | 10.80 |
| AC (KOH-treated) | 298/100 | 2800 | 0.85 |
| AC (KOH-treated) | 77/20 | 3190 | 7.08 |
| AC (Pt-doped) | 298/100 | 2033–3798 | 1.10 |
| AC (Pd-doped) | 298/80 | 2547 | 5.50 |
| CNT | 273–295/1.0 | 290–800 | ⩽ 1.0 |
| CNT (film) | 298/10 | — | 8.0 |
| SWCNT | 133/0.4 | — | 5–10 |
| MWCNT | 298/148 | — | 6.3 |
| CNT (Li-doped) | 653/1.0 | 130 (specific) | 20 |
| CNT (K-doped) | 343/1.0 | 130 (specific) | 14 |
| MWCNT (Ca-doped) | —(electrochemical) | — | 0.3 |
| MWCNT (Co-doped) | —(electrochemical) | — | 1.05 |
| MWCNT (Fe-doped) | —(electrochemical) | — | 1.5 |
| MWCNT (Ni-doped) | —(electrochemical) | — | 0.75 |
| MWCNT (Pd-doped) | - (electrochemical) | — | 0.4 |
| CNF | 298/120 | 51 | 6.54 |
| CNF (KOH-treated) | 77/40 | 1500–1700 | 3.45 |
| CNF (N-doped) | 298/100 | 870 (specific) | 2.0 |
| CNF (Ni-doped) | 298/100 | 1310 | 2.2 |

MOF-5 synthesized from zinc salt and 1,4-benzenedicarboxylic acid (BDC) resulted in $Zn_4O(BDC)_3$ and exhibited a hydrogen intake of 4.5 wt% at 78 K and 1.0 wt% at room temperature and pressure 20 bar. Poor moisture stability puts a limit on the performance of MOF-5. The change in structure and chemical linkage resulted in enhanced moisture, thermal, mechanical, and acid/base stabilities, as well as porosity. A series of isoreticular MOFs were subsequently synthesized with the substitution of linkers, metal ions, and functional groups for improved stability and capability. The introduction of ethynylene units into the p-phenylene and the carboxylic groups in MOF-5 resulted in novel MOFs with similar skeleton structures but with superior properties [10]. One such material NU-100 has a BET surface area of 6143 $m^2\,g^{-1}$ and a hydrogen uptake capacity of 10.0 wt% at 77 K and pressure 56 bar. Another material, NU-110, has a BET surface area of 7140 $m^2\,g^{-1}$ with a hydrogen capacity of 8.82 wt% at 77 K and pressure 45 bar. Several metal ions could be used to synthesize newer MOFs. One such example is Cu-MOF-5 ($Cu_3(BTC)_2$), which was synthesized using copper(II) ions and benzene-1,3,5-tricarboxylate(BTC). This material exhibited BET surface area of 1154 $m^2\,g^{-1}$ and a maximum hydrogen uptake of 3.6 wt% at 77 K and 0.35 wt% at room temperature and 65 bar.

A major limitation of MOF is the weak van der Waals interaction between hydrogen atoms and MOFs. The isosteric heat for hydrogen adsorption of MOFs is generally less than 10 kJ $mol^{-1}$ [10]. This interaction energy is strong enough for applying MOFs for hydrogen storage in a low-temperature environment. This capability of MOFs, however, decreases rapidly with increasing temperature. The isosteric heat needs to be increased to 15–20 kJ $mol^{-1}$ to stabilize the hydrogen atoms on MOF surfaces at room temperature. This can be achieved by the introduction of active metal sites, and several metals, e.g., Li, Na, K, Mg, Ca, Be, Ti, Pt, Pd, Cu, Fe, Co, Ni, and Zn, in elements or ion forms have been incorporated as clusters or on MOF decoration. Controlling the pore size and functionalization in MOFs can also be used to improve the isosteric heat. The varied design and synthesis from extended organic building blocks to functional MOFs are popular areas of research [10]. Some examples of MOFs and their hydrogen storage properties are summarized in table 12.2.

### 12.3.1.3 Covalent organic framework

Covalent organic frameworks (COFs) are carbon-based crystalline nanoporous organic polymers. They form with strong covalent linkages, e.g., B-O, C-O, B-C, C-C, and C-N, which leads two-dimensional and three-dimensional structures. These are relatively low-cost materials with large surface areas, and exhibit low density, high stability, and versatile structures. A two-dimensional material, COF-5, was prepared from 1,4-benzenediboronic acid and 2,3,6,7,10,11-hexahydroxytriphe-nylene, and exhibited a high BET surface area as 1590 $m^2\,g^{-1}$ and a 3.5 wt% hydrogen intake at 77 K and 80 bar. Another three-dimensional family of COFs, including COF-102, COF-105, and COF-108, also materialized, showing a larger surface area than two-dimensional COF-5 [7]. Among these COFs, COF-102 has a BET surface area of 3620 $m^2\,g^{-1}$, and its hydrogen uptake capacities are 7.2 wt% at 77 K and 35bar, and 10.0 wt% at 77 K and 100 bar. The hydrogen volumetric uptake

**Table 12.2.** Hydrogen storage properties of some selected MOFs and COFs. Reproduced from [10]. CC BY 4.0.

| Framework | Storage conditions Temperature (K)/pressure (bar) | BET surface area (m$^2$ g$^{-1}$) | Hydrogen capability (wt%) |
|---|---|---|---|
| MOF-5 | (a) 78/20 | 2500–3000 | (a) 4.5 |
| | (b) 298/20 | | (b) 1.0 |
| IRMOF-8 | 298/10 | 1801 | 2.0 |
| MOF-177 | (a) 78/70 | 4600 | (a) 7.5 |
| | (b) 298/100 | | (b) 0.62 |
| NU-100 | 77/56 | 6143 | 10.0 |
| NU-109 | 77/45 | 7010 | 8.30 |
| NU-110 | (a) 77/45 | 7140 | (a) 8.82 |
| | (b) 298/180 | | (b) 0.57 |
| MOF-399 | (a) 77/56 | 7157 | (a) 9.02 |
| | (b) 298/140 | | (b) 0.46 |
| Cr-MIL-53 | 77/16 | 1020 | 3.1 |
| Al-MIL-53 | 77/16 | 1026 | 3.8 |
| Cu-MOF-5 | (a) 77/65 | 1154 | (a) 3.6 |
| | (b) 298/65 | | (b) 0.35 |
| MOF-210 | (a) 77/80 | 6240 | (a) 17.6 |
| | (b) 298/80 | | (b) 2.7 |
| Be-MOF | (a) 77/ 1.0 | 4030 | (a) 1.6 |
| | (b) 298/95 | | (b) 2.3 |
| COF-1 | (a) 77/1.0 | 711 | (a) 1.7 |
| | (b) 77/70 | | (b) 3.8 |
| COF-5 | (a) 77/1.0 | 1590 | (a) 0.1 |
| | (b) 77/80 | | (b) 3.4 |
| COF-102 | (a) 77/1.0 | 3620 | (a) 0.5 |
| | (b) 77/100 | | (b) 10.0 |
| COF-102-3 | (a) 77/100 | — | (a) 6.5 |
| | (b) 300/100 | | (b) 26.7 |
| COF-105 | (a) 77/1.0 | 3472 | (a) 0.6 |
| | (b) 77/80 | | (b) 10.0 |
| COF-108 | (a) 77/1.0 | 4210 | (a) 0.9 |
| | (b) 77/100 | | (b)10.0 |
| CTC-COF | 77/1.1 | 1710 | 1.12 |
| COF-105 (Li-doped) | 298/100 | — | 6.84 |
| COF-108 (Li-doped) | 298/100 | — | 6.73 |
| COF-340-CoCl$_2$ | 298/250 | 7400 | 7.00 |

of COF-102 achieved 40.4 g L$^{-1}$, which is the best performance of these three-dimensional COFs. The phenylene groups in the COF-102 backbone can be further substituted by diphenyl, triphenyl, and naphthalene pyrene groups, giving COF-102-2, COF-102-3, COF-102-4, and COF-102-5, respectively. This modulation of the backbone and pore size further enhanced the hydrogen uptake capacity of COFs. COF-102-3 exhibited 26.7 wt% and 6.5 wt% hydrogen uptakes at 77 and 300 K under 100 bar, respectively [10]. The hydrogen uptakes of COF-105 and COF-108 at 77 K are 10.0 wt% at 80 bar for COF-105, and 10.0 wt% at 100 bar for COF-108.

As in the case of MOFs, the increase in temperature degrades the capacity of COFs. Metal doping with Li, Mg, Ti, and Pd has been tried to circumvent this problem and improve the hydrogen storage capacity in COFs under practical conditions. It was reported that the Li-doped COF-105 and COF-108 exhibited an improved hydrogen storage capacity of 6.84 wt% and 6.73 wt% at 298 K and 100. This increase of hydrogen uptakes is attributed to the bonds between positively charged Li-ions and hydrogen molecules. The effect of doping elements on several COFs with imine and hydrazide linkage has also been investigated. Such linkers in the COFs could chelate with Co(II), Cu(II), Fe(II), Mn(II), and Ni(II) anions. The synthesized COF-340 with Co(II)-doping exhibited the highest hydrogen uptake of 7.00 wt% at 298 K and 250 bar. Some examples of COFs and their hydrogen storage properties are summarized in table 12.2.

### 12.3.1.4 Porous aromatic framework

Porous aromatic frameworks (PAFs) are porous organic materials with a tetrahedrally diamond-like structure [10]. PAFs exhibit high porosity, large surface area, low mass densities, and high thermal and mechanical stability. In addition, the organic frameworks facilitate the postsynthetic functionalization and pore size modulation.

The first PAF, termed PAF-1, was synthesized in 2009. PAF-1 exhibited a BET surface area of 5600 $m^2$ $g^{-1}$ with a hydrogen uptake of 7.0 wt% at 77 K and 48 bar. In addition, PAF-1 showed a relatively low heat of adsorption as 4.6 kJ $mol^{-1}$, which suggested a weak interaction between hydrogen molecules and the surface of PAF-1. This results in the poor hydrogen intake capacity of PAF-1 at increased temperature or under ambient pressure.

A series of PAFs with a replaced quadrivalent atom (silicon (Si) or germanium (Ge)) in place of the carbon center were subsequently synthesized [10]. PAF-3 (PAF-4) with Si (Ge) centers exhibits BET surface area of 2932 $m^2$ $g^{-1}$ (2246 $m^2$ $g^{-1}$), 6.6 kJ $mol^{-1}$ (6.3 kJ $mol^{-1}$) heat of adsorption, and hydrogen uptake of 5.5 wt% (4.2 wt%) at 77 K and 60 bar. PAF-1 could be activated by potassium hydroxide followed by a carbonized process to increase its hydrogen gravimetric capacity to 3.06 wt% at ambient pressure and 77 K. It was also reported that PAF-4 carrying lithium tetrazolide doping had enhanced hydrogen storage capacities of 20.7 wt% (at 77 K and 100 bar) and 4.9 wt% (at 233 K and 100 bar), respectively. Apart from lithium, Mg- and Ca-doping were also successfully applied to increase the hydrogen storage capacities. Another series of PAFs were synthesized by diphenylacetylene derivatives as the linkers, and the resultant PAF-324 and PAF-334 exhibited reasonably high hydrogen uptakes of 6.32 wt% and 16.03 wt% at 298 K and 100 bar, respectively. Some examples of PAFs and their hydrogen storage properties are summarized in table 12.3.

**Table 12.3.** Hydrogen storage properties of some selected PAFs. Reproduced from [10]. CC BY 4.0.

| Framework | Storage conditions Temperature (K)/ pressure (bar) | BET surface area ($m^2$ $g^{-1}$) | Hydrogen capability (wt%) |
|---|---|---|---|
| PAF-1 | 77/48 | 5600 | 7.0 |
| PAF-3 | 77/60 | 2932 | 5.5 |
| PAF-4 | 77/60 | 2246 | 4.2 |
| PAF-1 (KOH-treated) | 77/1.0 | 1320 | 3.06 |
| PAF-1 (Li-doped) | 77/1.2 | — | 10 |
| PAF-4 (Li-doped) | 77/100 | 5525 | 20.7 |
| PAF-4 (Li-doped) | 233/100 | 5525 | 4.9 |
| PAF-Mg | 233/100 | 4479 (Langmuir) | 6.8 |
| PAF-Ca | 233/100 | 4479 (Langmuir) | 6.4 |
| PAF-324 | 298/100 | 5372 (specific) | 6.32 |
| PAF-334 | 298/100 | — | 16.03 |

### 12.3.1.5 Nanoporous organic polymers

Hypercrosslinked polymers (HCPs), conjugated microporous polymers (CMPs), and polymers of intrinsic microporosity (PIMs) have been used as hydrogen gas storage materials. They possess several advantages and unique properties for hydrogen storage applications [10]:

1. Variable polymer backbones and facile functionalization.
2. Tunable pose size and crosslinking density.
3. Lightweight and high surface area.
4. High processability for bulk, coating, and composite materials.
5. Low cost and accessibility for mass production.

The polystyrene-type HCPs have BET surface areas of 1930 $m^2$ $g^{-1}$ and capability of high uptakes of hydrogen as 1.5 wt% at 77 K and 1.2 bar. The polyphenylene-type HCP synthesized from para-dichloroethylene and 4,4 ′-bis (chloromethyl)-1,1′-biphenyl exhibits a surface area of 1904 $m^2$ $g^{-1}$ and a hydrogen uptake of 3.68 wt% at 77 K and 1.5 bar.

The poly(arylene-ethynylene) CMPs exhibit a large BET surface area of 1018 $m^2$ $g^{-1}$ and a hydrogen uptake of 1.4 wt% at 77 K and 1.0 bar. Using sterically demanding linkers with a trigonal or tetragonal geometry, specific surface areas as high as 1200 $m^2$ $g^{-1}$ have been achieved. Alkyne linkers gave CMPs with specific surface areas up to 842 $m^2$ $g^{-1}$ and a hydrogen intake of 131 $cm^3$ $g^{-1}$ at 77 K and 1.13 bar [10]. Nitrogen-rich CMPs have also been synthesized for hydrogen storage applications, and a CMP, PCZN-8, with a 20 mol% pyridine moiety in the backbone exhibited a BET surface area of 1126 $m^2$ $g^{-1}$ and 1.35 wt% hydrogen storage capability at 77 K and 1 bar. Improvement in the hydrogen storage of CMPs can be achieved with metal doping. It was found that the Li-doped CMP synthesized by homocoupling of 1,3,5-triethynylbenzene could enhance its hydrogen uptake value from 1.6 wt% for the undoped CMP to 6.1 wt% at 77 K and 1 bar.

Early research of PIMs for hydrogen storage reported that a cyclotricatechylene-(CTC-) based PIM-1 with a BET surface area of 760 m$^2$ g$^{-1}$ possess a hydrogen uptake capacity of 1.44 wt% at 77 K and 10 bar. Thermal treatment enhances the hydrogen storage capability of PIM-1. Subsequently, triptycene-based PIMs demonstrated an improved BET surface area up to 1990 m$^2$ g$^{-1}$ and a hydrogen uptake as 1.9 wt% at 77 K. A novel hexaazatrinaphthylene-(HATN-) based PIMs exhibits a surface area at 772 m$^2$ g$^{-1}$ and a hydrogen intake of 3.86 wt% at 77 K and nearly 120 bar.

The composite mixture of PIMs with high processability and PAFs with high surface area provided good processability for film casting and an improved hydrogen storage capability from 2.6 wt% (pure PIM film) to 4.1 wt% (PIM/PAF (=77.5/22.5 (wt%/wt%)) film). Similarly, PIM/AC and PIM/MOF composites have been used to prepare porous polymer-based composite membranes for mobile hydrogen storage applications [10]. It has been observed that the PIM with 60 wt% AC or 40 wt% MOF is useful for film casting processes. The synthesized films exhibited 1.6–2.5 times larger hydrogen intake capabilities. Such polymer/polymer and polymer/inorganic material composites offer advantages over powders in terms of safety, handling, and practical manufacturing for high-pressure hydrogen storage materials [10]. The hydrogen storage properties of selected HCPs, CMPs, and PIMs are summarized in table 12.4.

## 12.3.2 Chemisorption

In chemisorption, hydrogen is stored in solid materials by chemical bonding and is released through chemical reactions under specific conditions. Hydrogen can be combined with many metals to form hydrides, which will subsequently release hydrogen upon heating. Three types of hydrides are mainly studied for hydrogen storage [10, 11, 13]. The first type is metal hydride $MH_x$, where M stands for the main group or transition metal, such as Li, Na, Mg, Ca, and Ti, and X is the number of hydrogen atom. The second type is intermetallic hydrides $AB_xH_y$, where A is typically the hydriding metal and B is the non-hydriding metal. The third type $MEH_x$ is known as complex hydride or chemical hydride. A complex hydride contains a metal cation (M) and a hydrogen-containing polyatomic anion ($EH_x$). We have discussed the physical properties of different classes of hydrides in chapter 6.

### 12.3.2.1 Metal hydride

Metal hydrides are made up of metal atoms, which constitute a host lattice where hydrogen atoms are trapped in interstitial sites. The absorption of hydrogen leads to an increase in the size of lattices. The metal usually becomes powdered to prevent the decrepitation of metal particles [13]. There are two possible paths of hydriding a metal: (i) direct dissociative chemisorption, and (ii) electrochemical splitting of water. These reactions are expressed as [13]:

**Table 12.4.** Hydrogen storage properties of some selected HCPs, CMPs, and PIMs. Reproduced from [10]. CC BY 4.0.

| Polymer material | Storage conditions Temperature (K)/ pressure (bar) | BET surface area ($m^2\ g^{-1}$) | Hydrogen capability (wt%) |
|---|---|---|---|
| HCP (polystyrene) | 77/1.2 | 1930 | 1.5 |
| HCP (polystyrene) | 77/15 | 2920 | 3.04 |
| HCP (polyphenylene) | 77/15 | 1904 | 3.68 |
| HCP (polyaminobenzene) | (a) 77/1.2 | 384 | 0.97 |
| | (b) 273/90 | | 0.22 |
| HCP (polyphenylene-Pt) | 298/19 | 1399 | 0.21 |
| CMP (poly(arylene-ethynylene)) | 77/1.0 | 1018 | 1.4 |
| CMP (NCMP-0) | (a) 77/1.13 | 1108 | (a) 1.5 |
| | (b) 77/1.13 | | (b) 2.0 |
| CMP (E1) | (a) 77/1.13 | 1213 | (a) 1.33 |
| | (b) 77/8 | | (b) 2.66 |
| CMP (EOF-6) | 77/1.0 | 1380 | 1.29 |
| CMP (PCZN-8) | 77/1.0 | 1126 | 1.35 |
| CMP (Li-doped) | 77/1.0 | 834 | 6.1 |
| CMP (PTAT, Li-doped) | (a) 77/1.0 | 304 | (a) 7.3 |
| | (b) 273/1.0 | | (b) 0.32 |
| PIM-1 | (a) 77/1.0 | 760 | (a) 1.04 |
| | (b) 77/10 | | (b) 1.44 |
| PIM (STP-II) | 77/1.0 | 1990 | 1.9 |
| HATN-PIM | 77/120 | 772 | 3.86 |
| PIM (PAF mixture) | 77/100 | 1197 | 4.1 |
| PIM (AC (20 wt%) mixture) | 77/100 | 1130 | 3.7 |

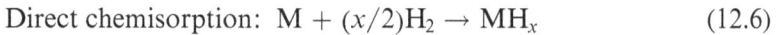

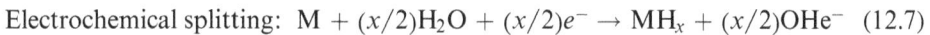

$$\text{Direct chemisorption: } M + (x/2)H_2 \rightarrow MH_x \qquad (12.6)$$

$$\text{Electrochemical splitting: } M + (x/2)H_2O + (x/2)e^- \rightarrow MH_x + (x/2)OHe^- \quad (12.7)$$

Here M represents the metal. A suitable catalyst such as palladium is needed for the electrochemical splitting process.

Figure 12.7 presents a schematic of the process of hydrogen chemisorption. The molecular hydrogen reaches a shallow potential minimum near the surface, and the atomic hydrogen has a deeper minimum almost at the surface. Hydrogen has a periodic potential minimum at the interstitial site in the metal lattice. A weak van der Waals force becomes operational when a hydrogen molecule arrives at the metal surface. This force draws the molecule closer, and it reaches the potential minimum. Large forces will be required to keep it in molecular form. However, the dissociation energy of the hydrogen molecule is exceeded by the chemisorption energy [13]. The hydrogen molecule gets dissociated and individual hydrogen atoms are attracted to

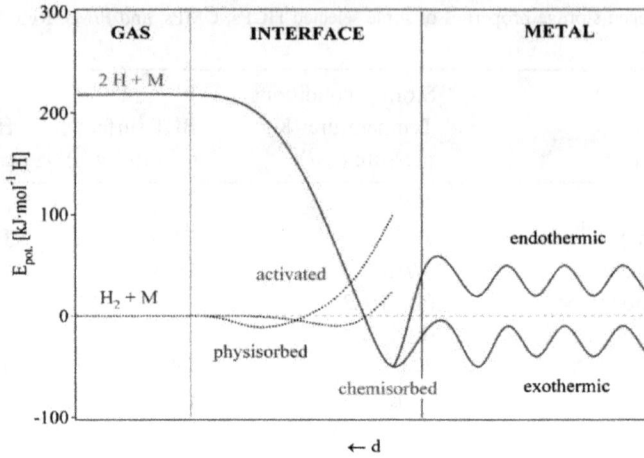

**Figure 12.7.** Schematic diagram of hydrogen chemisorption on metal. Reprinted from [13], Copyright (2017), with permission from Elsevier.

the surface by chemisorptive forces. At room temperature, even thermal energy is enough to increase the vibrational amplitude of hydrogen atoms, which can then reach and enter the metallic surface [13].

Two different kinds of hydrides are usually formed with metal and hydrogen: $\alpha$-phase and $\beta$-phase. In the $\alpha$-phase only some hydrogen is dissolved, whereas a stoichiometric phase is formed in the $\beta$-phase. This is elaborated here with the example of $Mg_2Ni$ with two hydride compositions $Mg_2NiH_{0.3}$ and $Mg_2NiH_4$. The $\alpha$-phase is formed initially, and then repeated cycling leads to the formation of the $\beta$-phase [13]:

$$Mg_2NiH_{0.3} + 3.7H \rightarrow Mg_2NiH_4 \tag{12.8}$$

The thermodynamics of metal hydride formation have been discussed in detail in chapter 6. The reaction is exothermic because the reaction enthalpy ($\Delta H$) of hydride formation is usually negative. The hydride formation releases energy as a result. Therefore, the dehydrogenation needs energy input. Most applications are at ambient temperature, or at least in the range 273–373 K. Thus, the reaction enthalpy needs to be small so that the hydride can take heat from the surroundings while releasing hydrogen. In some fuel cell systems, the hydride takes heat directly from the fuel cell. The stability of a hydride is also influenced by the reaction enthalpy because the pressure varies exponentially. The following are the essential requirements that need to be met by metal hydrides proposed for hydrogen storage application at a commercial level [13]:

1. High content of hydrogen storage.
2. Facile reversibility of formation and decomposition reactions. The hydride should be decomposable at moderate temperatures that can be provided from locally available heat sources such as solar, automobile exhaust, and waste heat sources.

3. Absorption-desorption kinetics that should be compatible with the charge-discharge requirements of the system.

4. Equilibrium dissociation pressure of the hydride at peak desorption rate should be compatible with the safety requirements of the hydride containment system. The hydride itself should have a high safety factor.

5. Sufficient chemical and dimensional stability of hydride to permit it to remain unchanged over a long number of charge-discharge cycles.

6. Minimal hysteresis in adsorption-desorption isotherms.

7. Reasonable resistance of hydride to deactivation by low concentrations of $O_2$, $H_2O$, $CO_2$, and CO.

8. Affordable total cost of hydride (raw materials, processing, and production) for the intended application. Long-term availability of raw materials (i.e., metal resources) must be ensured. The cost of the hydride system (which includes its containment) per unit of reversibly stored hydrogen should be as low as possible.

9. Moderate cost of the storage vessel, ancillary equipment, and fabrication/installation.

10. Low operating and maintenance costs and purchased energy requirements (that is, energy other than waste energy and energy extracted from ambient air) per storage cycle.

Magnesium hydride ($MgH_2$) is considered to be the most promising metal hydride material for hydrogen storage among the light-metal hydrides. The advantages here include the high gravimetric (7.6 wt% $H_2$) and volumetric (110 kgH m$^{-3}$) densities, natural abundance, low cost, lightweight, and chemical stability of $MgH_2$. The main problem with the practicality of $MgH_2$ is its high thermodynamic stability of Mg–H bonding ($\Delta H = 75$ kJ mol$^{-1}$ $H_2$), which causes difficulty in the release of hydrogen. To accelerate hydrogen sorption/desorption under ambient pressure, $MgH_2$ has to operate at a temperature above 300 °C. To overcome this limitation, other metals of metal oxides can be used as catalysts to improve the sorption kinetics of Mg and reduce the activation energies of hydrogenation/dehydrogenation processes. A variety of elements, e.g., V, Ti, Ni, Nb, Cr, and Fe, have been applied to prepare doped Mg through ball milling, melting, casting, and chemical processes to enhance the hydrogen storage/releasing performance [10]. The highest recorded hydrogen capacity for a metal hydride application for fuel cell electric vehicles is 2.0 wt%, which was recorded for a chromium–titanium–vanadium alloy at 313 K at ambient pressure [13].

Table 12.5 presents hydrogenation/dehydrogenation conditions, activation energy, and hydrogen capability of various metallic hydrides and intermetallic compounds.

### 12.3.2.2 Complex hydrides

Complex hydrides or chemical hydrides are of considerable interest because of their low molar mass and high gravimetric hydrogen capacity. Group I, II, and III metals, e.g., Na, Mg, B, and Al, result in a large variety of metal hydrogen compounds [13].

**Table 12.5.** Hydrogenation and dehydrogenation conditions, activation energy, and hydrogen storage capability of some metallic hydrides and intermetallic compounds. Reproduced from [10]. CC BY 4.0.

| Storage media | Storage conditions Temperature (K)/ pressure (bar) | $E_a$ (kJ mol$^{-1}$) | Hydrogen capability (wt%) | Dehydrogenation condition Temperature (K)/pressure (bar) |
|---|---|---|---|---|
| LiH | 1183/1.0 | 181.2 | 12.7 | —- |
| MgH$_2$ | 557/1.0 | 174 | 7.6 | 553/0.9 |
| MgH$_2$ (5 wt% V-doped) | 473/10 | 119 | 5.8 | 573/0.15 |
| MgH$_2$ (5 wt% Ni-doped) | 473/10 | 75 | 5.0 | 573/0.15 |
| MgH$_2$ (Ti–Nb-doped) | (a) 673/40 (b) 573/40 | 50.7 | (a) 6.8 (b) 5.7 | (a) 673/1.0 (b) 573/1.0 |
| MgH$_2$ (TiH$_2$-doped) | 573/20 | 16.4 | 4.6 | 573/0.01 |
| MgH$_2$ (Cr$_2$O$_3$-doped) | 573/1–2 | 86 | 6.4 | 573/1–2 |
| MgH$_2$ (Nb$_2$O$_5$-doped) | 573/8.4 | 95 | 7.0 | 573/vacuum |
| MgFeH$_6$ | 623/2.72 | 166 | 5.0 | 623/0.001 |

In contrast to metallic hydrides, the complex hydrides are ionic or covalent. A complex hydride contains a metal cation (M) and a hydrogen-containing polyatomic anion (EH$_x$). Examples of polyatomic anyons are alanates (AlH $_4^-$), borohydrides (BH $_4^-$), and amides (NH $_2^-$). Sodium alanate, for example, can decompose to give hydrogen through a series of reactions [13]:

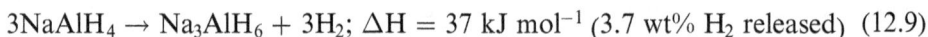

$$3NaAlH_4 \rightarrow Na_3AlH_6 + 3H_2; \quad \Delta H = 37 \text{ kJ mol}^{-1} \text{ (3.7 wt% H}_2 \text{ released)} \quad (12.9)$$

$$Na_3AlH_6 \rightarrow NaH + Al + 3/2H_2; \quad \Delta H = 47 \text{ kJ mol}^{-1} \text{ (1.9 wt% H}_2 \text{ released)} \quad (12.10)$$

These two reactions take place at two different temperatures, and a temperature greater than 373 K is required to complete the second step of decomposition. The total hydrogen capacity is 5.6 wt% [13]. The alanates have many advantages over conventional metal hydride systems as a reversible hydrogen storage system. A larger amount of hydrogen storage per unit weight of storage material is allowed due to the low molecular weight of the alanates. However, the alanates are relatively difficult to synthesize and the kinetics of the reaction used to store the elemental hydrogen are slow. These aspects somewhat lower the potential of alanates as hydrogen storage systems. It has been reported that doping sodium alanate (NaAlH$_4$) with a titanium catalyst is able to achieve reversible hydrogen storage of up to 4 wt% at 473 K with an enhanced kinetic release rate. The decomposition reaction temperature can also be lowered by ball milling, as well as doping with transition metals such as vanadium. Ball milling of sodium alanates in the presence

**Table 12.6.** Hydrogen storage capacity of some common complex hydrides.

| Complex hydride | Hydrogen storage capability (wt%) |
| --- | --- |
| $LiBH_4$ | 18.6 |
| $NaBH_4$ | 10.4 |
| $Al(BH_4)_3$ | 17.0 |
| $(NH_3BH_3, AB)$ | 19.6 |
| $LiAlH_4$ | 10.5 |
| $NaAlH_4$ | 7.5 |
| $KAlH_4$ | 5.7 |
| $Be(AlH_4)_2$ | 11.3 |
| $Mg(AlH_4)_2$ | 9.3 |
| $Ca(AlH_4)_2$ | 7.7 |
| $Ti(AlH_4)_4$ | 9.3 |

of carbon can enhance the reversible hydrogen storage capacities to 2.5–3 wt% at 353–413 K and 4.5–5 wt% at 423–453 K [13]. The required temperatures and reaction rates for the operation of sodium alanates and variants are not very appropriate for application and the exact mechanism of the catalyst is yet to be understood completely. Moreover, the full storage capacity is lost after the first cycle.

The reported hydrogen content of ammonia borane ($NH_3BH_3$, AB), $LiBH_4$, and $NaBH_4$ are 19.6 wt%, 18.6 wt%, and 10.6 wt%, respectively [10]. With high hydrogen capacity and moderate desorption temperature, ammonia borane is the most widely studied member of this group and has significant potential for onboard hydrogen storage. It contains protic (N-H) and hydridic (B–H) hydrogens, which show an opposite polarity. This can be simply regarded as dihydrogen bonding (DHB): $N–H^{\delta+} \cdots H^{\delta-}–B$, which enables hydrogen release under relatively mild temperature ($\approx$120 °C) with ultrahigh purity.

In summary, it can be said that despite the attractive advantages of hydrides as hydrogen storage media, the strong chemical binding in the hydride results in sluggish kinetics and inappropriate hydrogen desorption temperature. These aspects reduce their efficiency and suitability in mobile applications. Table 12.6 presents the hydrogen storage capacity of some common complex hydrides.

### 12.3.3 Liquid organic and circular carriers

Liquid organic molecules with conjugated $\pi$-bonds can be used for reversible hydrogen storage [14]. They are hydrogenated by saturating the $\pi$-bonds with hydrogen and are dehydrogenated in the reverse process. Hydrogen can also be stored and transported via circular carriers such as ammonia. Depending on the nature of the gas molecules, circular carriers in their hydrogen-rich state are liquid or can be liquefied with relative ease, and have similar desirable properties as liquid organics [14]. Liquid organics and circular carriers have been demonstrated to be promising for hydrogen storage, as well as transportation at the bulk level. We will talk about these chemicals in more detail in chapter 14.

Certain sections of text in this chapter have been reproduced with permission from [10]. CC BY 4.0. Copyright © 2021 Jie Zheng *et al.*

# References

[1] Nikolaidis P and Poullikkas A 2017 *Renew. Sustain. Energy Rev.* **67** 597

[2] Zohuri B 2019 *Hydrogen Energy* (Berlin: Springer)

[3] Sorensen B and Spazzafumo G 2018 *Hydrogen and Fuel Cells* (New York: Academic)

[4] Sheffield J W, Martin K B and Folkson R 2014 *Alternative Fuels and Advanced Vehicle Technologies for Improved Environmental Performance* ed R Folkson (Amsterdam: Elsevier)

[5] Zhou L 2005 *Renewable and Sustainable Energy Rev.* **9** 395

[6] Osman A I *et al* 2021 *Environ. Chem. Lett.* **20** 153

[7] Zheng J *et al* 2012 *Int. J. Hydrogen Energy* **37** 1048

[8] de Miranda P E V 2019 *Science and Engineering of Hydrogen-Based Energy Technologies* (New York: Academic)

[9] Aceves S M *et al* 2010 *Int. J. Hydrogen Energy* **35** 1219

[10] Zheng J *et al* 2021 *Research* **2021** 3750689

[11] Zacharia R and Rather S 2015 *J. Nanomater.* **2015** 914845

[12] Dillon A C *et al* 1997 *Nature* **386** 377

[13] Viswanathan B 2017 *Energy Sources Fundamentals of Chemical Conversion Processes and Applications* (Amsterdam: Elsevier) 185–212 pp

[14] Abdin Z *et al* 2021 *iScience* **24** 102966

**IOP** Publishing

Hydrogen
Physics and technology
**Sindhunil Barman Roy**

# Chapter 13

# Hydrogen safety and integrity

In the widespread use of hydrogen in its various forms, safety is a very important aspect. The potential risks of high-pressure gaseous hydrogen include vessel explosion, gas leakage, temperature rise in the fast-filling process, and hydrogen embrittlement of the container materials. The liquid phase is accompanied by the presence of hydrogen vapor, which is quite cold and dense. In the event of a spill of liquid hydrogen, it would initially give rise to the formation of a pool. The very low temperature of the fluid may also cause the solidification of the other air components such as nitrogen and oxygen. The particles of solid oxygen together with liquid hydrogen can form a potentially explosive mixture that could self-ignite. In the rest of this chapter, we will discuss various hazards and safety-related features of hydrogen.

## 13.1 Properties of hydrogen

Hydrogen in molecular form is colorless, odorless, and tasteless, and is about 14 times lighter than air. It diffuses faster than any other gas. The gaseous hydrogen has one of the highest heat capacities ($14.4$ kJ kg$^{-1}$ K$^{-1}$). Hydrogen is relatively nonreactive at ambient temperatures unless it is activated in some way. It becomes chemically very reactive at higher temperatures, and as a result, it is not found chemically free in nature.

At room temperature, atomic hydrogen can be a powerful reducing agent. It reacts with many metallic oxides and chlorides to produce free metals. It reduces some salts such as sodium and potassium nitrates to a metallic state. We have already studied in an earlier chapter that hydrogen reacts with several elements, both metals and nonmetals, to yield hydrides. Atomic hydrogen reacts with organic compounds to produce a complex mixture of products.

Hydrogen reacts violently with oxidizers such as nitrous oxide, halogens (more so with fluorine and chlorine), and unsaturated hydrocarbons (e.g., acetylene) with intense exothermic heat [1]. Hydrogen generates energy by reacting with oxygen either in a combustion or electrochemical conversion process. The resultant reaction product is water vapor. This reaction is immeasurably slow at room temperature but can be accelerated by catalysts such as platinum or by an electric spark.

Hydrogen is highly flammable and has a very wide range of flammability, with both lower and higher flammability limits of 4.1% and 74.8% (at a temperature of 25 °C and pressure of 1 atm), with the explosion range lying in between [1, 2]. The mixture of fuel and the air becomes flammable in the range of fuel concentration spanned by lower and higher flammability levels.

Hydrogen has a high autoignition temperature, which from a safety perspective indicates that hydrogen is relatively safe. However, the energy required to initially ignite hydrogen is very low compared to that of other fuels [2]. Moreover, hydrogen has an extremely low electro-conductivity rate. This means that both the flow and agitation of hydrogen have the potential to generate an electrostatic charge that might trigger the spark, whether in liquid or gaseous conditions. Thus, while dealing with hydrogen, the elimination of any potential ignition and heat sources (e.g., static electricity, hot objects, open flames, and electrical equipment), and proper electrical grounding of the devices are important. Hydrogen has an almost invisible, pale blue color on burning, which makes it difficult to detect. Hence, it requires very careful handling during ignition and combustion. The flame of a hydrogen fire is usually in the form of a torch or jet originating at the point of hydrogen discharge [2].

The speed of hydrogen vapor generation from a liquid state is significantly faster compared to any fossil fuel. This results in a very short period of hydrogen fire, which is very small compared to hydrocarbon-based fire for the same fuel volume. However, hydrogen combustion is relatively safe because the combustion product is water. Hydrogen is relatively sensitive to detonation and its wide range of oxygen mixtures potentially leads to convenient ignition and detonation [2]. This demands very careful attention during its storage.

A very rapid hydrogen burning rate leads to a reduction in the total energy that is radiated for equal volumes of fuels. As a result, the heat transferred to the object surrounding the flame through radiation becomes smaller. The mixture of hydrogen and air has a higher tendency to detonate in comparison to a mixture of other fuels with air. However, the rapid dispersion of hydrogen limits this detonation within a confined space.

Table 13.1 presents a selection of combustion properties of hydrogen and some hydrocarbon-based fuels at ambient conditions.

**Table 13.1.** Comparison of combustion properties among fuels at atmospheric conditions (temperature and pressure of 25 °C and 1 atm, respectively). Reproduced from [2]. CC BY 4.0.

| Properties | Hydrogen | Methane | Propane | Gasoline |
|---|---|---|---|---|
| Flammability limit (vol%) | 4.1–74.8 | 5.3–17.0 | 1.7–10.9 | 1.0–6.0 |
| Detonation limit (vol%) | 18.3–59.0 | 6.3–13.5 | 3.1–9.2 | 1.1–6.0 |
| Autoignition temperature (°C) | 585 | 540 | 490 | 246–280 |
| Minimum ignition energy (mJ) | 0.017 | 0.274 | 0.24 | 0.24 |
| Stoichiometric mixture (vol%) | 29.6 | 9.5 | 4.0 | 1.9 |
| Laminar burning velocity (cm $s^{-1}$) | 270 | 37 | 47 | 30 |

## 13.2 Hydrogen hazards

Hydrogen hazards are classified under three categories, namely physiological hazards, physical hazards, and chemical hazards [1].

### 13.2.1 Physiological hazards

When hydrogen or another nontoxic gas displaces oxygen and reduces the concentration below 19.5% by volume, it causes asphyxiation. Overpressure injury can take place due to blast waves from explosions. The threshold of lung rupture is 0.7 atm overpressure for 50 ms or 1.4–2 atm for 3 ms. Thermal burns can happen due to radiant heat emitted by hydrogen fire, which is directly proportional to exposure time, burning rate, the heat of combustion, size of the burning surface, and atmospheric wind and humidity [1]. A 0.95 W cm$^{-2}$ radiation can lead to skin burns in 30 s. Exposure to large liquefied hydrogen spills can cause cryogenic burns.

### 13.2.2 Physical hazards

Physical hazards include collision during transportation and mechanical failures of hydrogen equipment, such as safety devices, storage vessels, vents, and exhaust vaporization systems [1]. High-pressure high-purity hydrogen at room temperature can cause significant deleterious effects on the mechanical properties of metals. The invasion of hydrogen atoms inside metals promotes localized plastic processes and accelerates the crack propagation rate of the metal [3]. Thus, hydrogen embrittlement causes the mechanical properties of metals to degrade, which may ultimately result in leaks. This depends on environmental temperature and pressure, purity of metal, exposure time to hydrogen, and surface conditions [1]. Environmental hydrogen embrittlement can be controlled by oxide coatings, elimination of stress concentrations, additives to hydrogen, and alloy selection. We will talk about hydrogen embrittlement in the hydrogen damage section that follows.

In the case of liquid hydrogen, an inappropriate insulation of hydrogen equipment can provoke embrittlement. This is because at low temperature there can be a ductile-to-brittle transition and also elastic to plastic phase transformation in the crystalline structure of the containers or equipment [1].

### 13.2.3 Chemical hazards

The flammability limits are determined by the following factors: ignition energy, temperature and pressure diluents, and equipment size and configuration [1]. Liquid hydrogen and liquid or solid oxygen mixture can detonate when initiated by a shock wave. Therefore, buildings using hydrogen systems must properly isolate open flames, as well as electrical and heating equipment due to the low energy required for ignition (0.02 mJ).

### 13.2.4 Explosion phenomena

An explosion is caused by the sudden release of energy. Pressure waves are produced as a result. Conversely, deflagration is symbolized by the flame front, which propagates as a subsonic wave through the flammable mixture [1]. The flame front and shock wave are connected during detonation. They combine to generate a supersonic wave that moves 1000 times faster than the original reactions across a detonable combination. Accordingly, explosion has a far higher potential for harm and destruction than deflagration [1]. The concentration of hydrogen present in a detonable mixture determines the minimum ignition energy. The maximum pressure and temperature are roughly 1600 kPa and 3000 K, respectively. The stoichiometric hydrogen-air can give rise to 2/3 of the explosion energy of trinitrotoluene (TNT).

It can be finally summarized that [1]:

1. The combustion properties of hydrogen indicate that it is not easy to handle. But in terms of the energy offered, it is quite promising.
2. Hydrogen is very light. This causes trouble with leaking, along with the fact that it is highly flammable.
3. The flammability of hydrogen is a function of concentration level and is much greater than that of other fuels. The flame is also faster and much hotter.

## 13.3 Hydrogen integrity phenomena

Equipment used in hydrogen technology may lose its integrity due to specific physical and chemical processes. A thorough understanding of loss of integrity phenomena, such as hydrogen damage, can help to avert many crucial events, and consequently accidents, with serious repercussions. Other phenomena that can happen in components and instruments that deal with liquid nitrogen include thermal contraction and low-temperature embrittlement. Some of these phenomena will be covered in the subsections that follow.

### 13.3.1 Hydrogen damage

The term 'hydrogen damage' is used to categorize and characterize a variety of phenomena that may have an impact on the properties and integrity of the materials used in equipment for hydrogen technology. These phenomena can also influence applications where hydrogen is not directly employed. Three independent factors that influence hydrogen damage are shown schematically in figure 13.1. These three factors are: (i) the employment environment of the material (i.e., hydrogen amount, form, and processes); (ii) the field type (i.e., mechanical, electrochemical, and operating circumstances) where the material is employed; and (iii) the selected material itself [4].

There are three main forms of hydrogen damage [4]:

1. Blisters caused by hydrogen diffusion into the metal.
2. Hydrogen-assisted cracking, which includes different types of crack.
3. Metal hydride formation.

Figure 13.2 illustrates a schematic description of hydrogen attack, blistering, and metal hydride formation. When carbon and low-alloy steels are exposed to hydrogen over an extended period of time at high temperatures and pressures, a hydrogen

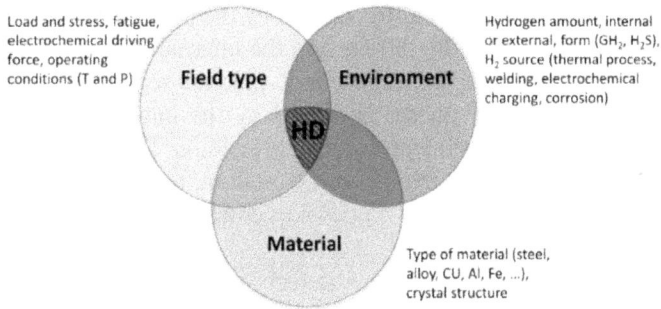

**Figure 13.1.** Schematic illustration of factors influencing hydrogen damages. Reproduced from [4]. CC BY 4.0.

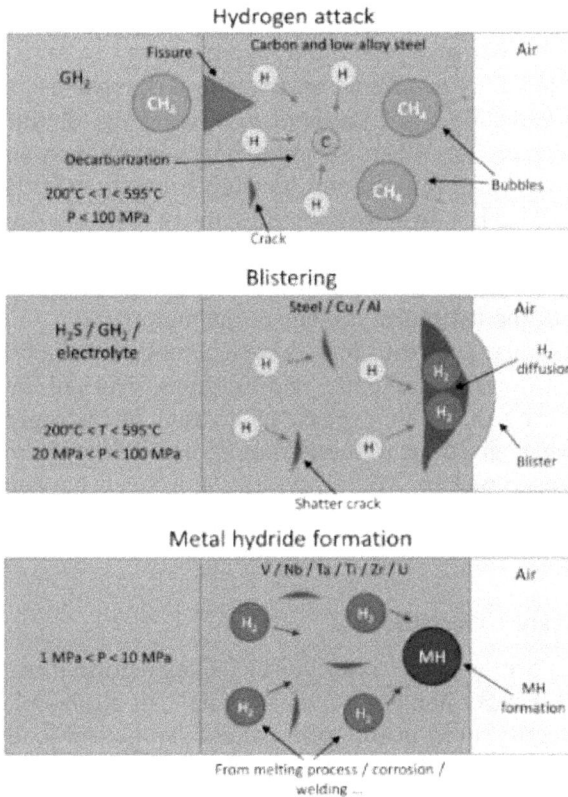

**Figure 13.2.** Schematic illustration of the phenomena of hydrogen attack, blistering, and metal hydride formation. Reproduced from [4]. CC BY 4.0.

attack occurs (see upper portion of figure 13.2). This results in steel decarburization due to methane formation from hydrogen and carbon contained in the material. Furthermore, the alloy may develop fissures and cracks, which weakens it. Blistering happens when low-strength alloys are exposed to a corrosive environment, such $H_2S$ [4].

Blisters are plastic deformations of alloys caused by the precipitation of hydrogen from atoms to molecules, which initially diffuse into the internal defects of the material (see the middle of figure 13.2). Thus, a localized high-pressure region is caused. Finally, the precipitation of metal hydride phases is caused by the introduction of hydrogen by welding, heat treatment, corrosion processes, or melting. These metal hydrides cause many metals, including titanium, magnesium, tantalum, niobium, vanadium, uranium, thorium, zirconium, and their alloys, to develop cracks and degrade their mechanical properties.

Another broad category of hydrogen defects is hydrogen embrittlement, which includes loss of tensile ductility, hydrogen stress cracking (HSC), and hydrogen environment embrittlement (HEE). HEE is typically connected to substances that operate in a hydrogen environment. This is more common in metals with low-hydrogen-solubility, such as metals with hexagon close-packed (hcp) and body-centered-cubic (bcc) lattice structures. HEE intensifies at a temperature of 20 °C, low strain rate, and high hydrogen purity and high pressure.

The second type of hydrogen embrittlement is known as internal hydrogen embrittlement, or HSC. This is a mechanism of cracking for ductile steels, where hydrogen can flow freely [4]. When the material is continuously subjected to a load that is between a specific threshold and its yield strength, a brittle fracture results from this process. The last hydrogen embrittlement phenomena show a reduced elongation and area in the tensile test for steels, stainless steels, nickel, aluminum, and titanium alloys exposed to hydrogen environment [4]. The decrease in ductility can be correlated to the hydrogen content of the material.

Similar to blistering, other forms of hydrogen damage include fisheyes, flakes, and shatter cracks. Castings, weldments, and forgings may all have these. These processes occur during the cooling phase as a result of hydrogen precipitating in material defects, which gives rise to these features, and during the melting operations as a result of hydrogen pickup. Micro-perforation in steels happens at nearly room temperature and under extremely high pressure. This may cause tiny fissures, and make the material permeable to gas and liquids.

### 13.3.2 Low-temperature embrittlement

Before fracturing, a brittle material exhibits no lasting deformation because it lacks a yielding area. On the other hand, a ductile material can display plastic deformation to maintain its integrity, even in situations where the applied load exceeds its yield stress. However, a lot of ductile metals are brittle at low temperatures. The onset of the brittle phase happens at the ductile-brittle transition temperature, which is metal specific. Metals with face-centered cubic crystal structures, such as aluminum, copper, nickel, some of their alloys, and austenitic stainless steel, are typically utilized at cryogenic temperatures.

### 13.3.3 Thermal contraction

Thermal strains are a common phenomenon for hydrogen equipment associated with liquid hydrogen technology. One property of the material that is dependent on operating temperature is the coefficient of thermal expansion (CTE). This is an important parameter to consider when designing systems and its component parts. The majority of the materials have a decreased volume when the temperature drops. The CTE is nonlinear in nature. The 90% of the total volume contraction typically occurs between room temperature and 77 K, the boiling point of nitrogen. We discuss the thermal stresses caused by dimensional change and thermal gradient in the following subsections.

### 13.3.3.1 Stresses cause by dimensional change

The source of these strains is thermal contraction brought on by exposure of the component materials to cryogenic temperatures. For instance, during the filling and discharging cycles, the interior vessel of a double-walled liquid hydrogen tank must be able to freely expand or contract. In any other case, internally generated stresses could lead to tank failures. Additionally, the suspension system and the interconnected pipework must be constructed to adapt to the dimensional variations of the inner shell after multiple cycles of contraction and expansion [4]. One major issue with liquid hydrogen tanks is the moving of the perlite powder in the vacuum jacket following thermal contraction of the inner vessel. The perlite powder is pressed by the expansion of the inner tank when the liquid hydrogen is released, and it cannot return to its previous position. This phenomenon is termed perlite compaction. A properly designed soft bat can be used on the outer surface of the inner tank or the support rod system to get around this problem [4]. It should also be possible for the liquid hydrogen vacuum-jacket pipe to deform appropriately.

### 13.3.3.2 Stresses caused by thermal gradients

Depending on whether the hydrogen equipment is in a steady state or is being filled and discharged, different sections are subjected to a thermal gradient. The supporting rods of the inner tank of a liquid hydrogen vessel serve as an illustration of the steady state gradient. The beam is linked to the inner shell at a cryogenic temperature on one end, while its other end is in contact with the outer vessel at room temperature.

The three factors that can affect the stresses caused by thermal gradients are the rate of cooling, thermal conductivity, and thickness of the cooled material. Inappropriate combination of these parameters may result in unanticipated stresses and eventual failure of the component. According to estimates for a stainless-steel flange, the thermal gradient produced by a liquid hydrogen flow can cause tension and compression in the interior and exterior of the component, respectively [4]. The thermal conductivity and thickness of the material are decided during the design stage. The cryogenic flow rate that is established during the operating phase determines the cooling rate.

During the cooldown of the long transfer line, the two-phase hydrogen flow regime inside the pipe creates a thermal gradient. The liquid phase flows through the bottom of the tube, which cools more quickly than the top of the tube and has a tendency to contract. The thermal gradient generated in the pipe section causes it to bow upward in the middle and generates unwanted stress, both in the pipe and its supports [4]. For the purpose of minimizing the vapor phase and preventing pipe bowing, the flow of liquid hydrogen must be greater than a critical lower limit.

## 13.4 Safety comparisons of hydrogen, methane, and gasoline

The thermo-physical, chemical, and combustion properties of hydrogen, methane, and gasoline are compared in table 12.1. Gasoline is the most straightforward and possibly safest fuel to store due to its higher boiling point, reduced volatility, and more limited flammability and detonability constraints. This remark is based on the previous discussions regarding hazards of fire and explosion. However, contemporary technology also allows for the safe storage of hydrogen and methane, the principal ingredients of natural gas.

Hydrogen is a very promising fuel for transport. As in any other situation, potential hazards here are frequently linked to fire, explosion, or poisoning. Because hydrogen is not poisonous, there are no toxicity risks associated with it or its combustion products. Fuel storage on a vehicle, fuel supply lines, or fuel cells are potential sources of fire and explosion when using hydrogen for transportation [1]. The fuel cell has the least risks among these options.

In conclusion, hydrogen, whether it is liquefied or compressed gas, may be utilized and stored securely for a considerable amount of time. Metal hydrides provide just as safe, if not safer, storage for hydrogen. Different applications of hydrogen might not pose a significant risk to public safety in the commercial and industrial spheres. However, the use of hydrogen in the transportation and residential fuel sectors requires more thorough safety evaluations.

## References

[1]  Nazzar Y S 2013 *Int. J. Hydrogen Energy* **38** 10716
[2]  Aziz M 2021 *Energies* **14** 5917
[3]  Zheng J *et al* 2012 *Int. J. Hydrogen Energy* **37** 1048
[4]  Ustolin F, Paltrinieri N and Berto F 2020 *Int. J. Hydrogen Energy* **45** 23809

**IOP** Publishing

Hydrogen
Physics and technology
**Sindhunil Barman Roy**

# Chapter 14

## Hydrogen transport and distribution

The transport of hydrogen is necessary to take clean energy to places where it is utilized. It is essential to transport hydrogen as easily as fossil fuels. Currently, there are four attractive options for hydrogen transportation. Three of these options involve shipping ammonia, liquid hydrogen, and liquid organic hydrogen carrier. The fourth option is transporting compressed hydrogen gas via pipelines. In the rest of this chapter, we will briefly discuss the various methods of hydrogen transport.

### 14.1 Hydrogen transport via ammonia

Ammonia, with a molecular formula of $NH_3$, has a hydrogen content of 17.6 wt% and contains no carbon [1]. It is possible to decompose it to a gas mixture of 75% $H_2$ and 25% $N_2$. Thus, it offers a high output and clean hydrogen generation with zero carbon emission. Ammonia has a boiling point of $-33\ °C$ at standard pressure. Therefore, it can be stored in a liquid form more easily than hydrogen with a boiling point of $-253\ °C$ and methane with a boiling point of $-160\ °C$. It can be also compressed to liquid at a mild pressure of 10 bar. Ammonia also has a relatively high volumetric energy density of 11.38 GJ $m^{-3}$. Furthermore, ammonia is the second most produced chemical in the world, with annual production up to about 180 million tons [1]. The infrastructure for ammonia storage and transport is already well-established and available worldwide. Thus, ammonia as an energy carrier for hydrogen is an attractive option for large-scale transportation of hydrogen. Ammonia ships have a wide range of size and distance covered, and are most attractive in this respect.

In the chemical process, hydrogen reacts with nitrogen to produce liquid ammonia. Ammonia is stored in liquid form in tanks, which is easy to transport. After reaching the destination, ammonia is cracked or broken down into its components, releasing hydrogen and nitrogen. The hydrogen is subsequently purified for use.

### 14.1.1 Synthesis of ammonia

Ammonia is typically synthesized by the Haber–Bosch process. In a green Haber–Bosch process, ammonia is synthesized by using renewable hydrogen from water electrolysis and nitrogen from an air separation unit as the feedstock. Typically, 27 kW of power is required for the synthesis loop to produce 1 tonne of ammonia per day, and the separation unit and mechanical compression require around 3.5 and 1.5 kW, respectively [2]. The ammonia synthesis process is exothermic, and is accompanied by a reduction of entropy. Low temperatures and high pressures are conducive for the reaction to proceed toward the product direction. However, the Haber–Bosch process is carried out at high temperatures and pressures (≈450 °C and ≈200 bar) because of the slow kinetics associated with the difficult dissociation of molecular nitrogen into atomic nitrogen [2]. The catalyst of choice for the Haber–Bosch process is usually iron promoted with main group element oxides. Ruthenium (Ru)-based catalysts are also often used. There are also newer methods for ammonia synthesis, including solid-state ammonia synthesis [2].

### 14.1.2 Ammonia decomposition

Ammonia decomposition usually takes place in the temperature range of 400–700 °C. However, for some catalysts such as Ni a temperature around 1000 °C is required [2]. The decomposition of ammonia takes place in a stepwise sequence of dehydrogenation reactions, including (1) adsorption of ammonia onto catalyst active sites, (2) successive cleavage of N-H bond on adsorbed ammonia to release hydrogen atom, and (3) re-combinative desorption of N and H atoms to form gaseous nitrogen and hydrogen molecules. This decomposition process is illustrated below by the following set of equations [1]:

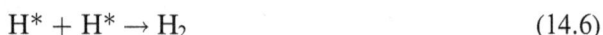

$$NH_3 + \phantom{}^* \rightarrow NH_3^* \tag{14.1}$$

$$NH_3^* + \phantom{}^* \rightarrow NH_2^* + H^* \tag{14.2}$$

$$NH_2^* + \phantom{}^* \rightarrow NH^* + H^* \tag{14.3}$$

$$NH^* + \phantom{}^* \rightarrow N^* + H^{**} \tag{14.4}$$

$$N^* + N^* \rightarrow N_2 \tag{14.5}$$

$$H^* + H^* \rightarrow H_2 \tag{14.6}$$

Ammonia decomposition or cracking is endothermic and only needs an appropriate catalyst that is highly active, durable, and scalable. A wide variety of materials are utilized for ammonia decomposition, among which elemental metal ruthenium catalyst is prominent. Binary and ternary catalysts such as Ni–Pt, NiMo, $MoN_x$, $NiMoN_x$, and $Co_3Mo_3N$ are also quite promising. The reactions of ammonia synthesis and decomposition follow the same fundamental principles, and hence are microscopically reversible. However, the optimal catalysts for ammonia decomposition are not the same as those for ammonia synthesis because the very different synthesis and decomposition reaction conditions, such as ammonia concentrations, need different optimal energy ranges for the critical step, namely nitrogen binding on the catalyst's surfaces [1].

Electric furnaces are used to decompose ammonia at a small scale to produce hydrogen and nitrogen gas mixture for various industrial applications. Reactor material choice is critical for ammonia decomposition because of the detrimental reaction environment, which may cause thermal stress, stress corrosion cracking, and hydrogen embrittlement in the reactor. A membrane reactor for ammonia decomposition has been demonstrated at the laboratory scale, in which chemical reactions and the selective separation of a product occur at the same time. Any downstream separation unit is not required here, which would help to increase efficiencies at lower operating temperatures. However, the usage of tantalum tubes as the membrane support, which exhibits selective hydrogen diffusion, would be quite costly for large-scale ammonia decomposition.

### 14.1.3 Advantages and disadvantages of ammonia

There are several advantages to using ammonia in hydrogen transportation. To start with, the infrastructure for ammonia synthesis is already in place because ammonia is used in various industries. Furthermore, ammonia can be stored in slightly refrigerated tanks at $-33\ ^{\circ}C$ or ambient temperatures under a pressure of 8–10 bar. Thus, storing and transporting ammonia is relatively straightforward and economically affordable.

There are also certain limitations in using ammonia as a hydrogen carrier. The cracking process requires high temperatures greater than $500\ ^{\circ}C$ for the production of high-purity hydrogen, which demands thermal energy of 4.2 $GJ_T$/ton $NH_3$ [3]. The conversion of ammonia to hydrogen needs to be conducted with a highly efficient purification and separation system. This causes additional consumption of energy, 0.5 $GJ_T$/ton $NH_3$. Furthermore, additional electrical energy of 2.0–4.3 $GJ_e$/ton $NH_3$ is required for the compression of hydrogen to 880bar for applications such as refilling fuel cell electric vehicles at 700 bar. Thus, the heat and electricity requirements for the cracking and compressing processes per unit of $NH_3$ demand $\approx$0.3 and 0.16–0.34 units of $NH_3$ (assuming a fuel cell efficiency of 60%), respectively [3]. The integration of an energy-intensive cracking reactor with a hydrogen compression system complicates the fueling and refilling processes on the consumer side. Figure 14.1 schematically presents the power-to-power energy consumption of ammonia as a hydrogen energy carrier. The limitations associated with ammonia further increase due to the intricacy of the cracking system and the performance and lifetime of catalysts in the presence of impurities [3].

## 14.2 Liquid organic hydrogen carrier

A liquid organic hydrogen carrier (LOHC) is a liquid that absorbs and subsequently releases hydrogen via a chemical reaction. The LOHC system is comprised of a pair of one hydrogen-lean organic compound (LOHC-) and one hydrogen-rich organic compound (LOHC+) [5]. Hydrogen is stored by converting LOHC- into LOHC+ in

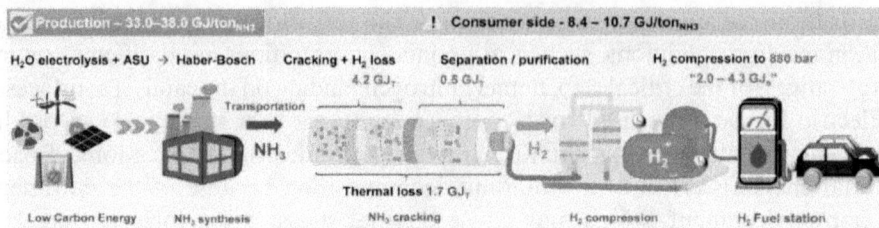

**Figure 14.1.** Power-to-power energy consumption of ammonia as a hydrogen energy carrier. Reprinted with permission from [3]. Copyright (2021) American Chemical Society.

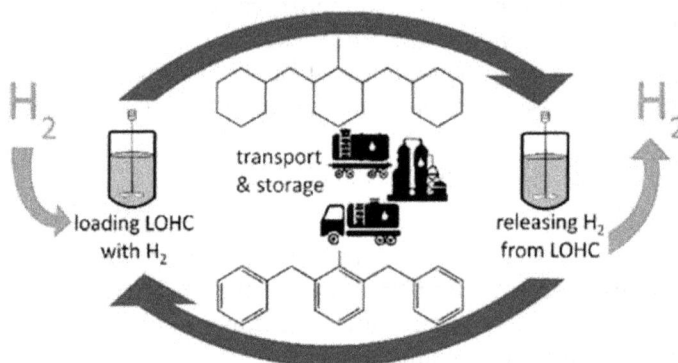

**Figure 14.2.** Schematic representation of the hydrogen storage and transport concept using a LOHC.

a catalytic hydrogenation reaction. Hydrogen can be released by converting LOHC+ into LOHC- in a catalytic dehydrogenation reaction.

It is possible to store hydrogen in a LOHC+ compound stably and without self-discharge for a long time. The hydrogen containing LOHC can be transported in atmospheric conditions. This enables seasonal energy storage and transportation to remote areas where dehydrogenation can be performed. This process is reversible and allows rapid hydrogenation and dehydrogenation without consumption of the compounds forming the LOHC system. Only hydrogen is released in the whole process, which is shown schematically in figure 14.2.

### 14.2.1 Characteristic properties of LOHC systems

A good LOHC system is characterized by as many of the following properties as possible [5].

1. Low melting point ($<-30$ °C) of all compounds involved and high boiling point ($>300$ °C) for simple hydrogen purification by LOHC condensation.
2. High hydrogen storage capacity ($>56$ kg m$^{-3}$ or $>6$ wt%)
3. Low heat of desorption (42–54 kJ mol$^{-1}$-H$_2$) to enable low dehydrogenation temperature against 1 bar hydrogen pressure (e.g., $<200$ °C).
4. Ability to undergo very selective hydrogenation and dehydrogenation for long life cycles of charging and discharging.

5. Compatibility with currently available infrastructure for fuels.
6. Low production costs and good technical availability.
7. Toxicological and eco-toxicological safety during transportation and use.

In order to supply high-purity hydrogen, purification technology with hydrocarbon condensers is frequently integrated into the system. The most researched LOHC compounds at the moment, both hydrogenated and unhydrogenated, are dibenzyltoluene and perhydrodibenzyltoluene (H0-DBT, H18-DBT), benzene and cyclohexane, toluene and methylcyclohexane, naphthalene and decalin, and N-ethylcarbazole and perhydro-N-ethylcarbazole (H0-NEC, H12-NEC) [5]. However, it should be mentioned that none of these LOHC systems can completely fulfill all the necessary characteristics. We will now briefly discuss the functions of a few of these LOHC systems. The readers are referred to the article by Modisha [5] *et al* for detailed information on LOHC systems.

### 14.2.1.1 Cyclohexane-benzene

At ambient temperature, the liquid LOHC compound cyclohexane-benzene (CHE) has a hydrogen storage capacity of ± 7.19 wt%. Because of its low boiling point of 81 °C and flammability, it is a challenging task to release high-purity hydrogen using CHE. Benzene, the reaction product, is also highly toxic. Figure 14.3 shows the hydrogenation reaction of benzene and the dehydrogenation of CHE to establish a LOHC system.

Various metals (e.g., Ni, Pt, Ag, etc) supported on activated carbon cloth (ACC) have been used as a catalyst for the dehydrogenation of CHE in a spray-pulsed reactor at 300 °C [5]. Experimental results showed that the bimetallic Ni/Pt activity (12.75 mmol $g_{cat}^{-1}$ min$^{-1}$) was 60 times higher than monometallic Pt/ACC activity (0.21 mmol g$^{-1}$$_{cat}$ min$^{-1}$) and 1.5 times higher than 20 wt% Ni/ACC activity (8.5 mmol $g_{cat}^{-1}$ min$^{-1}$). The hydrogen selectivity also improved from 98.8% to 99.7% when

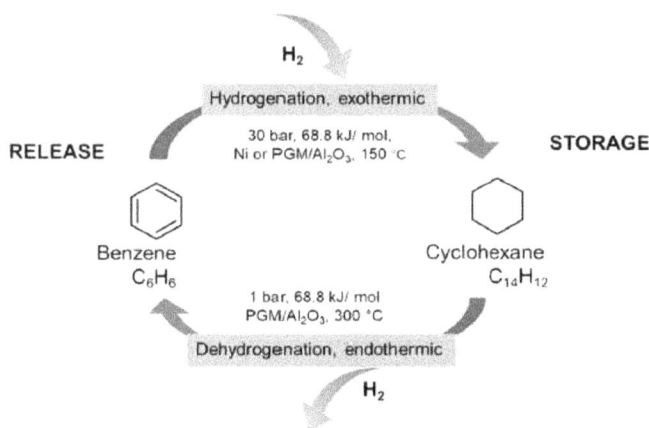

**Figure 14.3.** Schematic representation of the benzene/cyclohexane LOHC system. Reprinted with permission from [5]. Copyright (2019) American Chemical Society.

bimetallic Ni/Pt catalyst was used. With a similar reactor and reaction temperature, there can be an improvement in CHE dehydrogenation activity with 10 wt% Ag supported on ACC. The hydrogen production rate doubles in comparison to that achieved with the 10 wt% Ag (6.8 mmol $g_{met}^{-1}$ min$^{-1}$) catalyst [5] when the Pt-promoted catalyst 10 wt% Ag/ACC + 1 wt% Pt (14.2 mmol $g_{met}^{-1}$ min$^{-1}$) was used. A hydrogen production rate of 58 mmol h$^{-1}$ $g_{cat}^{-1}$ was obtained using a Ni–Cu/SiO$_2$ bimetallic catalyst (17.3 mol% Ni, 3.6 mol% Cu) for CHE dehydrogenation with a plug flow reactor system at 250 °C. CHE dehydrogenation experiments in a batch-type reactor at 300 °C with 2 wt% Pt/activated carbon and 3.82 wt% Pt/Al$_2$O$_3$ catalysts yielded 910 and 1800 mmol $g_{met}^{-1}$ min$^{-1}$, respectively [5]. It can be concluded from these studies that hydrogen productivity improves with an increase in metal loading, the application of bimetallic catalysts, and the use of noble metals as catalysts.

### 14.2.1.2 Methylcyclohexane-toluene

Methylcyclohexane (MCH) exists in a liquid state at ambient conditions and has a hydrogen storage capacity of 6.2 wt%. Figure 14.4 shows the equilibrium conversion of MCH as a function of temperature for various pressures. This picture is archetypical for many pure hydrocarbon LOHC systems. The temperature at which MCH may completely convert to toluene is directly related to pressure. Thus, for dehydrogenation, the lowest system pressure is ideal. However, this can cause adverse effects and catalyst deactivation, so it is not always necessarily beneficial for the catalyst.

Figure 14.5 shows the hydrogenation reaction of toluene and the dehydrogenation of MCH to establish a LOHC system. The dehydrogenation of MCH (with boiling point 100.9 °C) is a gas-phase reaction.

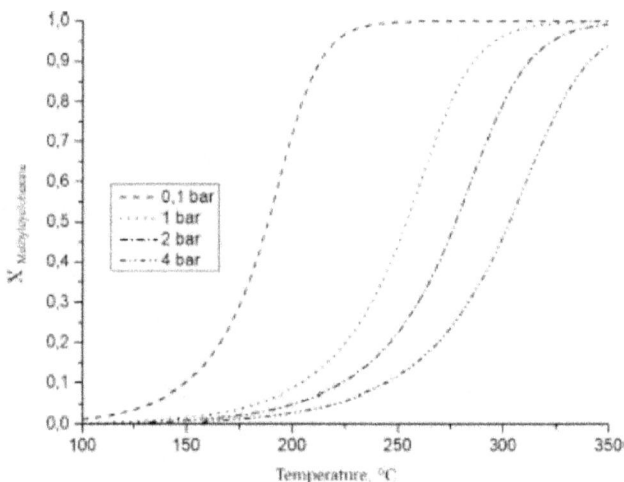

**Figure 14.4.** Equilibrium conversion as a function of system temperature for various system pressures in the dehydrogenation of MCH to toluene. Reprinted with permission from [5]. Copyright (2019) American Chemical Society.

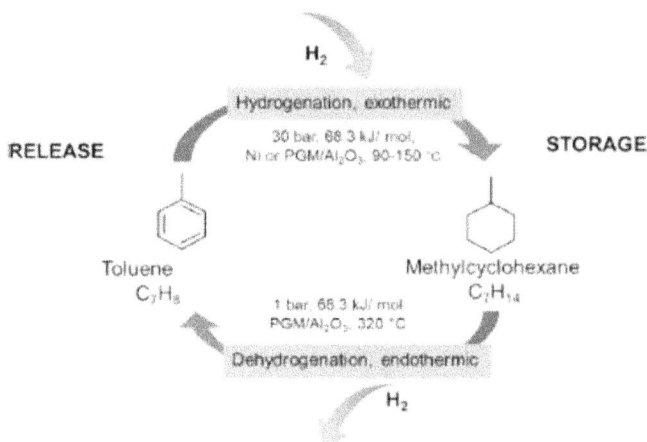

**Figure 14.5.** Schematic representation of the toluene and methylcyclohexane LOHC system. Reprinted with permission from [5]. Copyright (2019) American Chemical Society.

Okada *et al* [6] developed a dehydrogenation catalyst using a simple fixed-bed reactor that has a high stability and sufficient performance. They attempted to raise the catalytic activity by increasing the dispersibility of the platinum metal by using a pore-controlled γ-alumina carrier and optimizing the pH value for the impregnation of chloroplatinic acid. This catalyst could generate hydrogen from MCH: 99.9% with a conversion >95%, toluene selectivity >99.9%, hydrogen generation rate >1000 $Nm^3$ $h^{-1}$ $m^{-3}$cat at a reasonable temperature of 593 K and ambient pressure. Mixtures of different support materials like Ni on $Al_2O_3$ or $Al_2O_3$-$TiO_2$ hybrid composite have been used to effectively increase the catalyst performance. The performance of Ni/$Al_2O_3$-$TiO_2$ was found to be better with MCH conversion 99.9% at 400 °C [5].

Metal oxides $La_2O_3$, $ZrO_2$, $TiO_2$, $CeO_2$, $Fe_2O_3$, $Al_2O_3$, $MnO_2$, and perovskite ($La_{0.7}$ $Y_{0.3}$ $NiO_3$) catalyst supports have been tested for MCH dehydrogenation [7]. It was found that Pt supported on $La_{0.7}$ $Y_{0.3}$ $NiO_3$ exhibited high activity and 100% selectivity for the dehydrogenation of MCH. The productivity increased from 21.1 mmol $g_{met}^{-1}$ $min^{-1}$ (with Pt/$LaO_3$) to 45 mmol $g_{met}^{-1}$ $min^{-1}$ (with Pt/$La_{0.7}$ $Y_{0.3}$ $NiO_3$) [5]. This catalyst support screening was carried out using a spray-pulsed reactor at 350 °C.

Bimetallic Pt/Mo–$SiO_2$ catalysts with constant Pt loading of 5 wt% and Mo loading varying from 4.1 to 12.7 wt% were evaluated for the dehydrogenation of MCH in a down-flow fixed-bed reactor at 400 °C and 2.2 MPa [8]. The catalyst with a Mo loading of 8.0 wt% was found to be the most active, which delivered >80% selectivity toward toluene. This performance of the catalyst was correlated to high dispersion of the Pt on the $MoO_2$ phase.

## 14.3 Transport of gaseous and liquid hydrogen

Trucks, trains, and ships can be used for transporting hydrogen in compressed or liquid form. Another option for large-scale hydrogen transport is a hydrogen pipeline, more so for those regions where a natural gas network already exits.

### 14.3.1 Road and rail transportation of hydrogen

A very common method for the transport of hydrogen gas is filling it in pressure-proofed seamless vessels [9]. These vessels can either be industrial gas standard size (50–150 l in volume) or larger containers (>150 l) for transportation via tube trailers (figure 14.6). The small vessels are grouped in bundles and filled at a centralized production plant. They are then delivered to the final consumer where the empty set of bottles is replaced. A typical tube-trailer setup has nine tubes with a capacity of 2000 l each. This can be used to fill in distributed storage at the demand sites. This can also be used as stationary storage in refueling stations where a different hydrogen infrastructure such as a pipeline or a local production facility is not available. The hydrogen filling pressure of cylinders is usually around 18–25 MPa, although higher pressure is achievable and indispensable to improve the economic viability [9]. The need for improved levels of pressure has pushed cylinder manufacturers to opt for Type II, III, and some Type IV vessels. Figure 14.7 shows a Type IV pressure vessel for hydrogen. Such vessels come with a higher investment cost per unit, but the cost per MJ of hydrogen transported would come down if the internal pressure is increased to 40–70 MPa. Tubes or transport vessels for tube trailers are typically Type I cylinders with a filling pressure of 52 MPa to meet the economic feasibility [9].

The usual material used for the fabrication of transport cylinders/tubes is 34CrMo4 steel alloy, and the same material should be used for valves and safety devices [9]. Composite materials are also currently under investigation for tube construction.

**Figure 14.6.** Nine tube hydrogen trailer. Image source: City Machine & Welding, Inc.; https://cmwelding.com/configuration/hydrogen-h2-tube-trailer-9-tubes-dot-3aax-2400psi-40-ft.

**Figure 14.7.** A Type IV pressure vessel for hydrogen. Reprinted from [9], Copyright (2016), with permission from Elsevier.

**Figure 14.8.** A hydrogen tanker truck. Image source: The Chemical Engineer; https://www.thechemicalengineer.com/features/hydrogen-transport.

Liquid hydrogen is the main solution for transporting large quantities of hydrogen over medium/long distances. Tanker trucks have capacities that usually range from 20,000 to 50,000 l [9]. Figure 14.8 shows a typical tanker truck used for transporting liquid hydrogen. The pressure inside the tank is usually low (0.6–1 MPa). Liquid hydrogen can also be transported via rail and the tank has the same size as that used when attached to a trailer [9].

The critical point during the transport of the liquid hydrogen from the tank to the final destination is the extremely high difference in temperatures between the ambient and the internal environment. Flashing may occur, which can cause injuries and damage to the surroundings where workers and equipment operate [9]. The transport of liquid hydrogen in tanker trucks is controlled by regional, national, and international regulations concerning dangerous goods transportation. The use of trucks for the delivery of liquid hydrogen is monitored with particular care, and it is usually recommended to avoid routing through populated areas.

A physiological boil-off occurs in all transportation phases of liquid hydrogen. This leads to a net loss in terms of payload. About 20% of the hydrogen is lost when it is transferred from one small dewar vessel to another. The loss when hydrogen is transferred from a trailer is said to be around 10%.

### 14.3.2 Ocean transportation of hydrogen

Ocean transportation of hydrogen is an attractive possibility when a large quantity of liquid hydrogen is to be transported over a long distance. This option has been studied intensively since the late-1980s. The first high-level study in this field was carried out within the framework of the Euro-Quebec Hydro-Hydrogen Pilot Project. This project was jointly funded and carried out by the European Commission and the government of Quebec, with the support of a multitude of industrial partners, including many German industries [9]. The basic idea was the exploitation of the excess of cheap but difficult-to-store hydroelectricity produced in Quebec, which was planned to be converted into an easier-to-manage intermediate energy vector such as hydrogen and then shipped to one of the most energy-demanding areas of the world, namely Germany. However, the project was stopped before the construction of the liquid hydrogen-carrying vessel due to the economic aspects of the operation.

The second study was promoted by the Japanese government and focused on the assessment of the viability of hydrogen as an energy carrier for Japan. The World Energy Network (WENET) research program was addressed in many phases and one of them dealt with the design of liquid hydrogen tankers for overseas transport [9]. The researchers proposed different storage architectures (spherical and prismatic vessels) and also different tanker designs. Liquid hydrogen as cargo has a conflicting requirement: large cargo volume and small cargo weight [10]. In the case of hydrogen cargo with a specific gravity lighter than water, the carrying capacity of a tanker depends on cargo capacity and not on cargo weight. The heating value of hydrogen per capacity is as small as 10.1 $GJm^{-3}$. This value is only 42% of liquefied natural gas (LNG) (23.7 GJ $m^{-3}$) and 22% of oil (45 GJ $m^{-3}$). This would suggest the requirement of five liquid hydrogen tankers to transport the same amount of energy by one oil tanker. On the other hand, the heating value per weight is 142 kJ $g^{-1}$, which is 3.1 times higher than that of oil [10]. Thus, it is clear while liquid hydrogen has an advantage as fuel for spaceships, it does not have similar merit in the case of tanker transportation. The specific gravity of liquid hydrogen is 1/14 of water. As a result, liquid hydrogen ships have a small displacement despite their huge capacity. Thus, these ships require ballast and a double hull to guarantee stability.

Additionally, liquid tankers need to maintain ultralow temperature ($-253$ °C) which is much lower than LNG tankers ($-163$ °C).

The ultralow temperature of liquid hydrogen leads to the following problems [10]:

**Figure 14.9.** A Japanese hydrogen carrier ship developed by Kawasaki Heavy Industries, Ltd. Reprinted from [9], Copyright (2016), with permission from Elsevier.

1. Large energy is lost at liquefaction (over 20%).
2. Large amount of boil-off gas is lost at the loading and unloading, and during navigation.
3. An expensive and large liquid hydrogen tank has to be constructed at the ports of origin and arrival.
4. A spherical tank needs to be adopted because of its high-performance insulation. Consequently, 1/3 body of the ship becomes dead space and this makes the tanker much bigger.

Figure 14.9 shows a hydrogen tanker ship that was developed as a part of the Japanese liquid hydrogen carrier project.

### 14.3.3 Pipeline transportation of hydrogen

The alternative for hydrogen transport over short or medium distances is pipelines [9, 11]. The viable distances are determined by the ease and cost of establishing pipelines, not only usually over land but also conceivably in some cases offshore. The technology is already developed for natural gas transport, and only modest alterations are assumed for containing the smaller hydrogen molecules with acceptable leakage rates. The polymer pipes used for the transportation of natural gas at pressures under 0.4 MPa have sufficient strength to carry hydrogen. The diffusion losses arise mainly from pipe connectors and have been estimated to be three times higher for hydrogen than for natural gas [11]. Steel pipes with welded connections are used for the transmission of natural gas at pressures up to 8 MPa. However, most metals can develop brittleness after absorption of hydrogen, especially at locations contaminated with foreign substances such as dirt or $H_2S$. These effects can be minimized by adding a little oxygen (about $10^{-5}$ by volume). Auxiliary components such as pressure regulators and gauges are not expected to cause problems, although lubricants used should be checked for hydrogen tolerance [11]. Spark ignition is a potential hazard, as in the case of natural gas pipelines. If hydrogen is distributed to individual buildings, then gas detectors must be installed and outdoor emergency venting devices should be considered.

## References

[1] Wan Z et al 2021 Energy Convers. Manage. **228** 113729
[2] Abdin Z et al 2021 iScience **24** 102966
[3] Chatterjee S, Parsapur R K and Huang K W 2021 ACS Energy Lett. **6** 4390
[4] Chen X et al 2021 ACS Sustainable Chem. Eng. **9** 6561
[5] Modisha P M et al 2019 Energy Fuels **33** 2778
[6] Okada Y et al 2006 Int. J. Hydrogen Energy **31** 1348
[7] Shukla A et al 2010 Int. J. Hydrogen Energy **35** 4020
[8] Boufaden N et al 2016 J. Mol. Catal. A: Chem. **420** 96
[9] Gerboni R 2016 Compendium of Hydrogen Energy, Volume 2: Hydrogen Storage, Transportation and Infrastructure (Woodhead Publishing Series in Energy) (Cambridge: Woodhead Publishing) pp 283–99
[10] Sano H Energy Carriers and Conversion Systems, Volume II: Ocean Transportation of Hydrogen (UNESCO-EOLSS) (Oxford: Eolss Publishers)
[11] Sorensen B and Spazzafumo G 2018 Hydrogen and Fuel Cells (New York: Academic)

**IOP** Publishing

Hydrogen
Physics and technology
**Sindhunil Barman Roy**

# Chapter 15

# Hydrogen energy conversion technologies

Energy carriers are a convenient form of stored energy [1, 2]. Electricity is a very prominent energy carrier, which can be produced from many sources, transported over large distances, and distributed to the end user. Hydrogen is another prominent energy carrier. We already know that it can be produced from clean sources, and it constitutes a clean, efficient, and versatile energy carrier that can supplement electricity well.

In the earlier chapters we have studied that hydrogen has some unique properties, and together these make hydrogen an ideal energy carrier. We summarize these properties below:

1. Hydrogen can be produced from electricity and solar energy at relatively high efficiencies.
2. The raw material for hydrogen production is water, which is abundantly available.
3. Hydrogen is a completely renewable fuel. The product of hydrogen utilization is pure water or water vapor.
4. It can be stored in gaseous form, liquid form, or in the form of metal hydrides.
5. Hydrogen can be transported over large distances through pipelines or via tankers. In most of cases, the transport processes are more efficient and economical than electricity.
6. Hydrogen production, storage, and transportation, and also its end use does not produce any pollutants (except small amounts of $NO_x$ if hydrogen is burned with air at high temperatures [1]), greenhouse gases, or cause any other harmful effects on the environment.

In this chapter, we will discuss how hydrogen can be converted into other forms of energy in more ways and more efficiently than other fuels. We will see that in addition to flame combustion (as in other fuels) hydrogen may also be converted through catalytic combustion, electrochemical conversion, and metal hydriding.

## 15.1 Flame combustion

Burning or combustion is the basic chemical process of releasing energy from a fuel and air mixture. The internal combustion engine (ICE) is the most common type of heat engine. The ignition and combustion of the fuel occurs within the ICE itself to perform work inside the engine.

The working principle can be understood in terms of the ideal gas law: $pV = nRT$. A rise in the temperature of a gas increases the pressure, which causes the gas to expand. In an ICE chamber, the fuel added is ignited to raise the temperature of the gas. The system with added heat forces the gas inside to expand. In a standard ICE piston engine, this causes the piston to rise. The piston is attached to a crankshaft, and thus the engine converts a part of the energy input to the system into useful work. In an intermittent ICE, the gas is exhausted to compress the piston. The system is kept running at a constant temperature by attaching it to a heat sink. Figure 15.1 presents the schematic diagram of an intermittent ICE.

There are two kinds of ICEs, which are very common: the spark ignition gasoline engine and the compression ignition diesel engine. Most are so-called four-stroke cycle engines, which produce one power stroke for every two cycles of the piston, and the basic steps/processes are described as follows:

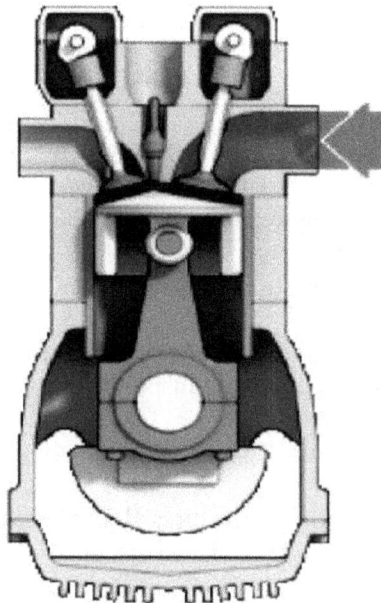

**Figure 15.1.** Schematic diagram of an intermittent ICE. This 4StrokeEngine Ortho 3D image has been obtained by the author(s) from the Wikimedia website where it was made available under a CC BY-SA 3.0 licence. It is included within this book on that basis. It is attributed to Zephyris.

1. Fuel is injected into the ICE chamber.
2. The fuel combusts, i.e., catches fire.
3. This fire pushes the piston, thus causing useful motion. This in turn rotates the crankshaft and ultimately drives the wheels of the vehicle through a system of gears.
4. To compress the piston in an ICE, the engine pushes out the gas containing the waste chemicals. This is mostly water vapor and pollutants such as carbon dioxide. There may be additional pollutants such as carbon monoxide from incomplete combustion.

Four-stroke ICEs are used in various automobiles, including cars, trucks, and some motorbikes. The spark ignition gasoline and compression ignition diesel ICEs differ in how the fuel is supplied and ignited. The fuel is mixed with air and then inducted into the cylinder in a spark ignition engine during the intake process. The spark ignites the fuel-air mixture to cause combustion after the piston compresses the mixture. The expanding combustion gases then push the piston during the power stroke. In a diesel engine, only air is inducted into the engine and then compressed. The fuel is sprayed subsequently into the hot compressed air at a suitable, measured rate, to cause ignition.

A second class of ICE uses continuous combustion, including gas turbines, jet engines, and most of the rocket engines. In gas turbines, hot air is forced into the turbine chamber and turns the turbine. Figure 15.2 presents a schematic diagram of a gas turbine engine.

A piston engine is extremely responsive and more fuel-efficient at low outputs. These engines are ideal for use in vehicles, and they also start up relatively quickly. A gas turbine engine has a better power-to-weight ratio in comparison to a piston engine. The design of a gas turbine engine is also more appropriate for continuous high outputs. A turbine engine also works better than a naturally aspirated piston engine at high altitudes and cold temperatures [3]. With its lightweight build, reliability, and high altitude capability, gas turbine engines are very suitable for airplanes. Gas turbine engines are also commonly used at power plants for electricity generation.

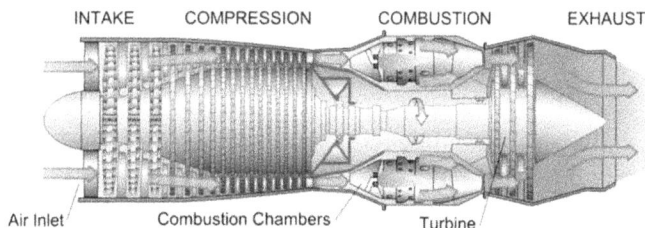

**Figure 15.2.** Schematic diagram of a gas turbine engine. This Jet engine image has been obtained by the author(s) from the Wikimedia website where it was made available under a CC BY-SA 4.0 licence. It is included within this book on that basis. It is attributed to Jeff Dahl.

The conventional ICE with gasoline as fuel is not a promising technology for the future because of its low efficiency and polluting emissions. It is, however, possible to increase the efficiency of ICE substantially when operated with hydrogen [1, 2].

### 15.1.1 Hydrogen ICE

Hydrogen is a very good fuel for ICEs. It has a wide flammability range and high autoignition temperature. These make hydrogen particularly suitable for combustion. The wide flammability range means that it can be used with a lower temperature creating fewer pollutants, and the high autoignition temperature means that less energy is lost. The only problem is that oxides of nitrogen form in the high-temperature oxidation of hydrogen fuel. These oxides, i.e., NO and $NO_2$, are collectively known as NOx, but this is a tractable problem.

ICEs can operate on different fuels. From economic and practical points of view, this 'flex-fuel' ability is an advantage for hydrogen vehicles [2]. First, this enables a gradual buildup of a hydrogen fueling infrastructure. Second, this can mitigate the onboard storage challenge, with a second fuel, e.g., gasoline, thus serving essentially as a "range extender'. The hydrogen-fueled ICE is also less costly compared to a fuel cell (this will be discussed later on in this chapter). This lower cost is not due to only the ICE itself, but also to the fuel. The ICE can handle relatively low-purity hydrogen without any problems [2].

In comparison to the standard gasoline engines, hydrogen-powered ICEs are on average about 20% more efficient. The ideal thermal efficiency of an ICE is [1, 2]:

$$\eta = 1 - \left(\frac{1}{r}\right)^{k-1} \tag{15.1}$$

Here $r$ is the compression ratio and $k$ is the ratio of specific heats ($C_p/C_v$).

Equation (15.1) indicates that the thermal efficiency in hydrogen engines can be improved by increasing either the compression ratio or the specific heat ratio. Both of these ratios are higher in hydrogen engines than in a comparable gasoline engine. This is due to the relatively low self-ignition temperature of hydrogen and its ability to burn in lean mixtures.

The use of hydrogen in ICEs, however, results in loss of power due to lower energy content in a stoichiometric mixture in the cylinder of the engine [1]. An externally premixed stoichiometric mixture of gasoline and air, and gaseous hydrogen and air, occupies ≈2% and 30% of the cylinder volume, respectively. Thus, the energy of the hydrogen mixture is only 85% of the hydrogen mixture. This results in about 15% reduction in power. Therefore, the same engine running on hydrogen will have ≈15% less power than when operated with gasoline. The power output of a hydrogen engine can be improved by employing more advanced fuel injection techniques or liquid hydrogen. The amount of hydrogen that can be introduced in the combustion chamber can be increased significantly if liquid hydrogen is premixed with air.

Hydrogen engines emit far fewer pollutants than comparable gasoline engines. Water vapor and small amounts of nitrogen oxides (NOx) are the only byproducts of hydrogen combustion in air. Hydrogen has a wide flammability range in air (5–75 vol%), and therefore high excess air can be utilized more effectively [1]. With the introduction of an excess amount of air, the NOx formation in hydrogen/air combustion can be minimized. The NOx emissions can also be reduced by cooling the combustion environment using techniques such as water injection, exhaust gas recirculation, or liquid hydrogen [1]. In comparison to the emissions from comparable gasoline engines, the emissions of NOx in hydrogen engines are typically one order of magnitude lower. Some small amounts of unburned hydrocarbons, $CO_2$, and CO can also be traced in hydrogen engines, originating from the lubrication oil.

The low ignition energy and fast flame propagation of hydrogen can have problems with preignition and backfire. These problems can be minimized by adding hydrogen to the air mixture at the point where and when the conditions for preignition are less likely. This can be done by delivering the fuel and air separately to the combustion chamber, and/or injecting hydrogen under pressure into the combustion chamber before the piston is at the top dead center and after the intake air valve has been closed. The other techniques to control premature ignition in hydrogen engines are water injection and exhaust gas recirculation. The research on hydrogen combustion in ICEs so far has been conducted mostly with modifications of existing engines designed to burn gasoline. Redesigning the combustion chamber and coolant systems to accommodate the unique combustion properties of hydrogen could be the most effective method of solving the problems of preignition and knocking in hydrogen engines.

### 15.1.2 Turbines and jet engines

Hydrogen can be used in turbines and jet engines in a very similar way to conventional jet fuel. Liquid or gaseous hydrogen is burnt in a gas turbine engine to generate thrust. The problems of sediments and corrosion on turbine blades can be avoided by using hydrogen. This prolongs life and reduces the maintenance costs in turbines and jet engines. The overall efficiency can be increased by pushing the gas inlet temperatures beyond normal gas turbine temperatures of 800 °C. The only pollutants from the use of hydrogen in turbines and jet engines are nitrogen oxides [1].

In late-2022, Rolls-Royce of Britain successfully tested an aircraft engine running on hydrogen fuel. This is a first in the world of aviation that marks a major step towards proving that gas could be key to decarbonizing air travel [4]. The ground test was performed with a converted Rolls-Royce aircraft engine, using green hydrogen created by wind and tidal power.

## 15.2 Steam generation by hydrogen/oxygen combustion

The combustion of pure hydrogen and pure oxygen in a stoichiometric mixture generates superheated steam (i.e., $H_2O$) with the temperature in the flame zone reaching around 3000 °C. Every two moles of hydrogen react with one mole of oxygen in the combustion chamber to produce steam at high temperature [1, 5]:

$$2H_2 + O_2 \longrightarrow 2H_2O + \text{heat} \tag{15.2}$$

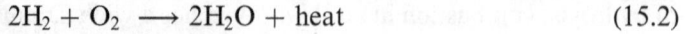

This reaction is highly exothermic. Therefore, additional liquid water is injected into the combustion chamber to reduce the steam temperature and utilize the steam in a steam turbine.

A steam generator is a device to burn a stoichiometric mixture of hydrogen and oxygen in a combustion chamber to produce high-temperature superheated steam [5]. Figure 15.3 presents a schematic of a hydrogen–oxygen steam generator device. It consists of the propellent injection or ignition, combustion, water injection, and evaporation chambers. A combustible mixture of hydrogen and oxygen at a low oxidant/fuel ratio is ignited using a spark plug in the ignition chamber [1, 6]. In the combustion chamber, the rest of the oxygen is introduced to make the oxidant/hydrogen fuel ratio exactly the stoichiometric mixture. The water injection system cools the hot gases of the combustion chamber to a temperature that is suitable for the turbine operation. This cooling process needs to be as smooth as possible. This is to avoid quenching of the hot gases, which may cause non-tolerable high concentrations of unburned propellants in the exhaust steam [6]. The complete cross section of the steam generator needs to be filled with a specific amount of water to achieve a homogeneous temperature distribution. This is needed to avoid hot and cold spots in the steam, which are sources of possible turbine blade damage. Finally, the hot gases pass through a vaporization zone, which is long enough to ensure the mixing and vaporization of the injected water [6]. This vapourization part of the steam generator performs two tasks. First, it enhances the turbulence to promote the mixing of the two-phase flow. Second, it increases the residence time to ensure complete vaporization of big droplets. The outlet of the vapourization part is designed to hold sensors for temperature and steam quality determination. These facilitate online control of the steam properties.

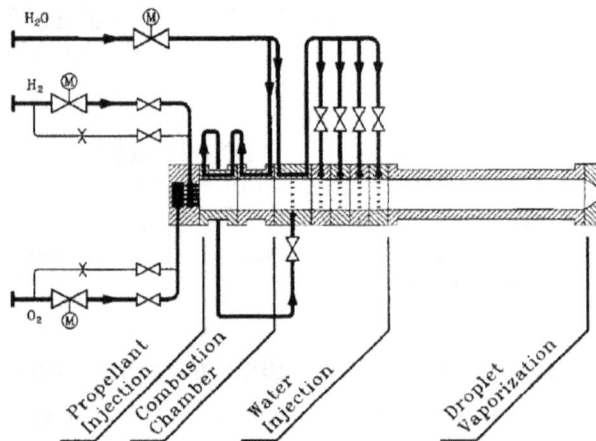

**Figure 15.3.** Schematic representation of an oxygen/hydrogen-fuel steam generator. Reprinted from [6], Copyright (1998), with permission from Elsevier.

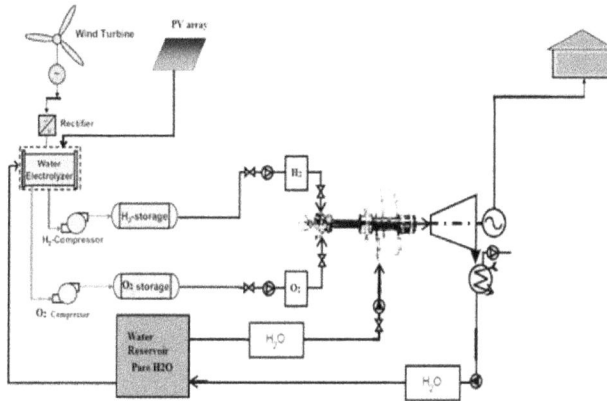

**Figure 15.4.** Schematic representation of an oxygen/hydrogen-fuel steam turbine power unit integrated with a renewable energy source. Reprinted from [5], Copyright (2012), with permission from Elsevier.

A hydrogen steam generator can be used to generate steam for spinning reserve in power plants, for peak load electricity generation, for industrial steam supply networks, and as a micro-steam generator in medical technology and biotechnology [1]. Figure 15.4 presents a schematic diagram of the oxygen/hydrogen fuel steam turbine power unit integrated with renewable energy resources. It consists of a renewable energy source (solar or wind, or a combination of both), electrolyzer (alkaline or PEM), electrolyzer auxiliary unit (water pumps, gas drier, and deoxidizer), hydrogen compressor, oxygen compressor, hydrogen storage system, oxygen storage system, hydrogen and oxygen supply system, hydrogen–oxygen steam generator, steam turbine, electric generator, condenser, and a water reservoir [5]. Renewable energy resources provide electric power for the hydrogen and oxygen production unit (electrolyzer, auxiliary unit, compressors, and storage system). Subsequently, hydrogen and oxygen are supplied to the oxygen/hydrogen fuel steam generator for steam production. Steam is fed to the turbine to produce work for electricity generation. The steam exhaust is then passed through the condenser to the main water reservoir. The system operates as a closed cycle.

## 15.3 Catalytic hydrogen combustion

When hydrogen gas is converted on a catalyst surface by an oxidizer, the process is called catalytic hydrogen combustion (CHC) [7]. The oxidizer in combustion systems typically is air. Catalytic combustion of hydrogen can take place at low temperatures, which minimizes pollutant emissions of nitrogen oxides $NO_x$ and improves flame stability to a greater extent than conventional combustion. Moreover, a lean hydrogen-air mixture can be reacted outside the limit of flammability.

CHCs can be classified into hybrid and low-temperature catalytic combustions, according to their reaction temperature (figure 15.5). The low-temperature catalytic combustions are entirely heterogeneous because the reaction temperature is generally

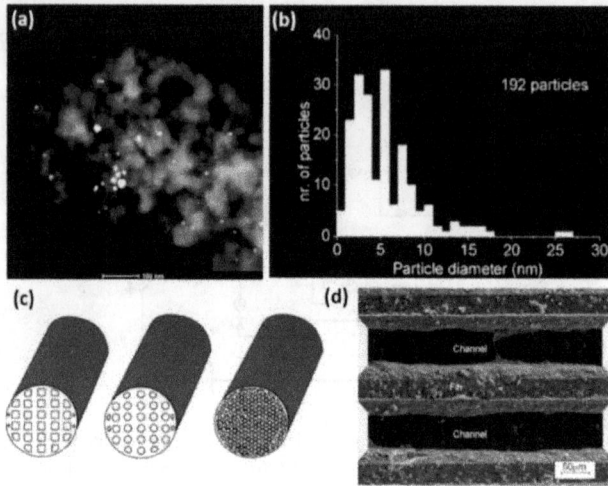

**Figure 15.5.** Various morphologies of catalysts: (a) high-angle annular dark-field image of SiC-supported platinum (Pt) catalyst obtained with a scanning transmission electron microscopy (white spots are Pt nanoparticles) and (b) particle-size distribution of Pt catalyst; (c) different shapes of channels in monolithic catalysts; and (d) nanosized catalyst wash-coated onto microchannels. Reprinted from [7], Copyright (2021), with permission from Elsevier.

below the autoignition temperature of hydrogen (585 °C) [7]. The low-temperature catalytic combustion of hydrogen finds usage in domestic and commercial heating systems, central boilers, and cooking appliances. This is due to its lower operating temperatures than those used for flame combustion. The advances in such CHC systems require the development of inexpensive and highly reactive catalysts and catalytic reactors with optimized heat and mass transfer for superior thermal efficiency. Hybrid catalytic combustion, on the other hand, involves both heterogeneous and homogeneous combustion. This can drive a gas turbine for power generation and aircraft applications. This can also find usage in portable power generators, which involve a micro-combustor to combust the gas to drive a micro-gas turbine, providing electricity for portable electronic devices, laptops, and cellular phones.

Catalysts are the core components of CHC systems. They affect hydrogen conversion and reaction rates. They are mainly classified into noble metal and metal-oxide catalysts, which include non-noble metal oxides and perovskites. The physicochemical properties of catalysts are influenced by multiple factors, namely synthesis method, composition, and heat treatment. Therefore, the catalysts exhibit varying properties depending on their type. A fairly recent survey of such CHC catalysts, including their synthesis, chemical structures, physical properties, and the analytical techniques for their characterization, can be found in the article by Kim *et al* [7, 8].

The catalysts used in CHC reactors can have various morphologies, including powder catalysts with nanosized particles and wash-coated monoliths and microchannels. Figure 15.5 presents such morphologies of the catalysts. A high-angle annular dark-field scanning transmission electron microscopy image of a SiC-supported 0.5 wt% Pt catalyst and its particle-size distribution with 5 nm of the average size of the nanoparticles is shown in figure 15.5(a) and (b), respectively. Figure 15.5(c) shows the channels of various shapes (i.e., square, round, triangle) that are used in monolithic reactors. Here, triangular interconnected channels can achieve a higher specific surface area (400 cells cm$^{-2}$) with enhanced mass and heat transfer [7]. In a microchannel reactor, a catalyst is commonly wash-coated. The typical shape of such a microchannel is shown in figure 15.5(d).

A high conversion rate of hydrogen over a range of temperatures is generally desired in a reactor for CHC. This necessitates many aspects to be considered when designing such a catalytic reactor with optimal performance, including the type of reactor, catalyst, and support [7]. Catalytic reactors need more surface area than conventional flame burners. Thus, the catalyst is typically spread across a porous structure. The reaction rate and resulting temperature are adjusted by controlling the hydrogen flow rate. The reaction takes place in a reaction zone of the porous catalytic sintered cylinders or plates in which hydrogen and oxygen are mixed by diffusion from opposite sides [1]. A combustible mixture forms in the reaction zone and burns at low temperatures in the presence of a catalyst. In the catalytic combustion of hydrogen, the only product is water vapor. There is no formation of $NO_x$ due to the relatively low temperatures of combustion. The reaction does not migrate into the hydrogen supply because there is no flame and the hydrogen concentration is above the higher flammable limit [1]. We briefly introduce the different types of CHC reactors in the following subsections.

### 15.3.1 Fixed-bed reactor

The catalysts used in fixed-bed reactors are generally oxides of transition metals or supported noble metals. This type of reactor is employed for determining catalytic activities and kinetic studies to evaluate performance of the catalysts. Figure 15.6 shows such a fixed-bed quartz reactor loaded with catalyst, and heated in an electric furnace to measure the NO oxidation reactivity of the catalysts with different compositions and test $SO_2$ poisoning [8]. Fixed-bed reactors are commonly used in

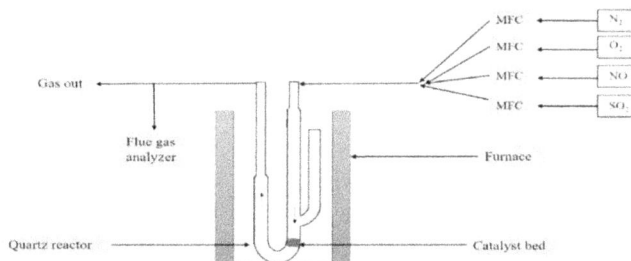

**Figure 15.6.** Schematic diagram for a CHC fixed-bed reactor. Reprinted from [7], Copyright (2021), with permission from Elsevier.

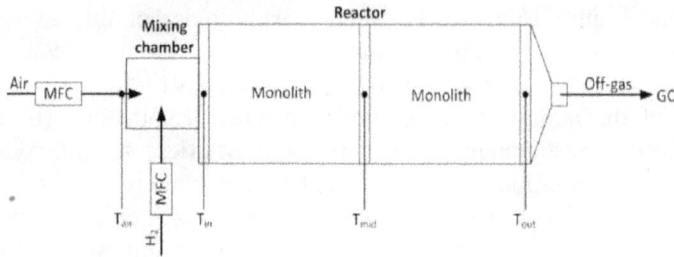

**Figure 15.7.** Schematic diagram of a monolithic reactor for a hydrogen-air mixture. Reprinted from [9], Copyright (2018), with permission from Elsevier.

laboratory-scale experiments and industrial applications due to the flexibility in catalyst packing [7].

### 15.3.2 Monolithic reactor

In monolithic reactors, catalysts are fixed at particular locations inside the reactor, and this makes them hydrodynamically superior to other reactor types [7]. Such reactors contain several narrow, parallel channels, which dictate the wall thicknesses and cell densities. The usage of wash-coated monolithic catalysts is beneficial in these reactors because of the high surface areas of such catalysts, which ensure small pressure drops within the system. Figure 15.5(c) shows various internal channels of monolithic catalysts, including square, round, and triangular. The highest mass and heat transfers take place in the triangular monolith due to a higher specific surface area and a greater number of cells per unit area. Figure 15.7 shows a schematic diagram of a monolithic reactor for a hydrogen-air mixture. This consists of a ceramic-coated mixing chamber and reactor with catalyst-coated honeycomb monoliths made with cordierite and ceramic $Al_2O_3$ [9]. In this horizontal reactor, the exhaust pipe also has a horizontal orientation. A high-temperature insulation material covers the entire burner. Hydrogen and air are introduced into the reactor by the assigned mass flow controller. The air inlet, gas inlet, and exhaust outlet are also suitably insulated to avoid heat losses. The catalysts used in monolithic reactors are only required to be packed inside a specific part of the reactor. Hence, this type of reactor is widely used in industrial applications. Furthermore, their high thermal stability enables their usage in gas turbines at high temperatures for power generation [7].

### 15.3.3 Microchannel reactor

Figure 15.8 shows a configuration of a microchannel reactor used for generating hydrogen from steam reforming of ethylene glycol [10]. Stainless steel was used for the manufacturing of this microchannel reactor. The parallel microchannels were etched on the stainless-steel sheet by a micromilling technique. The microchannel reactor has three zones: an inlet triangular manifold zone for uniform flow distribution, a parallel microchannel zone for catalytic reactions, and an outlet manifold zone. The microchannels are sealed with a copper gasket to prevent

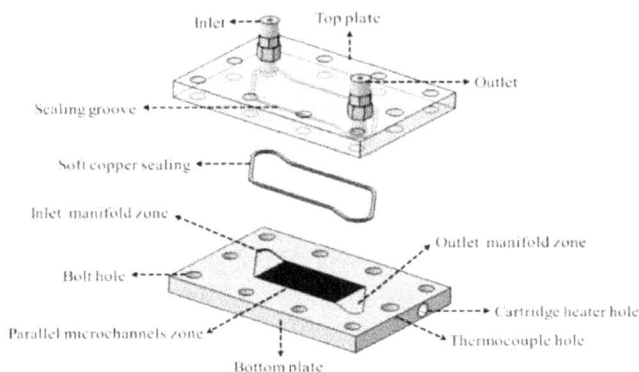

**Figure 15.8.** Schematic presentation of the configuration of a microchannel reactor. Reprinted from [10], Copyright (2018), with permission from Elsevier.

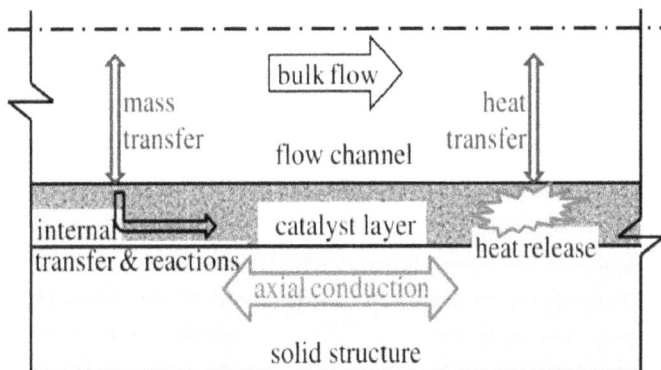

**Figure 15.9.** Schematic presentation of various processes within a catalytic microreactor for combustion. Reprinted from [7], Copyright (2021), with permission from Elsevier.

leakage of fluid. There are embedded holes for cartridge heaters on both end-sides of the bottom plate of the microchannel reactor. A thermocouple is used in the bottom plate of the microchannel reactor to monitor the reaction temperature.

Figure 15.9 shows a schematic of the various processes in a catalytic microreactor for combustion. The operating conditions, including the energy input and residence time, are decided from the gaseous-phase flow conditions, containing the reactants for catalytic combustion [7]. The reactants are absorbed into the catalytic surfaces from the bulk flow. This leads to the formation of products on the catalyst surfaces. The produced gas species are then desorbed and released back into the bulk flow. Microchannel reactors have a modular and compact design, which can be efficiently utilized for industrial applications. Large-scale reactors can easily be constructed by stacking multiple microchannel reactors.

## 15.4 Electrochemical conversion

Hydrogen can be combined with oxygen without combustion in an electrochemical reaction and produce electricity [1]. This electrochemical reaction happens in a device called the electrochemical fuel cell, or just a fuel cell. One can generally define the fuel cell as an electrochemical device for the conversion of chemicals into direct-current (dc) electricity [11, 12]. A fuel cell to some extent resembles a primary battery. However, all the chemicals necessary for operation in a battery are normally confined within a sealed container. Thus, the capacity of a battery, measured in ampere-hours, is determined by the quantity of chemicals that a battery can hold [11]. In a fuel cell, the chemicals are supplied from external sources. The capacity of the fuel cell is controlled only by the available supply of reactant chemicals.

A typical fuel cell consists of two electrodes, an anode, and a cathode, separated by an electrolyte, which can conduct ions but not electrons. Figure 15.10 shows the general operation of a fuel cell. The fuel is oxidized at the anode with the release of electrons, which then pass through the external circuit and reduce oxygen at the positive electrode. The flow of electrons is balanced by a flow of charged ions in the electrolyte, and overall an electric current is generated in the closed circuit.

In the absence of any current flow, the reversible voltage $V_r$, produced by the fuel cell is expressed as [11]:

$$V_r = -\Delta G/nF \qquad (15.3)$$

Here $\Delta G$ is the free energy of the cell reaction (joules per mole), $n$ is the number of electrons involved in the reaction, and $F$ is the Faraday constant (9.648 53 $\times 10^4$

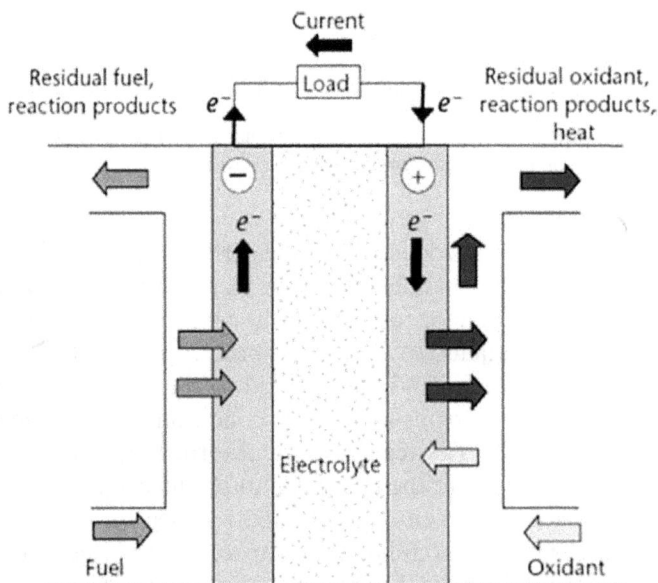

**Figure 15.10.** Schematic representation of the operation of a fuel cell. Reprinted from [11], Copyright (2014), with permission from Elsevier.

coulombs per mole). This voltage $V_r$ is also termed sometimes as the 'open-circuit' voltage. The relationship expressed in equation (15.3) is also applicable to the reverse reaction, which is also known as 'electrolysis'. In the case of hydrogen fuel, the reversible voltage $V_r$ under standard conditions is $V^0 = 1.229$ V at 25 °C, just as for an electrolyzer [11]. The voltage developed when the current is being drawn from the cell is, however, much lower than $V_r$, typically 0.6–0.8 V, depending on the current density. As a result, a stack is usually formed by electrically connecting the fuel cells in series to build up the voltage to the desired level.

A typical plot of cell voltage ($V$) versus current ($I$) in a fuel cell is shown in figure 15.11(a). The voltage falls with the current being drawn from the cell. Various factors that cause this degradation in performance are shown in different regions of the $V$ - $I$ curve. The initial fall in voltage is due to electrokinetic limitations at the electrodes. The 'ohmic loss' arises from resistance to the flow of ions in the electrolyte. This ohmic loss is reflected in the central linear portion of the curve. Finally, at high current densities mass-transport limitations resulting from gas consumption outstripping supply become significant. Additionally, there are also electrical-resistance losses in the collectors and conductors of the current, and side-reactions and leakages of ionic current between series-connected cells. All of these may also contribute to the voltage drop. All such effects collectively limit the working cell voltage well under 1 V. Figure 15.11(b) shows an example of the power performance of a cell. The cell voltage declines with the rise in current density. The power density is the product of the voltage and the current density, and this passes through a maximum, which in this example is at around 1 A cm$^{-2}$.

Historically, the discovery of fuel cell operation is attributed to the German scientist Christian Friedrich Schönbein, who discovered this process in 1839 [13]. While Schönbein conducted mainly theoretical research, a Welsh scientist, William Robert Grove, experimentally investigated the effects of hydrogen–oxygen combination. Grove developed the first fuel cell, which he called 'voltaic battery gas'. This battery contained pairs of platinum electrodes immersed in sulfuric acid in a closed system containing hydrogen and oxygen. Grove made two important observations: (i) the flow of current when the electrodes were connected in series, indicating the generation of electricity, and (ii) the containers containing gases were filled with water. The term 'fuel cell' was adopted around 1890 by chemists Ludwig Mond and Charles Langer in an attempt to construct the first functional gas battery device composed of air and industrial coal gas [13]. In the 1930s, Francis Bacon at the University of Cambridge introduced newer concepts on fuel cells by changing the electrolyte from acid to alkaline (potassium hydroxide). Nowadays, there are various classes of fuel cells depending on the type of electrolyte used. We will provide a brief introduction to such fuel cells. The readers are referred to the various review articles [1, 11, 12] for a detailed discussion on such fuel cells.

1. Alkaline fuel cells (AFC): These fuel cells use concentrated (85 wt%) potassium hydroxide (KOH) as the electrolyte for high-temperature operation ($\approx$250 °C) and less concentration (35–50 wt%) for lower temperature operation (<120 °C). The electrolyte is stored in a matrix-like asbestos. A wide range of electro-catalysts can be used for the operation, such as Ni, Ag,

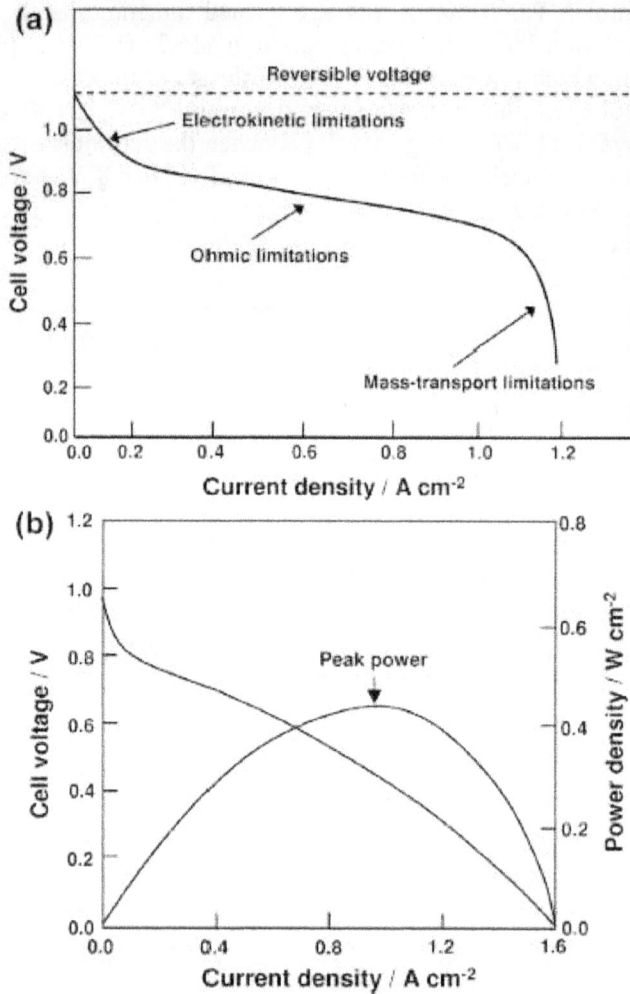

**Figure 15.11.** Typical performance characteristics of fuel cells: (a) voltage vs current curve; (b) power output as a function of current density. Reprinted from [11], Copyright (2014), with permission from Elsevier.

metal oxides, and noble metals. This fuel cell is intolerant to the presence of $CO_2$, either in the fuel or in the oxidant.

2. Polymer electrolyte membrane or proton exchange membrane fuel cells (PEMFC): These fuel cells use a thin polymer membrane, e.g., perfluorosulfonated acid polymer, as the electrolyte. The membranes can be as thin as 12–20 $\mu$m, which are excellent proton conductors. The catalyst used is typically platinum with loadings of about 0.3 mg cm$^{-2}$. If the hydrogen fuel contains minute amounts of CO, then Pt–Ru alloys are used. These fuel cells usually operate at temperatures below 100 °C, and more typically between 60 °C and 80 °C.

3. Phosphoric acid fuel cells (PAFC): These fuel cells use concentrated phosphoric acid ($\approx 100\%$) as the electrolyte. A silicon carbide (SiC) matrix is usually used to store the acid. The electrocatalyst in both the anode and cathode is platinum black. These fuel cells operate at temperatures typically between 150 °C and 220 °C.

4. Molten carbonate fuel cells (MCFC): The electrolyte in these fuel cells is composed of a combination of alkali (Li, Na, K) carbonates, which is stored in a ceramic matrix of $LiAlO_2$. They operate in temperatures between 600 °C and 700 °C. The carbonates form a highly conductive molten salt, and carbonate ions facilitate ionic conduction. Noble metal catalysts are usually not required at such high operating temperatures [1].

5. Solid oxide fuel cells (SOFC): These fuel cells use a solid, nonporous metal oxide, usually $Y_2O_3$-stabilized $ZrO_2$ as the electrolyte. The fuel cell operates at 900–1000 °C, where ionic conduction by oxygen ions takes place [1].

Among the various classes of fuel cells, PEMFCs are highly efficient. Depending on the operating temperature, the efficiency ranges between 50% to 70%. These efficiency values are almost unaffected by the size of the fuel cells, which makes PEMFCs suitable for a variety of applications. PEMFCs can deliver electrical output for portable systems (from microwatts to a few watts), APU and backup units (from a few watts up to a few kilowatts), transportation (all sectors, from a few kilowatts up to several hundred kilowatts), and stationary systems (from few kilowatts up to several megawatts) [12]. PEMFCs are the most practical candidate for general road-transport applications [11]. We will now study PEFMCs in more detail. The electrolyte in PEFMC is a thin membrane of a copolymer that is highly conductive to hydrated protons ($H_3O^+$) [12]. The standard membrane material is sold under the trade name Nafion, which is produced by the DuPont Corporation in the USA. The backbone consists of polytetrafluoroethylene (PTFE or 'Teflon'). The side chains of perfluorinated vinyl polyether are bonded to this via oxygen atoms. Sulfonic acid groups (- $SO_3H$) at the ends of the side chains can exchange with protons to enable their passage through the membrane [12]. Nafion has two notable characteristics [12]: (i) the PTFE backbone is water-repellent, and thereby prevents flooding by-product water; and (ii) the sulfonic acid group attracts water molecules, which form aqueous micelles within the polymer, which, in turn, serve to conduct the protons across the membrane.

Figure 15.12 presents a schematic of the basic operation of a PEMFC. The overall reaction is split into two semi-reactions, anodic (negative pole) and cathodic (positive pole) [12]:

$$2H_2 \rightarrow 4H^+ + 4e^- \tag{15.4}$$

$$O_2 + 4H^+ + 4e^- \rightarrow 2H_2O \tag{15.5}$$

Hydrogen is introduced at the anode. It then diffuses into the anodic catalytic layer and gets oxidized into protons and electrons. Electrons move into the external electrical circuit from the anode to the cathode, thus providing electricity. Protons

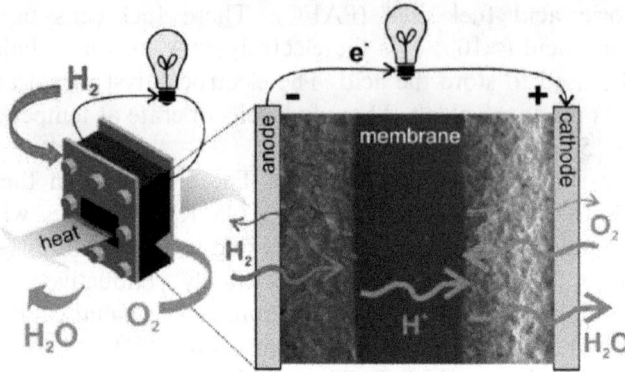

**Figure 15.12.** Schematic presentation of the operation of a PEMFC. Reprinted from [12], Copyright (2018), with permission from Elsevier.

cross the electrolyte to reach the cathode. They then diffuse into the cathodic catalytic layer, and, along with the electrons, reduce oxygen to water, which is the sole product of reaction [12].

## 15.5 Energy conversions involving metal hydrides

We have learned in chapter 6 that hydrogen can form metal hydrides, which can be used for the storage of hydrogen. The same metal hydrides can also be employed for energy conversions. The process of metal hydride generation is exothermic. The chemical combination of hydrogen with a metal leads to heat generation. On the other hand, heat needs to be supplied to release hydrogen from a metal hydride. These processes can be expressed in terms of the following chemical reactions [1]:

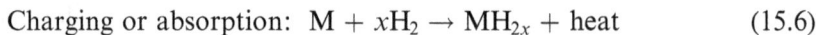

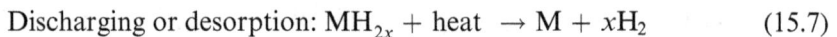

$$\text{Charging or absorption: } M + xH_2 \rightarrow MH_{2x} + \text{heat} \qquad (15.6)$$

$$\text{Discharging or desorption: } MH_{2x} + \text{heat} \rightarrow M + xH_2 \qquad (15.7)$$

The hydriding substance here is M, typically a metal in elemental form or an alloy. An increase in the surface area enhances the rate of these reactions. Thus, a powdered form of the hydriding substances is, in general, used to speed up the reactions. Suitable hydriding substances have metal atoms with unfilled electron shells or subshells. The hydrogen atoms form chemical compounds by sharing their electrons with the unfilled subshells of the metal atom.

It is ideal for the processes of charging, or absorption, and discharging, or desorption, to happen isobarically (or at a constant pressure) at a specific temperature. However, the pressure fluctuates in practice. The pressures needed for hydrogen charging surpass those needed for hydrogen to discharge at a specific temperature. The amount of heat generated during the charging process and the amount of heat needed for discharging are determined by the hydriding substance, the hydrogen pressure, and the temperature at which the heat is delivered or extracted [1]. Different alloys made of different metals have varied hydriding properties. Consequently, it is possible to synthesize hydriding substances that are more suited for a given application, such as hydrogen purification, power generation, waste heat storage, isotope separation, and pumping [1].

There are two ways to store electrical energy by employing metal hydrides [1]:

1. DC electricity is used to electrolyze the water, and the hydrogen produced is stored in a hydriding substance. When needed, the hydrogen is retrieved from the hydriding substance by adding heat. This is subsequently used in a fuel cell to produce DC electricity. The heat from the fuel cell can be utilized to release hydrogen from the metal hydride.
2. One electrode in an electrolyzer can be covered with a hydriding substance (e.g., titanium nickel alloy). The hydrogen produced on the surface of the electrode during the electrolysis of water is then absorbed by this hydride coated electrode. When power is required, the electrolyzer functions like a fuel cell in reverse and generates energy by using hydrogen released from the metal hydride.

Building heating and cooling technologies may result from the combination of hydrogen and hydriding materials. A graphical representation of the operation of one such potential system is shown in figure 15.13. Four hydride tanks, a heat source or solar collector, and several heat exchangers make up the system. Hydrogen can flow from hydride tank numbered as 1 to hydride tank 3 through a hydrogen pipe connection between the two tanks. The connections between the hydride tanks numbered 2 and 4 are identical. The hydriding material $CaNi_5$ is present in tanks 1 and 2, while tanks 3 and 4 contain another hydriding material $LaNi_5$. Water-carrying pipe circuits or loops, furnished with a set of switches and valves, connect the heat exchangers and the hydride tanks [1]. With this configuration, one hydride tank in a given water loop can be swapped out for another hydride tank.

In addition, the system can function as a heater. In this instance, heat is transferred to tank 1 at roughly 100 °C from a heat source, such as a solar collector. Heat moves hydrogen from tank 1 to tank 3, where it absorbs and forms a hydride. At 40 °C, heat is released. The water loop then transports this heat to the heat exchangers, where it warms the air in the building. Simultaneously, the water in the other loop draws heat from the surroundings and transfers it to tank 4. The hydrogen in tank 4 is driven out of tank 2 by this heat. After that, in tank 2, a hydride is created and heat is produced at 40 °C. It takes about two minutes to complete the whole operation of driving hydrogen from tanks 1 and 4 to tanks 3 and 2 [1]. The hydride tanks are switched from one loop to the other in cycle II at the end of this cycle (see figure 15.13). The hydrogen in tanks 2 and 3 is now driven off to tanks 4 and 1, respectively, using the heat from the Sun and the surrounding air. The building is heated by the heat generated during the absorption processes in tanks 1 and 4, after which the cycles are repeated. Additionally, this system functions as an air conditioner. Next, the heat exchangers inside the building are positioned in the 8 °C water loop, and the heat exchangers outside are positioned in the 40 °C water loops. The operation proceeds in two cycles as described above.

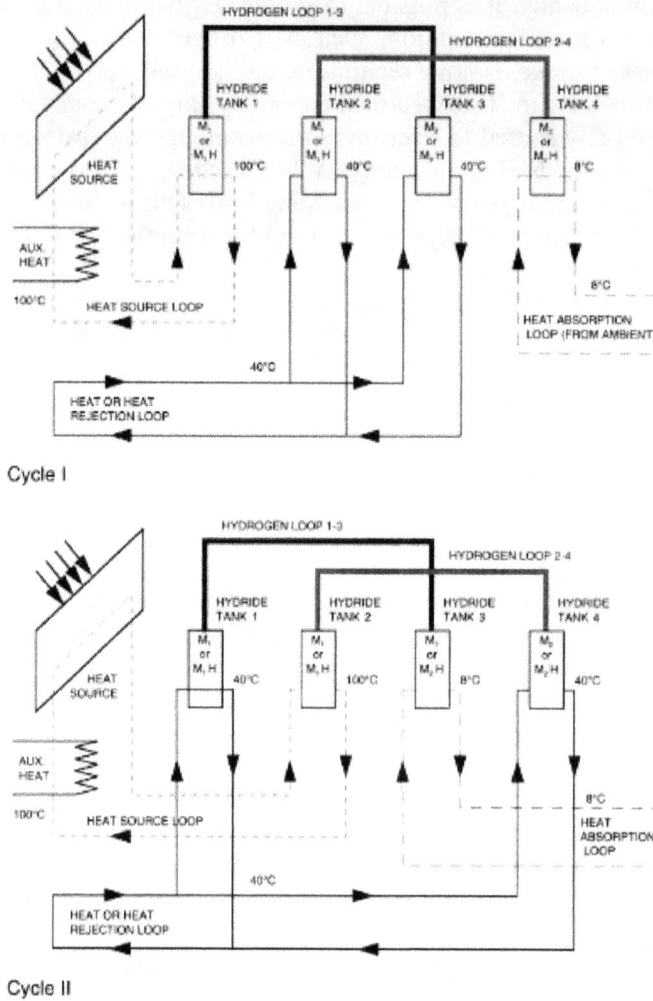

**Figure 15.13.** Schematic representation of a hydrogen/hydride heating-cooling system. Reprinted from [1], Copyright (2005), with permission from Elsevier.

If hydrogen is pushed via a turbine or expansion engine while it moves from one hydride tank to the next, it has the potential to provide both mechanical and electrical energy. Such a system, which is quite similar to the one proposed for heating and cooling, is shown in figure 15.14. This specific system is made up of three tanks that hold LaNi$_5$ as a hydriding material. Using heat from the Sun or any other source, hydrogen is forced out of the tank designated as tank 1, or the desorption tank, in the first cycle. After that, it goes through the expansion turbine, which generates both electrical and mechanical energy. The hydriding material in tank 2 then absorbs hydrogen at a reduced pressure. After going through the turbine,

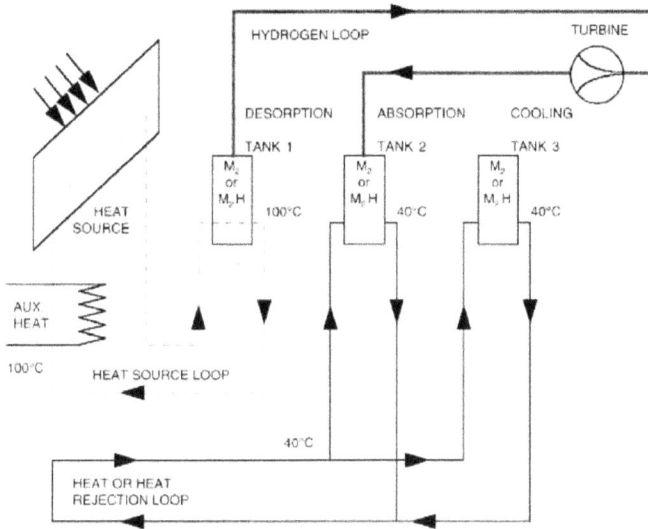

**Figure 15.14.** Schematic representation of electricity generation via hydrogen and hydrides. Reprinted from [1], Copyright (2005), with permission from Elsevier.

hydrogen is at a lower pressure, and in this instance heat is produced at a temperature that is lower (≈40 °C) than the desorption temperature. The water-cooling system rejects the heat generated in tank 2, also known as the absorption tank, to the outside world. Since the cooling tank (tank 3) served as the desorption tank in the previous cycle, it is likewise cooled using the same water-cooling system, from 100 °C to 40 °C. In the second cycle, a switch and valve system moves the tanks one step to the right in the diagram. The cooling tank becomes the absorption tank, the absorption tank becomes the desorption tank, and the desorption tank becomes the cooling tank. After that, the cycles are repeated. This technique can be used to convert low-quality heat into electricity.

## References

[1] Nejat Veziroglu T, Sherif S A and Barbir F 2005 *Environmental Solutions* (New York: Academic) pp 143–80
[2] Verhelst S *et al* 2013 *Renewable Hydrogen Technologies* (Amsterdam: Elsevier) ch 16
[3] Energy Education, https://energyeducation.ca/encyclopedia/Internal_combustion_engine
[4] https://www.reuters.com/business/aerospace-defense/rolls-royce-successfully-tests-hydrogen-powered-jet-engine-2022-11-28/
[5] Alabbadi S A 2012 *Energy Procedia* **29** 12
[6] Haidn O J *et al* 1998 *Int. J. Hydrogen Energy* **23** 491
[7] Kim J *et al* 2021 *Int. J. Hydrogen Energy* **46** 40073
[8] Rahaman S M *et al* 2020 *Energy Fuels* **34** 6052
[9] Van Nhu Nguyen *et al* 2018 *Int. J. Hydrogen Energy* **43** 17520

[10]  Kiadehi A D and Taghizadeh M 2018 *Int. J. Hydrogen Energy* **43** 4286
[11]  Dell R M, Moseley P T and Rand D A J 2014 *Towards Sustainable Road Transport* (New York: Academic) p 260
[12]  Coralli A *et al* 2018 *Science and Engineering of Hydrogen-Based Energy Technologies* (Amsterdam: Elsevier) p 39
[13]  dos Santos M C *et al* 2017 *Reference Module in Materials Science and Materials Engineering* (Amsterdam: Elsevier); https://doi.org/10.1016/B978-0-12-803581-8.09263-8

**IOP** Publishing

Hydrogen
Physics and technology
**Sindhunil Barman Roy**

# Chapter 16

## Hydrogen nuclear fusion technology

Hydrogen nuclear fusion technology involves combining two nuclei of hydrogen or its isotopes into a heavier one. This process releases substantial energy. The hydrogen fuel in a fusion reactor at the temperature required for fusion will be in the form of plasma, which is a fully ionized gas. Methods of generating energy production from fusion generally use thermal reactions where electron temperature $T_{el}$ and ion temperature $T_{ion}$ are similar. To achieve substantial reactions, the hydrogen fuel must be at extremely high ion temperatures, which are of the order of several KeV or $10^7$ K. The main difficulty faced in hydrogen fusion is that while the plasma is self-heated in the fusion reactions, cooling mechanisms such as bremsstrahlung x-ray radiation can lead to rapid energy losses at these temperatures.

The most promising path to a hydrogen fusion reactor is the fusion of two heavier isotopes of hydrogen, namely deuterium (D) and tritium (T) [1]. This is because it has a higher yield of 17.6 MeV at a relatively modest temperature $\approx 4.3$ keV, where self-heating power in fusion just can balance the losses from bremsstrahlung x-ray radiation. There can be additional energy loss mechanisms, such as heat conduction. This also cools the plasma and raises the requirement of temperature higher than 4.3 keV to achieve a self-heating condition where all loss mechanisms are balanced out. The temperature then rises through a positive thermodynamic instability, which results in an increasing fusion burn. This happens in nature in celestial bodies such as novae and type-1a supernovae, and on the Earth in thermonuclear weapon explosions.

The deuterium–tritium reaction is expressed as:

$$^3_1H + {}^2_1H \rightarrow {}^4_2He + {}^1_0n + 17.6 \text{ MeV} \tag{16.1}$$

Tritium can be found in seawater but it is too small for economic extraction. However, tritium can be produced from two isotopes of natural lithium by bombarding them with neutrons:

$$\frac{6}{3}\text{Li} + \frac{1}{0}n \rightarrow \frac{3}{1}\text{H} + \frac{4}{2}\text{He} \tag{16.2}$$

$$\frac{7}{3}\text{Li} + \frac{1}{0}n \rightarrow \frac{3}{1}\text{H} + \frac{4}{2}\text{He} + \frac{1}{0}n \tag{16.3}$$

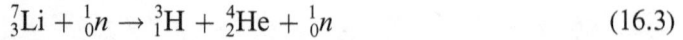

Tritium can in fact be made during the fusion reaction itself with the introduction of a lithium blanket to absorb the neutrons generated during the operation of a fusion reactor. Liquid lithium circulating between the fusion reactor and steam generator can also be used to extract the evolved energy.

Three conditions are necessary for the successful operation of a fusion reactor:

1. A high temperature of $10^8$K or more, so that the nuclei are moving fast enough to come together for collisions despite the repulsion due to their positive electric charges.
2. A fairly high density of the nuclei to ensure a high frequency of such collisions.
3. The confinement of reacting nuclei for a long enough time to generate more energy than that required for the operation of the fusion reactor.

Sustaining a high enough particle density $n$ for a long time $\tau$ in a hot plasma needed for a net energy gain is a difficult proposition. Apart from the interior of stars, the necessary combination of temperature, density, and confinement time happens during the explosion of fission bombs. The introduction of the fuels for fusion in a fission bomb leads to an uncontrolled fusion process, and the result is a devastating weapon, namely a hydrogen bomb. In a controlled thermonuclear hydrogen fusion reactor, appropriate isotopes of light elements are made to undergo nuclear fusion, with the result of controlled production and extraction of useful quantities of energy over the energy required to operate the reactor.

In a controlled thermonuclear reaction, an alpha particle and a neutron are generated from the fusion reaction $D + T \rightarrow {}^4\text{He} + n$ between deuterium (D) and tritium (T), and these particles carry the released energy away in the form of kinetic energy. The main source of heating in thermonuclear reactors is alpha particles (${}^4\text{He}$ nuclei), which emerge with an energy of $E_\alpha \approx 3.5$ MeV. In a controlled fusion on Earth, the goal of achieving energy production depends on two factors:

1. The created alpha particles remain in the plasma and heat the fusion fuel to keep the reactions going.
2. The kinetic energy of neutrons can go out of the plasma and get converted into electrical energy.

The confined alpha particle must efficiently transfer their energy to the plasma particles during slowing down in the plasma to provide the power for a self-sustained D–T burning. This self-sustaining condition is termed ignition. It is, however, very difficult to know what is going on inside the fuel. So, scientists use a more practical definition of ignition, which is based on the outgoing energy from fusion being greater than the input energy from the external heating sources.

## 16.1 Magnetic confinement fusion

In the controlled release of the fusion energy, the earliest and still active approach is to contain the reactive plasma with the help of a strong magnetic field. A solid vessel for containing plasma is not quite suitable here because of the contamination and cooling of the plasma by the vessel wall.

There are two classes of magnetic-field geometry, which are used for wall-free confinement of hot plasma. The one geometry uses the concept of a 'magnetic bottle', which consists of two magnetic mirrors. A magnetic mirror has a configuration of static magnetic field, which within a localized region has a shape such that any approaching charged particle is repelled back along its path of approach.

A magnetic field **B** is usually visualized as a distribution of nearly parallel nonintersecting field lines, the direction of which determines the direction of the magnetic field, and the density determines magnetic field strength. The magnetic force on a charge particle moving with a velocity **v** is perpendicular to both **B** and **v**. So, the magnetic force will have both a backward and an inward component. Thus, the charged particles, here for example electrons, would move through the space along a helical path around a magnetic-field line (see figures 16.1 and 16.2). If the magnetic-field lines along the path of the electron converge, the electron would enter

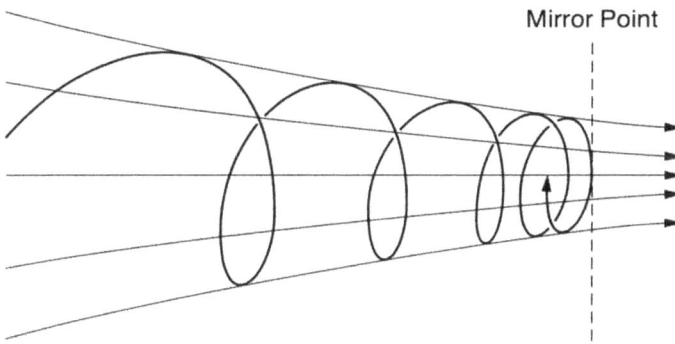

**Figure 16.1.** Schematic representation of the concept of a magnetic mirror.

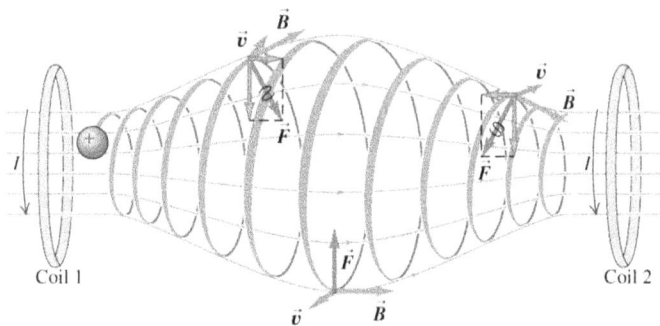

**Figure 16.2.** Schematic representation of the concept of a magnetic bottle. Reprinted from [3], Copyright (2020), with permission from Elsevier.

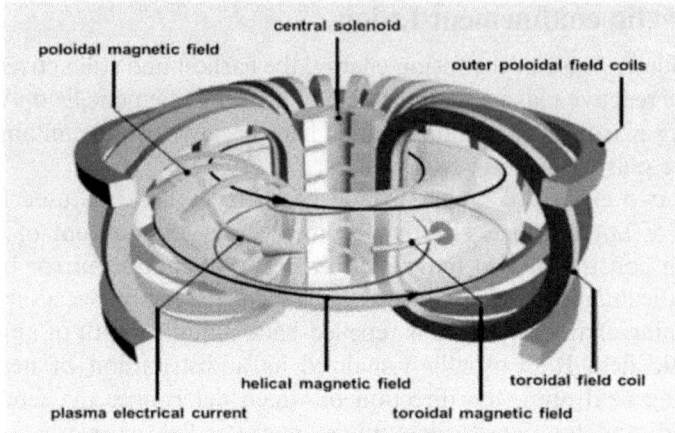

**Figure 16.3.** Schematic representation of the basic tokamak components, including the toroidal field coils (in blue), the central solenoid (in green), and poloidal field coils (in gray). Image source: DOE, Office of Sciences; https://www.energy.gov/science/doe-explainstokamaks.

a region of a stronger magnetic field. The electron will continue circling the magnetic-field line, but if the backward force is powerful enough its forward motion will be damped until it is stopped and finally repelled back along its original path. This mirroring effect takes place at a particular location, which depends only on the initial pitch angle describing the helical path of electron (figure 16.1). Two such magnetic mirrors are used to form a magnetic bottle that can trap charged particles in the middle (see figure 16.2). In a magnetic bottle the homogeneous magnetic field $\mathbf{B}_0$ in the central part is smaller compared to the magnetic field strength $\mathbf{B}_1$ at the neck of the magnetic bottle, i.e., $|\mathbf{B}_0| < |\mathbf{B}_1|$. In such a configuration, charge particles are trapped when their pitch angle $\theta_0$ at the center satisfies the relation $\sin^2 \theta < B_0/B_1$ [3].

The other magnetic confinement scheme uses a modified toroidal or doughnut-shaped magnetic-field configuration (see figure 16.3). This design originated in Russia and is popularly known as a tokamak. In a tokamak, magnetic-field coils confine plasma particles to allow the plasma to achieve the conditions necessary for fusion. One set of magnetic coils generates an intense toroidal field, directed the long way around the torus. A central solenoid (a magnet that carries electric current) creates a second magnetic field directed along the short way around the torus, i.e., poloidal direction. A current setup in the plasma itself can also create such a poloidal field. The two field components result in a twisted magnetic field that confines the particles in the plasma. A third set of field coils generates an outer poloidal field that shapes and positions the plasma. The operation of the fusion reactor relies on keeping the hot fusing plasma sustainably contained by having the plasma ions and electrons spiral along magnetic-field lines in this doughnut-shaped vessel. Once the plasma discharge is generated by resistive heating, external power sources such as radio-frequency antennae provide additional plasma heating as the plasma is brought to fusion conditions.

The first tokamak became operational in Russia in the late-1950s. Further advances in this field led to the construction of the Tokamak Fusion Test Reactor (TFTR) at Princeton Plasma Physics Laboratory and Joint European Torus (JET) in England. These facilities achieved record fusion power in the 1990s. These successes motivated multiple nations including Japan, several European countries, and India to collaborate on the superconducting tokamak in the International Thermonuclear Experimental Reactor (ITER). ITER is expected to generate 400 MW from deuterium–tritium reactions. Superconducting magnets will keep the reacting ions in a doughnut-shaped region whose volume is that of a large house. The neutrons that are produced during the fusion reaction, carry away about 80 % of the energy released during the fusion reaction. These neutrons are absorbed by lithium pellets in tubes that surround the fusion chamber. The resulting heat is carried away by circulating water, and this heat can be used to power steam turbines connected to electric generators in a commercial thermonuclear fusion reactor.

A key step before practical hydrogen nuclear fusion energy technology becomes a reality is the creation of a 'burning-plasma', in which the fusion reactions themselves supply most of the heating needed to sustain the plasma at the temperatures necessary for achieving fusion. A burning-plasma state signifies a transformational change to the energy and power balance in the D–T plasma, which opens up the potential for rapidly increasing performance [4]. To provide power in a fusion reactor for a self-sustained D–T burning plasma, the confined alpha particles need to transfer their energy efficiently to the plasma particles during the slowing-down process.

Alpha-particle heating in magnetically confined fusion machines is a high-priority subject and has been studied in the largest tokamaks with D–T plasma capabilities, TFTR and JET [2]. It is indeed a big challenge to identify alpha-particle heating in modern fusion devices because it is rather a small amount relative to other heating effects. In recent D–T plasma experiments performed at JET, the alpha-particle self-heating effects were observed in high-performance hybrid discharges with power modulation of the neutral D and T beams [2]. The core electron temperature in the D–T discharge was found to be about 30% higher than that in the reference deuterium discharge. After the neutral beam injection power was turned off, the alpha particles continued to transfer their kinetic energy to plasma electrons while slowing down.

## 16.2 Inertial confinement fusion

Inertial confinement fusion (ICF) involves the rapid implosion of a millimeter-sized capsule filled with a thermonuclear fuel mixture of deuterium and tritium. The capsule is heated with x-rays generated by high-power lasers, which turn the capsule into a plasma. This plasma accelerates inward, like a collapsing star, and compresses the deuterium–tritium (D–T) fuel in the capsule into a tiny sphere with a temperature exceeding 100 million degrees centigrade and a pressure more than 100 billion times greater than that of Earth's atmosphere. Hydrogen atoms in the fuel fuse under such conditions and release energy. The operation relies on the inertia of the assembled material to keep it together for a time long enough for fusion reactions to propagate through the fuel.

There are two principal approaches using lasers to generate the energy flux and pressure required to drive an ICF implosion [6]. In the direct-drive approach, the laser beams are aimed directly at the target (see figure 16.4(a)). The beam energy is absorbed by electrons in the outer, low-density corona of the target. That energy is transported by electrons to the denser shell material to drive the ablation and the resulting implosion. In the indirect-drive approach (see figure 16.4(b)), the laser energy is absorbed and converted to x-rays by high-$Z$ material inside the hohlraum that surrounds the target. In radiation thermodynamics, a hohlraum is a cavity whose walls are in radiative equilibrium with the radiant energy within the cavity. This idealized cavity is approximately realized in practice by making a small perforation in the wall of a hollow container of any opaque material [7]. Indirect drive is generally less efficient at coupling energy to a capsule than direct drive [6] because of the x-ray conversion and transport step. Ablation driven by electron conduction is, however, generally less efficient and more hydrodynamically unstable than ablation driven by x-ray. Direct-drive targets are also quite sensitive to intensity variations within individual beams. Hydrodynamic instability further amplifies the perturbations caused by such beam intensity variations. Calculations for current target designs indicate that with adequate beam uniformity direct-drive targets have about the same ignition threshold as indirect-drive targets, and there is a factor of two higher gain than the best indirect-drive-based designs, depending on a tolerable hydrodynamic instability growth [6].

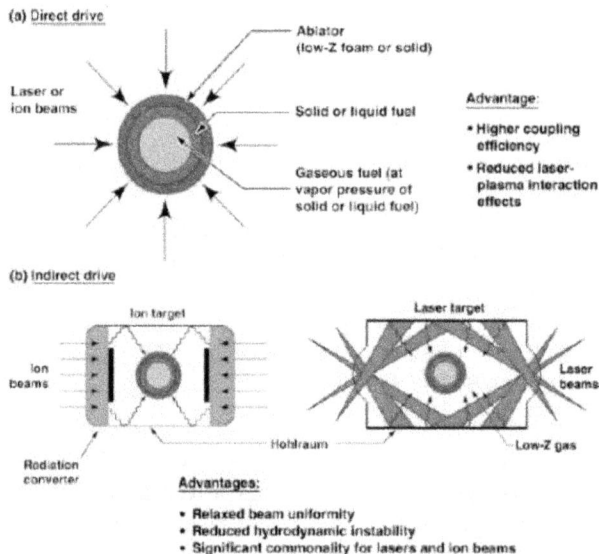

**Figure 16.4.** ICF process to produce a high shell-ablation pressure to drive an implosion using: (a) direct drive involving electron conduction, (b) indirect-drive with x-rays Reprinted from [6], with the permission of AIP Publishing.

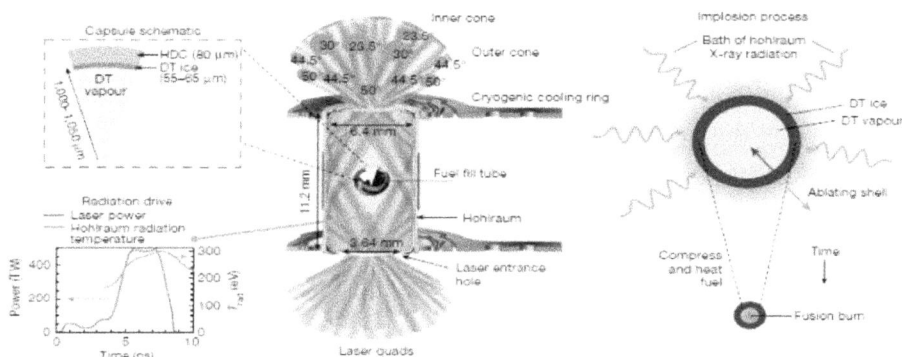

**Figure 16.5.** Schematic of the indirect-drive inertial confinement approach to achieve burning-plasma state in ICF at NIF, USA. Reproduced from [4]. CC BY 4.0.

In the US National Ignition Facility (NIF), inertially confined fusion plasmas use a rocket-like ablation effect to compress millimeter-sized capsules filled with D–T fuel to hundreds of billions of times the pressure of Earth's atmosphere [5]. This is the condition required for fusion and subsequent alpha heating. At the NIF facility, 192 lasers deliver up to 1.9 MJ of frequency-tripled light into a high atomic number (Z) 'hohlraum' (see figure 16.5), which serves the purpose of an x-ray converter generating a nearly Planckian x-ray bath [4]. The incident laser beam is designed to generate a specific radiation temperature history (see the bottom left-hand of figure 16.5) inside the hohlraum, with sufficient uniformity to match the specifics of target geometry and desired final plasma state. The exposed surface of a capsule at the center of the hohlraum absorbs approximately 10–15% of the x-rays. This results in the ionization of the outer edge of the capsule, generation high pressures of the order of hundreds of Mbar, and expansion away from the capsule. A shell of cryogenic D–T fuel is layered against the inside surface of the ablator capsule, which is in partial-pressure equilibrium with D–T vapor in the center of the capsule (see the top left-hand of figure 16.4). The inwardly directed acceleration caused by the ablation drives the capsule and D–T fuel, causing an implosion, as shown schematically at the right-hand side of figure 16.4, with enormous acceleration. The fuel is accelerated towards the center of the D–T gas core at extreme implosion velocities ($v_{imp}$) of nearly 400 km s$^{-1}$. The kinetic energy of the imploding shell and D–T fuel during stagnation is converted into internal energy in a dense fuel layer surrounding a central lower-density 'hot spot', where most of the fusion reactions occur. It is essential to have a symmetric compression of the D–T fuel surrounding the hot spot for providing inertial confinement and time for the alpha particles to redeposit their energy before the system explodes and rapidly cools as it expands. It is also required to achieve adequate areal densities for sufficient alpha deposition. This redeposition of alpha-particle energy back into the 'hot spot' creates further fusion reactions and amplified neutron yield [5]. Recently performed experiments in this configuration resulted in fusion powers of 1.5 petawatts, which was greater than the input power of the laser, and 170 kJ of fusion energy [4, 5]. Radiation hydrodynamics simulations show energy deposition by alpha particles as the dominant term in the hot-spot energy balance. This was an indication of a burning plasma state in ICF [4, 5].

# References

[1] Post R F 1956 *Rev. Mod. Phys.* **28** 338

[2] Kiptily V J *et al* 2023 *Phys. Rev. Lett.* **131** 075101

[3] Tjus J B and Merten L 2020 *Phys. Rep.* **872** 1

[4] Zylstra A B *et al* 2022 *Nature* **601** 542

[5] Kritcher A L *et al* 2022 *Nat. Phys.* **18** 251

[6] Lindl J *et al* 2004 *Phys. Plasmas* **11** 339

[7] Wikipedia; https://en.wikipedia.org/wiki/Hohlraum

**IOP** Publishing

Hydrogen
Physics and technology
**Sindhunil Barman Roy**

# Chapter 17

# Hydrogen in semiconductor technology

Hydrogen is extensively used in semiconductor technology, namely in the production of semiconductors, displays, LED, photovoltaic applications, etc. Hydrogen has excellent heat transfer capabilities, and is an efficient reducing and etching agent. The major ways of the usage of hydrogen in the semiconductor industry are as follows:

1. Annealing: Silicon (Si) wafers are heated to temperatures over 1000 °C in a hydrogen atmosphere to repair their crystal structure. Hydrogen transfers the heat uniformly to the wafer to assist in the reconstruction of the crystal structure in the final surface layers. Hydrogen also reacts to remove any oxide formation in the silicon.

2. Epitaxy: In many chemical reactions, hydrogen serves as a reducing agent, especially when new crystalline films are deposited. This is termed as epitaxy and is often used to create a starting silicon surface for semiconductor manufacturing. The newly cut and polished silicon wafers are reacted with trichlorosilane ($SiHCl_3$) in an epi-house. The gas-phase chlorine atoms are reduced by hydrogen, and the HCl byproduct is removed from the reactor as a gas. Materials such as strained silicon, silicon–germanium, and germanium are also grown using hydrogen-mediated epitaxy.

3. Deposition: Hydrogen is often directly incorporated into Si thin films to make them less crystalline. Interstitial monatomic hydrogen is electrically active and it tends to counteract the intrinsic conductivity of the semiconductor. As a result, the silicon layer is more electrically insulating.

4. Plasma etch: By reacting directly with the semiconductor wafer surface, hydrogen and hydrogen-containing plasmas can be employed to clean or remove undesirable thin films. This works particularly well for the removal of undesired fluorocarbon deposits from silicon oxides.

5. Passivation: Hydrogen can react and remove native oxides on silicon surfaces and facilitate the reconstruction of silicon-silicon bonds in the final

layers of the crystal. When hydrogen binds itself to existing defects or other impurities in silicon, it frequently stops the electrical activity of the impurities. This phenomenon is termed passivation. Such passivation of defects at the $Si/SiO_2$ interface is necessary for the reliable operation of complementary metal-oxide-semiconductor devices.

6. Ion implantation: In a semiconductor thin film, protons produced from hydrogen gas can be implanted to specific depths and concentrations with the help of ion implanters. Hydrogen atoms in higher doses and implantation energies can also be used to cleave silicon wafers, apart from the modification of a thin film.

7. Carrier gas: In a reaction chamber, hydrogen find use as a carrier gas to entrap and transport less volatile chemicals. This is done by heating and bubbling hydrogen through the liquid chemicals. The mass of hydrogen is very light compared to entrapped chemical vapor. Thus, specialized mass flow controllers can be used to sense, measure, and precisely control the amount of chemical vapor dispensed.

8. Deuterium defense against hot-carrier injection: Deuterium can also be reacted with semiconductor wafers at high temperatures in a pressurized furnace similar to regular hydrogen annealing and passivation. In environments with high levels of short-wavelength radiation, e.g., semiconductor components in a spacecraft, hydrogen bonds to silicon at thin-film interfaces are susceptible to breakage by hot-carrier injection, which are electrons and holes caused by short-wavelength radiation such as x-rays and gamma rays. Deuterium-passivated thin films have different bond strengths and a considerably lower probability of cleavage due to hot carriers.

9. Stabilization: The addition of hydrogen helps to extend the shelf life of important electronic chemicals such as diborane ($B_2H_6$) and digermane ($Ge_2H_6$). These chemicals usually decompose slowly.

## 17.1 A brief history of hydrogen in semiconductors

The study of hydrogen in semiconductors started in early-1950s when it was reported that the exposure of ZnO to hydrogen caused a significant increase in the conductivity of the ZnO [1]. In the 1980s it was observed that hydrogen in germanium could passivate electrically active impurities and at the same time activated isoelectronic impurities that were normally inactive electrically [2, 3]. Further experimental works soon attributed the passivation of boron in metal-oxide semiconductors transistors to hydrogen released from the aluminum gate [4, 5]. Hydrogen in boron-doped silicon was behaving as a donor because it was passivating the acceptor [6]. It was subsequently demonstrated that hydrogen could also passivate phosphorus donors in Si. This indicated that hydrogen could also behave as an acceptor in n-type semiconductors. Thus, hydrogen behaves as a donor ($H^+$) in p-type semiconductors and as an acceptor ($H^-$) in n-type semiconductors. Hydrogen consistently opposes the existent conductivity of semiconductor materials. This behavior of hydrogen was validated and extensively explained by a range of experimental techniques, conducted in tandem with theory and computing [6]. In general, the theory demonstrated the relationship between atomic and electronic structure of hydrogen and elucidated the physics of its interactions with semi-conductors. In a compound semiconductor, the $H^+$ prefers to stay close to the anion. The Coulomb attraction between the proton and the electronic charge density surrounding the more electronegative anion provides an explanation for this. Conversely, the $H^-$ has a tendency to remain close to the cation.

We shall now describe some of the mechanisms that govern interactions between hydrogen and semiconductors, including monatomic hydrogen, hydrogen molecules, and hydrogen-related complexes.

## 17.2 Monoatomic hydrogen

As we have already indicated, hydrogen in its positively charged state, $H^+$, is expected to optimize its Coulomb interaction with the host electrons in order to reduce its energy. Indeed, $H^+$ is located at the bond center in silicon, a strongly covalent semiconductor. In contrast, in GaN with partially ionic nature, $H^+$ sites on a spherical shell around the anion are all rather close in energy. In the negatively charged state of hydrogen, $H^-$ seeks to be close to the minimum in the electronic charge density. In silicon, $H^-$ is placed at the tetrahedral interstitial site. In GaN, $H^-$ is found at the center of the hexagonal channel, thus maximizing its distance from the six surrounding N atoms [6]. Hydrogen diffusion paths can be mapped by performing first-principles calculations for a specific charge state in several different lattice locations, and then tracing a path through the resulting total-energy surface [6]. In general, the migration barrier for $H^-$ is much higher than for $H^+$ [6].

### 17.2.1 Electronic structure and transition levels

We have already stated, hydrogen can behave either as a donor or an acceptor in most of the semiconductors. We shall now illustrate this behavior, taking the example of interstitial hydrogen in GaN [6]. The same treatment applies qualitatively to silicon and also to other semiconductors, such as GaAs and ZnSe. Figure 17.1(a) shows formation energy obtained with density functional theory (DFT)—local density approximation (LDA) calculations for an interstitial hydrogen atom in GaN as a function of the Fermi level for $H^+$, neutral hydrogen $H_0$, and $H^-$ (solid lines). The formation energy $E_H^f$ is expressed as [6, 8]:

$$E_H^f(q) = E_H^{tot}(q) - E_{bulk}^{tot} - \mu_H - qE_F \qquad (17.1)$$

Here $E_H^{tot}(q)$ is the total energy of $H^q$ in GaN, $E_{bulk}^{tot}$ is the total energy of the corresponding bulk supercell, and $\mu_H$ is the chemical potential for hydrogen. It is assumed that $\mu_H$ is fixed at the energy of a free hydrogen atom as a reference. $E_F$ is the Fermi level and it is set to zero at the top of the valence band. The formation energy is referenced to the energy of a free hydrogen atom.

It can be seen from figure 17.1(a) that $H^+$ is the energetically most stable species for Fermi energies below $\approx 2.1$ eV. $H^-$ is more stable for Fermi levels higher in the gap [8]. It can also be concluded from figure 17.1(a) that the solubility of hydrogen is

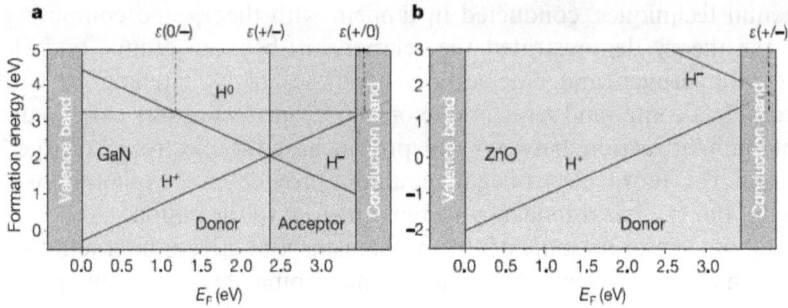

**Figure 17.1.** Formation energy as a function of the Fermi level for $H^+$, $H_0$, and $H^-$ in (a) GaN (b) ZnO. Reprinted from [7], Copyright (2003), with permission from Springer Nature.

relatively higher under p-type conditions than under n-type conditions. Hydrogen is most stable as $H^+$ in p-type semiconductor. This explains the experimentally observed passivation of acceptors. In a similar manner, in n-type semiconductors $H^-$ passivates donors. Neutral hydrogen is always higher in energy than either $H^+$ or $H^-$. This causes the donor level $\varepsilon(+/0)$ to lie well above the acceptor level $\varepsilon(0/-)$ [6]. As a result, the electronic behavior is governed by the transition level $\varepsilon(+/-)$ between the $+$ and $-$ charge states. The positively charged state is stable when the Fermi level is below the transition level. When the Fermi level lies above the transition level, the negatively charged state becomes stable. The instability of the neutral hydrogen state $H_0$ is characteristic of a 'negative-U' system. It was found from the $+/0$ and $0/-$ transition levels that $U = E^{0/-} - E^{+/0} = -2.4$ eV [8]. A negative-U behavior was also found for H in Si and GaAs, but the value found here for GaN is unusually large. For a detailed discussion on the origin of the large negative-U value, the readers are referred to the article by Neugebauer and Van de Walle [8].

Detailed knowledge about the electrical activity of hydrogen in a semiconductor is extremely important. Many semiconductor growth techniques employ hydrogen in the growth environment. This causes hydrogen to be easily and sometimes unintentionally incorporated into the semiconductor. Thus, to avoid unexpected device performance, the presence of hydrogen in the semiconductor, irrespective of whether it acts as an electrically active dopant or as a passivating agent, needs to be carefully monitored and controlled.

### 17.2.2 Alignment of hydrogen levels

First principle calculations show that hydrogen can exist only as $H^+$ in ZnO, and the other charge states of hydrogen are not stable [7, 9]. Figure 17.1(b) shows that in ZnO, for any Fermi-level positions, $H_0$ and $H^-$ lie higher in energy than $H^+$. Thus, hydrogen in ZnO acts as a source of doping, rather than just causing a reduction of the conductivity. Various experiments have confirmed these theoretical results (see reference [6] and references therein). GaN and ZnO have almost the same band gap and have quite similar physical properties, hence the qualitative difference between these two semiconductors is rather surprising. Hydrogen reduces the existing conductivity in GaN, whereas it behaves as a doping source in ZnO. The $\varepsilon(+/-)$ level lies within the band gap in GaN, which leads to its amphoteric behavior. In contrast, the $\varepsilon(+/-)$ level in ZnO resides above the conduction-band minimum (CBM). This leads to hydrogen behaving solely as a donor. This difference in behavior is explained by the very different positions of the band structures of ZnO and GaN on an absolute energy scale. The valence band and conduction band in ZnO lie 1.3 eV lower in energy than the corresponding bands in GaN [6]. After taking into account this band alignment, the electronic level $\varepsilon(+/-)$ for interstitial hydrogen lies at virtually the same energy. It is constant on an absolute energy scale.

### 17.2.3 Effects on doping and passivation

We now understand that in most semiconductors and insulators hydrogen can exist in both positive and negative charge states. As can be seen in figure 17.2, the $\varepsilon(+/-)$ level is located well within the band gap in the majority of semiconductors and insulators. Using GaN as an example, we will now investigate the consequences of the amphoteric behavior of hydrogen. Figure 17.1(a) shows the formation energy of hydrogen in its different charge states in GaN. The stable charge state directly shifts from positive (for $E_F$ below 2.4 eV) to negative (for $E_F$ above 2.4 eV) when Fermi energy $E_F$ goes through the band gap. In p-type GaN with $E_F$ near the valence-band maximum (VBM), this suggests that $H^+$ is preferred [6]. In n-type GaN, Fermi energy $E_F$ lies close to the CBM. Here, $H^-$ is stable. As a result, in both cases, hydrogen counteracts the existing conductivity. This also implies that hydrogen cannot act as a source of doping in GaN [6]. Let us consider a situation where hydrogen is the only impurity present in GaN and there is no compensation for other defects or impurities. Here if one starts with the intrinsic material (i.e., Fermi energy $E_F$ residing near midgap) and slowly adds hydrogen, it will first function as a donor

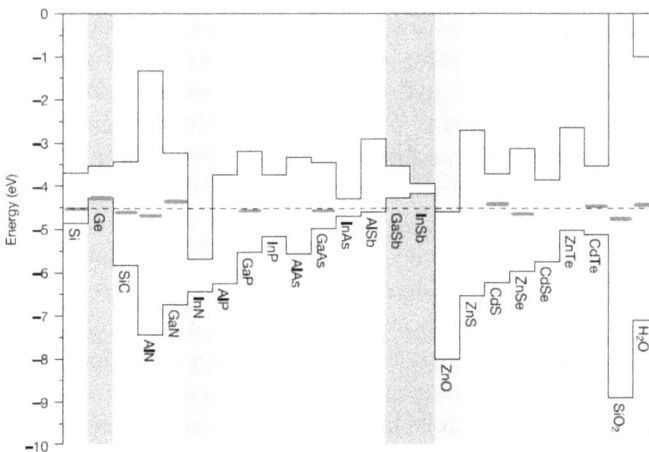

**Figure 17.2.** Band line-ups and position of the $\varepsilon(+/-)$ level for a range of semiconductors and insulators. Reprinted from [7], Copyright (2003), with permission from Springer Nature.

($H^+$). This will bring the Fermi level closer to the conduction band, and hydrogen will begin to behave as an acceptor ($H^-$) when it reaches $\varepsilon(+/-)$. The Fermi level will be generally pushed down as a result of this. Hydrogen thus effectively gets self-compensated, and $E_F$ ends up pinned at the $\varepsilon(+/-)$ level. Only when the $\varepsilon(+/-)$ level is very close to or above (below) the CBM (VBM) does hydrogen doping succeed in leading the Fermi level close to the corresponding band edge and have a visible effect on electron or hole concentrations.

$H^+$ is the sole stable charge state in ZnO since the $\varepsilon(+/-)$ level is above the CBM (see figure 17.1(b)) [6, 7]. The concentration of free electrons will always rise as hydrogen releases its electron. Since $H^-$ is not formed in this scenario, self-compensation does not occur. In ZnO, the level of hydrogen that may be observed experimentally corresponds to a shallow donor level [6]. Since the $\varepsilon(+/-)$ level lies above the CBM, hydrogen's electron can reduce its energy by transferring to the CBM. This electron can, however, still remain bound to the ionized donor (i.e., the proton) in a 'hydrogenic effective mass state' [6]. Hydrogen does not form a hydrogenic shallow dopant state in the majority of materials, but it does in ZNO.

Band line-ups and the location of the $\varepsilon(+/-)$ level for a variety of semi-conductors and insulators are shown in figure 17.2. The position of the VBM, CBM, and $\varepsilon(+/-)$ level determined within DFT local density approximation (LDA) with respect to the VBM for each material are indicated by the lower line, the upper line, and the thick red line, respectively. Figure 17.2 lists several materials in which hydrogen exhibits electrically active dopant behavior. For instance, it is anticipated that hydrogen in InN will only function as a donor, and experimental evidence supports this hypothesis [6]. Hydrogen could potentially be a source of p-type doping, as seen in figure 17.2. In Ge and GaSb, hydrogen can act as a shallow acceptor [6].

## 17.3 Hydrogen molecules and molecular complexes

Hydrogen molecules are quite hard to detect experimentally because it is challenging to utilize vibrational spectroscopy to observe free hydrogen molecules since they lack a dipole moment. Putting a hydrogen molecule within a semiconductor causes it to have a weak dipole moment. The hydrogen molecule can be observed then through infrared absorption. The first direct observations of hydrogen molecules in semiconductors were recorded in the-late 1990s. This was first made with Raman spectroscopy in GaAs, and then in silicon first with infrared absorption and followed by Raman spectroscopy [6]. Numerous other issues regarding the observation of spectral lines connected to ortho-hydrogen and para-hydrogen were then resolved. All these experimental results showed remarkably large shifts in the vibrational frequency when compared to that of the free hydrogen molecule. Computational investigations confirm that strong interactions exist between hydrogen molecules and the semiconductor. This results in a considerable increase in the bond length and lowers the vibrational frequency.

There can be another type of diatomic hydrogen complex $H_2^*$ in semiconductors in addition to hydrogen molecules [6]. This complex was first proposed on the basis of calculations in Si [11] and diamond [10]. The hydrogen complex has $C_{3v}$ symmetry and consists of one hydrogen in a bond-center-like position and one hydrogen in an antibonding-type position. The original Si–Si bond is broken because the Si atom between the two hydrogens is substantially relaxed from its equilibrium state. The resultant dangling bond is secured by the hydrogen of the bond-center type [6]. Since the $H_2^*$ complex has a lower activation barrier for migration than molecular hydrogen, it is anticipated to be involved in hydrogen diffusion in semiconductors [11]. The $H_2^*$ complex has since been observed experimentally [12].

### 17.3.1 Complexes with impurities and point defects

It is possible to have a formation of strong bonds between hydrogen and host atoms. This can happen when there is a deviation from the perfect crystal structure of the solid. There are such deviations at a crystalline surface, at an interface, in polycrystalline or amorphous material, near a point defect, and in an impurity in the bulk crystal. The last type of interaction with defect structures, particularly with dopant impurities, can be understood straightforwardly based on the discussion presented earlier on monatomic hydrogen. We have studied that interstitial hydrogen prefers to be in the positive charge state, $H^+$ in p-type semiconductors. The dopant impurities that made the semiconductor to be p-type to start with are acceptors that exist in a negative charge state $A^-$ by contributing a hole to the valence band. The tendency for hydrogen to be in the $H^+$ state in p-type material counteracts the characteristic conductivity of the semiconductor. In other words, hydrogen compensates the acceptors. It may be noted that the term compensation refers to a general effect on the conductivity of the layer and makes no assumptions on the spatial location of the compensating center [6]. In reality, the negatively charged acceptors attract $H^+$. The interstitial hydrogen atom will diffuse toward the acceptor and become attached to it because of the high diffusivity of hydrogen. Passivation is the term used to describe the resulting configuration with a complex formation. Hall effect measurements would reveal distinctly different results in the two cases, demonstrating the unambiguous contrast between passivation and compensation [6]. $H^+$ and $A^-$ appear in distinct locations in the case of compensation, and they contribute independently to ionized impurity scattering. The measured mobility is lower than in the absence of hydrogen [6]. An electrically neutral $(AH)^0$ complex is formed in the case of compensation. This results in a decrease of ionized impurity scattering, which in turn increases mobility. The microscopic location of hydrogen can be inferred by using this difference between compensation and passivation as an experimental tool.

Defects in the bulk crystal lattice, such as point defects, can also lead to complex formation. Passivation of dangling bonds at interfaces, grain boundaries, or vacancies is often assumed to be the cause of the creation of strong bonds between hydrogen and host atoms [6]. Hydrogen passivation can, however, occur at both over- and under-coordinated atoms. For instance, the formation energies of self-interstitials in crystalline Si are comparable to those of vacancies. In addition to under-coordinated atoms, over-coordination defects may also occur in amorphous silicon [13]. It is known that self-diffusion, impurity diffusion, surface reconstructions, planar interstitial defects, and nucleation of dislocation are all influenced by the self-interstitials in crystalline silicon. One or two hydrogen atoms are attracted to these self-interstitials. The calculated binding energy for such complexes is large enough for the complexes to be stable at room temperature, although it is smaller than that for hydrogen interacting with a vacancy. When the self-interstitial attaches itself to two hydrogen atoms, it becomes electrically neutral.

### 17.3.2 Defect and impurity engineering

The incorporation of dopants and the suppression of compensating point defects are significantly impacted by the presence of hydrogen during the growth of a semiconductor. Thus, in defect and impurity engineering it might be possible to beneficially use the capacity of hydrogen to influence these variables. However, the hydrogen-enhanced doping mechanism only functions under the following specific conditions [6]:

1. The predominant compensatory defect must be hydrogen. In other words, its formation energy needs to be comparable to the formation energy of the dopant impurity and less than that of all native defects.
2. Simple process of hydrogen removal from the doped layer after growth. High-temperature annealing is limited to a certain temperature range. Simultaneously, the temperature must be high enough to facilitate all three processes necessary to eliminate hydrogen as a compensating center, namely complex dissociation, migration, and overcoming of any additional barriers (e.g., surface barriers) toward activation.

Condition 2 is met in p-type GaN. Hydrogen in p-type ZnSe acts as a donor, and this gives rise to N–H complexes that are electrically neutral. The strength of the N–H bond in Zn–Se makes it more stable than the Zn–Se bonds of the host crystal. Thus, the removal of the hydrogen from the nitrogen-doped layer by annealing becomes quite difficult. This explains why metal–organic chemical vapor deposition (MOCVD) generally fails to produce p-type ZnSe of device-quality [6].

## 17.4 Hydrogen on semiconductor surfaces

Hydrogen is present in high concentrations in widely used semiconductor growth techniques, such as MOCVD, gas-source molecular beam epitaxy (MBE), and hydride vapor-phase epitaxy. Utilizing the example of GaN, we will now talk about the significant effects of hydrogen on semiconductor growth using such techniques. An increase in the growth by a factor of two has been observed when hydrogen was added during the MBE growth of GaN using an RF-plasma source [14]. Then, using time-of-flight scattering and recoiling spectrometry, low energy electron diffraction (LEED), and thermal decomposition mass spectrometry, the composition and structure of GaN(0001) surfaces produced via MOVPE were examined [15]. It was concluded that hydrogen was present on the surface, and it removed surface states and facilitated passivation of the partially occupied dangling bonds. Hydrogen

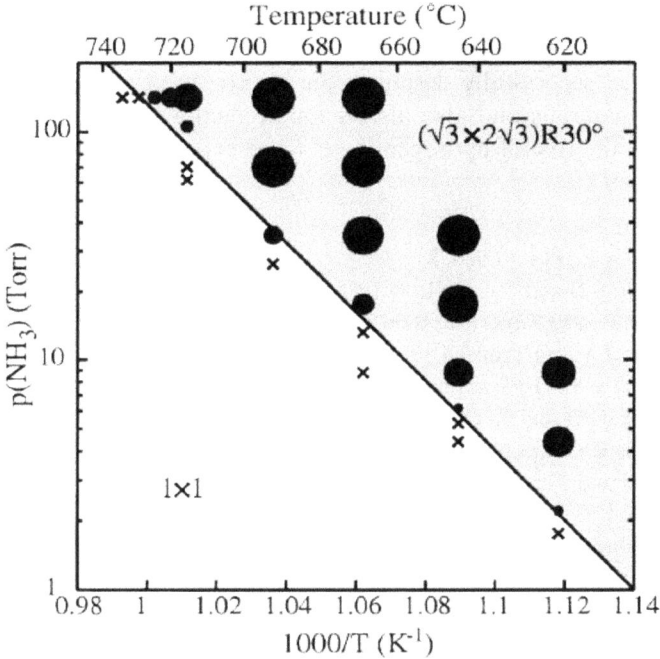

**Figure 17.3.** Equilibrium phase diagram of the GaN(0001) surface under MOCVD conditions. Shaded area denotes conditions where the reconstruction with a periodicity $\sqrt{3} \times 2\sqrt{3} - R30^0$ is observed. Reprinted with permission from [16], Copyright (1999) by the American Physical Society.

desorption was also experimentally investigated [6]. Experiments with the help of electron energy loss spectroscopy and LEED revealed that hydrogen desorption from Ga sites occurred between 250 °C and 450 °C. However, all such studies were performed in conditions far away from realistic growth.

In situ grazing incidence x-ray scattering can provide direct information about the structure of the surface during MOCVD growth [16]. A surface phase diagram as a function of temperature and ammonia partial pressure $p(NH_3)$ revealed a transition between the two phases. Figure 17.3 represents the equilibrium phase diagram of a clean GaN surface. At high temperatures, a $1 \times 1$ reconstruction of the unit cell was observed. A different reconstruction with a periodicity of $\sqrt{3} \times 2\sqrt{3} - R30^0$ was found at lower temperatures and sufficiently low $NH_3$ pressures. This $\sqrt{3} \times 2\sqrt{3} - R30^0$ phase has a novel 'missing row' structure with 1/3 of the surface Ga atoms being absent. An activation energy of $3.0 \pm 0.2$ eV was calculated from the temperature dependence of p(NH3) at the transition.

## 17.5 Summary

In this chapter, we have discussed the major uses of hydrogen in the semiconductor industry. To understand the science behind such usage, we described the behavior of hydrogen in the bulk and on the surface of semiconductors in general. It is shown that a combination of first-principles theory and experimental investigations can provide a very detailed picture of the behavior of hydrogen in semiconductors as an isolated impurity, and its interactions with impurities or point defects. A 'universal alignment' model successfully describes the electronic behavior of hydrogen in a wide range of materials and also allows for predictions for newer semiconductor materials where the role of hydrogen is yet to be explored.

## References

[1]  Mollwo E 1954 *Z. Phys.* **138** 478
[2]  Haller E E, Joos B and Falicov L M 1980 *Phys. Rev.* B **21** 4729
[3]  Kahn J M *et al* 1983 *Phys. Rev.* B **36** 8001
[4]  Sah C T, Sun J Y and Tzou J J 1983 *Appl. Phys. Lett.* **43** 204
[5]  Pankove J I *et al* 1983 *Phys. Rev. Lett.* **51** 2224
[6]  Van De Walle C G and Neugebauer J 2006 *Annu. Rev. Mater. Res.* **36** 179
[7]  Van De Walle C G and Neugebauer J 2003 *Nature* **423** 626
[8]  Neugebauer J and Van de Walle C G 1995 *Phys. Rev. Lett.* **75** 4452
[9]  Van de Walle C G 2000 *Phys. Rev. Lett.* **85** 1012
[10]  Briddon P, Jones R and Lister G M S 1988 *J. Phys.* C **21**
[11]  Chang K J and Chadi D J 1989 *Phys. Rev.* B **40** 11644
[12]  Holbech J D *et al* 1993 *Phys. Rev. Lett.* **71** 875
[13]  Pantelides S T 1986 *Phys. Rev. Lett.* **57** 2979
[14]  Yu Z *et al* 1996 *Appl. Phys. Lett.* **69** 2731
[15]  Sung M M *et al* 1996 *Phys. Rev.* B **54** 14652
[16]  Munkholm A *et al* 1999 *Phys. Rev. Lett.* **83** 741

# Chapter 18

## Road towards hydrogen economy

Summarizing the presentations and discussions in this book so far, we can say that hydrogen energy technology involves the use of hydrogen and hydrogen-containing compounds to generate energy. Hydrogen has the potential for perennial and sustainable energy production, and can be utilized to fulfill various needs of modern civilization. Hydrogen technology can utilize a wide spectrum of raw materials and an energy source, such as electricity, heat, or mechanical work, to generate fuel. This, in turn, can be used in energy-efficient devices to generate the same energy elements, such as electricity, heat, or mechanical work, with little noise and no toxic waste [1].

The adoption and implementation of hydrogen energy introduces the possibility of the creative and innovative concept of a circular economy, namely hydrogen economy, which is restorative by nature [1]. The term 'hydrogen economy' was introduced first by electrochemist John Bockris during a meeting at the General Motors Technical Centre in 1970 [2]. The concept of a hydrogen economy envisages hydrogen replacing carbon-containing fuels, and thus eliminates air pollution and accumulation of carbon dioxide in the atmosphere (see [3] and references therein). The process involves the usage of intermittent renewable energy along with sufficient control of seasonal energy storage. Currently, the world is mainly dependent on the fossil fuel-based energy system, which has a few clear drawbacks: (i) the fuel ownership is confined in selective geographical locations, (ii) the growing consumption leads to steady depletion of the known fuel reserves, (iii) it causes serious harm to the environment, and (iv) its utilization involves energy-inefficient and pollutant engineering procedures and technologies. The transition to a hydrogen economy will enable the creation of a system that is based on renewable energies, such as hydroelectric, solar, wind, geothermal, and oceanic energies, and is also based on a host of raw materials as a source of the energy carrier, which includes water and virtually any type of biomass [1]. The selection of primary energy and raw material can be adapted for local conditions and the generated fuel can be easily and widely distributed throughout the world. This will reduce concentrated ownership for market control. The hydrogen economy will also introduce the utilization of the most efficient energy converter, known as the fuel cell. This will avoid the sequential and inefficient conversion of energy forms used by heat engines, turbines, and motors, which would be replaced by the direct, unique, and highly efficient electrochemical energy conversion of the chemical energy contained in the fuel into electric energy and heat, with water as the only byproduct [1].

The transition to the era of the hydrogen economy is concomitant with the growing interest in renewable energies. The inherently intermittent production of renewable energies is nicely complemented by the production and storage of hydrogen [1]. This leads to a perennial renewable and circularly sustainable cycle. In an earlier chapter of this book, we have seen that as an energy carrier hydrogen is versatile, clean, and safe. It can generate electricity, heat, and power, and can also be used in many applications as a raw material for industry. Hydrogen can be stored and transported with high energy density in liquid or gaseous forms. Green hydrogen can be produced by water electrolysis technologies using renewable energies. Fuel cells that use hydrogen as fuel and oxygen from the air as oxidant constitute the most energy-efficient devices known to date to generate electricity [1]. They find usage in all kinds of electrical energy applications, with particular interest in the distributed stationary generation of electricity, heavy-duty vehicles, and automobiles.

In summary, hydrogen energy technology uses hydrogen and hydrogen-containing compounds to generate energy to be supplied to all needed practical uses with high energy efficiency, overwhelming environmental and social benefits, as well as economic competitiveness [1]. The arrival of the era of hydrogen energy provides a revolutionary circular path concerning energy production and use. This in turn entails convenient and beneficial utilization of the huge amount of waste resulting from the lifestyle in modern civilization, decarbonizing different areas of intense energy consumption, implementation of large-scale production of renewable energy, and uniformity in the distribution and access of energy across the world. However, this transition from the fossil fuel-based economy to the hydrogen energy economy requires new approaches, which are presented schematically in figure 18.1. These include [1]:

1. Circular, clean, and beneficial path for energy production and use.
2. Widespread use of renewable energies, including production and storage of hydrogen to stabilize the delivery of electric energy, regulating the inherent intermittence associated with renewable energies. Use of hydrogen to act as a buffer to increase resilience in the energy system of a country or region.
3. Conversion of urban and rural organic wastes and sewage to produce hydrogen and hydrogen-rich gases and compounds.
4. Use of hydrogen to decarbonize activities in sectors such as industry: (i) supplying electrical and thermal energies, and (ii) renewable feedstock produced by conveniently reacting hydrogen with biomasses. Energy supply, as combined heat/cooling and power to buildings and households, introduces the distributed generation of electrical and thermal power. Hydrogen can be

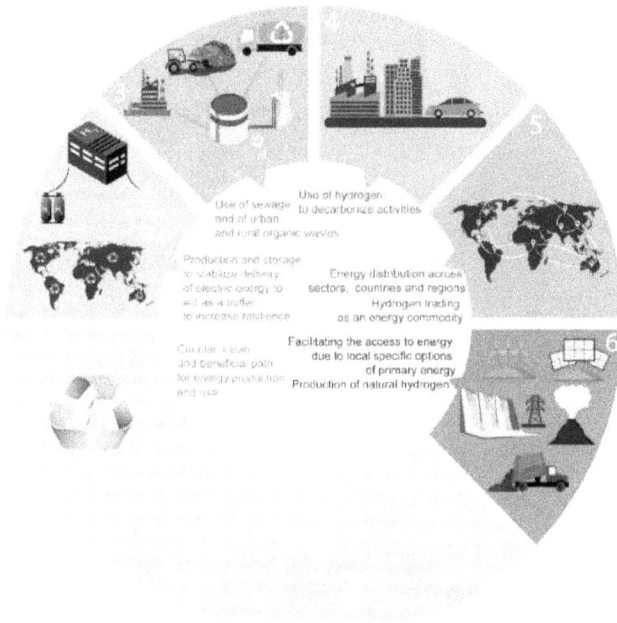

**Figure 18.1.** The various characteristics and possibilities related to hydrogen energy technology application. Reprinted from [1], Copyright (2019), with permission from Elsevier.

used in transportation, including light-duty and heavy-duty vehicles and automobiles for terrestrial, nautical, and aeronautical applications.

5. Energy can be distributed across sectors, countries, and regions using hydrogen and hydrogen-rich gases and compounds as carriers, and also trading hydrogen as an energy commodity.

6. Facilitating access to energy in different countries or regions because of local specific options of primary energy sources and raw materials to produce hydrogen, and also local production of natural hydrogen.

The road to a hydrogen economy naturally involves some amount of financial risk [4]. This can be minimized if the energy industry adds hydrogen technology in small increments that closely follow demand. Companies can, for instance, build power plants that generate both electricity and a small stream of hydrogen for early fuel-cell cars [4]. The most viable option in the near term is production of hydrogen from natural gas, either in centralized reformers that supply fueling stations by delivery truck or in smaller on-site reformers located at the stations (see figure 18.2). These fueling stations can also use electricity from the power grid to make hydrogen by water electrolysis. However, these processes cause greenhouse gas emissions with the assumption that fossil fuels are used to make electricity. In the long term (see figure 18.3), (i) advanced power plants can produce hydrogen from coal and bury the carbon dioxide deep underground, (ii) wind turbines and other renewable energy

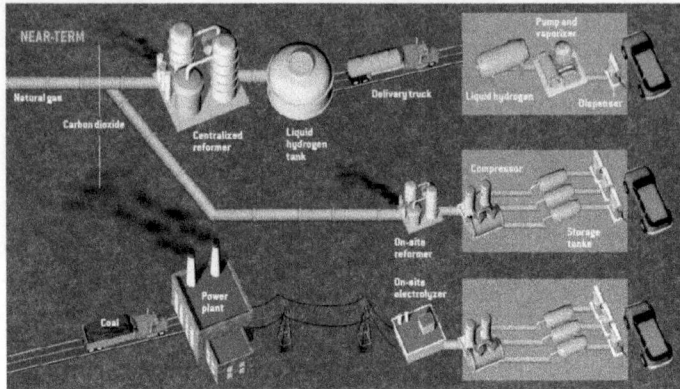

**Figure 18.2.** The near-term options for hydrogen infrastructure. Reprinted from [4], Copyright (2006), with permission from Springer Nature.

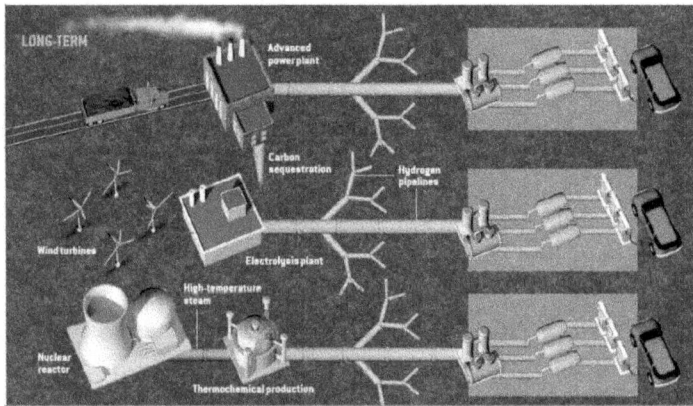

**Figure 18.3.** The long-term options for hydrogen infrastructure. Reprinted from [4], Copyright (2006), with permission from Springer Nature.

sources could provide the power for electrolysis, and (iii) high-temperature steam from nuclear reactors can generate hydrogen through the thermochemical splitting of water [4].

The progression to a hydrogen transportation system consists of several parallel tracks. The first important step is improving fuel economy. Developing lightweight cars, more efficient engines, and hybrid electric-driven trains can greatly reduce carbon emissions and oil consumption over the next few decades [4]. Hydrogen and fuel cells will build on this technical development, taking advantage of the efficiency improvements and the increasing electrification of vehicles. While the transition may take place over several decades, hydrogen fuel-cell vehicles could ultimately help protect the global environment and reduce the world's reliance on fossil fuels.

# References

[1] de Miranda P E V 2019 *Science and Engineering of Hydrogen-Based Energy Technologies* (Amsterdam: Elsevier)
[2] Bockris J O M 1972 *Science* **176** 1323
[3] Bockris J O M 1911 *Materials* **4** 2073
[4] Ogden J 2006 *Sci. Am.* **295** 94

**IOP** Publishing

# Hydrogen
Physics and technology
**Sindhunil Barman Roy**

# Appendix A

## Schrödinger wave equation for the hydrogen atom

### A.1 Stationary state wave functions and eigenvalues

The Schrödinger equation is the wave equation for de Broglie waves, and its solution lies at the core of quantum mechanics. In general, the time-independent Schrödinger equation for a particle of mass $m$ moving in one-dimension in a potential $U(x)$ is expressed as:

$$-\frac{\hbar^2}{2m}\frac{\mathrm{d}^2\psi(x)}{\mathrm{d}x^2} + U(x)\psi(x) = E\psi(x) \tag{A.1}$$

Here the first term in equation (A.1) is associated with the kinetic energy, $U(x)$ is the potential energy function, and $E$ is the total energy. The relations expressed in equation (A.1) can be expressed in the more formal language of quantum mechanics in terms of *operators* and *eigenvalue equations*. An operator is an instruction, which carries out an operation on a function. In equation (A.1) the differential operator $-\frac{\hbar^2}{2m}\frac{\mathrm{d}^2}{\mathrm{d}x^2}$ operates on the wave function $\psi(x)$. This differential operator takes the second derivative of the wave function $\psi(x)$ with respect to $x$ and then multiplies the result by $-\frac{\hbar^2}{2m}$. This operation regenerates the same function $\psi(x)$ with a numerical multiplier, which in this case is the kinetic energy $E_{\mathrm{Kin}}$. The result can be formally written as:

$$-\frac{\hbar^2}{2m}\frac{\mathrm{d}^2\psi(x)}{\mathrm{d}x^2} = E_{\mathrm{Kin}}\psi(x) \tag{A.2}$$

Equation (A.2) is an example of an eigenvalue equation, an operator acts on a function $\psi(x)$, called an eigenfunction, to produce the same function $\psi(x)$ with a numerical multiplier $E_{\mathrm{Kin}}$, which is called an *eigenvalue* of the kinetic energy

doi:10.1088/978-0-7503-5172-0ch19 © IOP Publishing Ltd 2024

operator $-\frac{\hbar^2}{2m}\frac{d^2}{dx^2}$. In the framework of quantum mechanics, various physical observables, such as kinetic energy, total energy, linear momentum, angular momentum, etc, have an associated operator with each such variable, and they form parts of an eigenvalue equation. The allowed values of the observables are given by the eigenvalues obtained by the eigenvalue equation. If an eigenfunction representing an energy state of the system is simultaneously an eigenfunction of many such operators, then the eigenvalues associated with these operators represent the outcomes of relevant measurements carried out on the system, when residing in this state.

The total energy operator or Hamiltonian operator is the sum of the operators for kinetic energy and potential energy:

$$\hat{H} = -\frac{\hbar^2}{2m}\frac{d^2}{dx^2} + U(x) \tag{A.3}$$

The eigenvalue we are looking for here is the total energy $E$, and the eigenvalue equation is expressed as:

$$\hat{H}\psi(x) = E\psi(x) \tag{A.4}$$

In the case of a hydrogen atom, the problem is three-dimensional and the energy eigenfunctions are represented by $\psi(x, y, z)$. The potential energy function $U(x)$ is the Coulomb potential energy, which depends only on the distance of separation $r = \sqrt{x^2 + y^2 + z^2}$ between the electron and the nucleus of the hydrogen atom. It is assumed here that the nucleus of a hydrogen atom consisting of one proton is massive in comparison to the electron, and the motion of the nucleus is negligible and for all practical purposes the nucleus is at rest. The Hamiltonian operator for the hydrogen atom problem can now be written as:

$$\hat{H} = -\frac{\hbar^2}{2m_e}\left[\frac{\partial^2}{\partial x^2} + \frac{\partial^2}{\partial y^2} + \frac{\partial^2}{\partial z^2}\right] - \frac{e^2}{4\pi\epsilon_0 r} \tag{A.5}$$

The eigenfunctions $\psi(x, y, z)$ and energy eigenvalues $E$ are then found by solving the three-dimensional time-independent Schrödinger equation:

$$-\frac{\hbar^2}{2m_e}\left[\frac{\partial^2}{\partial x^2} + \frac{\partial^2}{\partial y^2} + \frac{\partial^2}{\partial z^2}\right]\psi(x, y, z) - \frac{e^2}{4\pi\epsilon_0 r}\psi(x, y, z) = E\psi(x, y, z) \tag{A.6}$$

An important aspect to note here is the nature of the Coulomb potential energy, which depends only on the distance from the origin where the nucleus of a hydrogen atom is located. The Coulomb potential energy does not depend on the orientation of the direction from the nucleus to the electron. Thus, it is quite convenient not to use the Cartesian coordinates $(x, y, z)$ as shown in figure A.1(a), and rather use the spherical polar coordinates $(r, \theta, \phi)$ as shown in figure A.1(b). The Coulomb potential energy is then only a function of the radial coordinate $r$, and does not

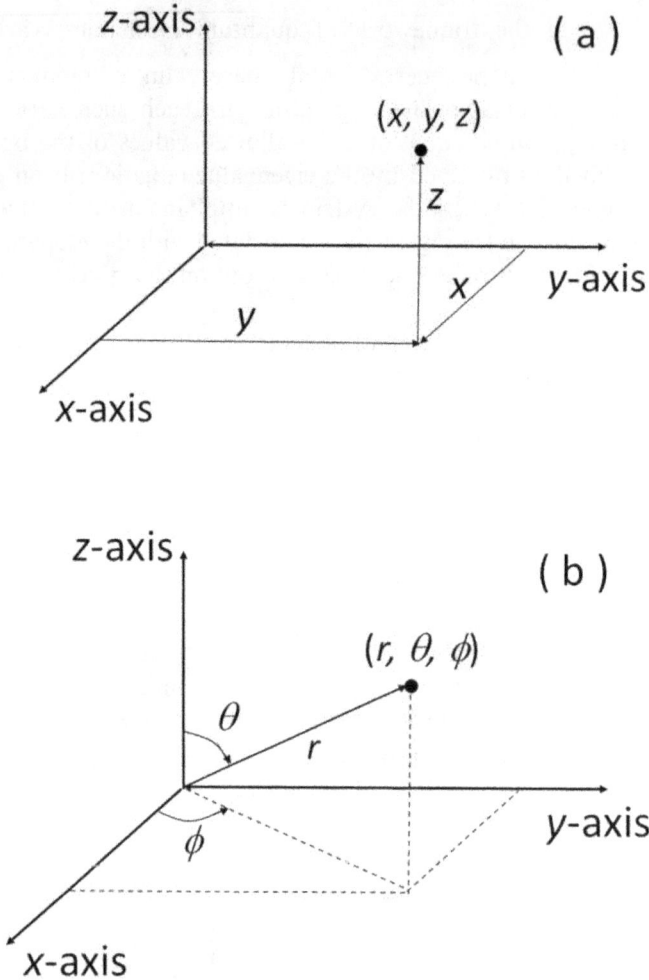

**Figure A.1.** The location of a point in (a) Cartesian coordinates, (b) spherical polar coordinates.

depend not at all on the angular coordinates $\theta$ and $\varphi$. Such a potential $U(r)$, which depends only on $r$ and not on $\theta$ and $\varphi$, is known as a *central potential*.

The Hamiltonian operator in the spherical polar coordinate is expressed as:

$$
\begin{aligned}
\hat{H} &= -\frac{\hbar^2}{2m_e}\left[\frac{1}{r^2}\frac{\partial}{\partial r}\left(r^2\frac{\partial}{\partial r}\right) + \frac{1}{r^2}\frac{1}{\sin\theta}\frac{\partial}{\partial\theta}\left(\sin\theta\frac{\partial}{\partial\theta}\right) + \frac{1}{r^2}\frac{1}{\sin^2\theta}\frac{\partial^2}{\partial\phi^2}\right] - \frac{e^2}{4\pi\epsilon_0 r} \\
&= -\frac{\hbar^2}{2m_e}\left[\frac{1}{r^2}\frac{\partial}{\partial r}\left(r^2\frac{\partial}{\partial r}\right)\right] + \frac{\hat{L}^2}{2m_e r^2} - \frac{e^2}{4\pi\epsilon_0 r}
\end{aligned}
\tag{A.7}
$$

Here $\hat{L}^2$ is the operator for the square of the orbital angular momentum. We also define $\hat{L}_z$ as the operator for the z-component of orbital angular momentum.

A-3

We recall here that the commutator of two operators $\hat{A}$ and $\hat{B}$ is defined by $[\hat{A}, \hat{B}] = \hat{A}\hat{B} - \hat{B}\hat{A}$, and they are said to be commutating if $[\hat{A}, \hat{B}] = 0$. When two operators commute, they are compatible and the measurement of one does not affect the value of the other in any way. We note here that the Hamiltonian expressed in the above equation (A.8) contains three terms. The first term is a function of $r$ only and involves multiplication and division by $r$ and differentiation with respect to $r$. So, this term will commute with $\hat{L}^2$ and $\hat{L}_Z$ because both $\hat{L}^2$ and $\hat{L}_Z$ do not involve $r$. The second term contains $\hat{L}^2$ and $r^2$. Now $\hat{L}^2$ commutes with $\hat{L}^2$ and $\hat{L}_Z$, and $r^2$ also commutes with $\hat{L}^2$ and $\hat{L}_Z$. Hence, the second term in equation (A.8) commutes with $\hat{L}^2$ and $\hat{L}_Z$. The third term only involves $r$, hence it commutes with $\hat{L}^2$ and $\hat{L}_Z$. Thus, the Hamiltonian expressed in equation (A.8) commutes with $\hat{L}^2$ and $\hat{L}_Z$. This result is valid for any central potential, which depends only on $r$ and not on $\theta$ and $\varphi$. We also note the result here that an observable whose operator commutes with the Hamiltonian is a conserved quantity and is known as a constant of motion [1].

Since $\hat{H}$, $\hat{L}^2$ and $\hat{L}_Z$ commute with each other, it is possible to write the eigenfunction of $\hat{H}$ from the common eigenfunction of $\hat{H}$, $\hat{L}^2$, and $\hat{L}_Z$. It is from the quantum mechanics of angular momentum that the eigenfunctions of $\hat{L}^2$ and $\hat{L}_Z$ can be expressed as [1]:

$$\psi(r, \theta, \phi) = f(r) Y_l^{m_l}(\theta, \phi) \tag{A.8}$$

The function $f(r)$ is an arbitrary function of $r$, and the function $Y_l^{m_l}(\theta, \phi) = P_l^{m_l}(\cos\theta)e^{i\phi}$ with proper normalization constant is called *spherical harmonics*. The normalization constant is fixed by the requirement [1]:

$$\int_{\phi=0}^{2\pi} \int_{\theta=0}^{\pi} |Y_l^{m_l}(\theta, \phi)|^2 \sin\theta d\theta d\phi = 1 \tag{A.9}$$

The function $P_l^{m_l}(\cos\theta)$ is known as an *associated Legndre function*. Here $l$ is a positive integer representing orbital angular momentum quantum number and $m_l$ is an integer with constraints $|m_l| \leqslant l$.

The Schrödinger equation for the hydrogen atom in spherical polar coordinate involving a central potential like Coulomb potential can be expressed as [1]:

$$\hat{H}\psi(r, \theta, \phi) = E\psi(r, \theta, \phi)$$

$$or, -\frac{\hbar^2}{2m_e}\left[\frac{1}{r^2}\frac{\partial}{\partial r}\left(r^2\frac{\partial}{\partial r}\right)\right]f(r)Y_l^{m_l}(\theta, \phi) + \frac{f(r)}{2m_e r^2}\hat{L}^2 Y_l^{m_l}(\theta, \phi) - \frac{e^2}{4\pi\epsilon_0 r}f(r)Y_l^{m_l}(\theta, \phi) = Ef(r)Y_l^{m_l}(\theta, \phi)$$

$$or, -\frac{\hbar^2}{2m_e}\left[\frac{1}{r^2}\frac{d}{dr}\left(r^2\frac{df(r)}{dr}\right)\right]Y_l^{m_l}(\theta, \phi) + \frac{f(r)}{2m_e r^2}l(l+1)\hbar^2 Y_l^{m_l}(\theta, \phi) - \frac{e^2}{4\pi\epsilon_0 r}f(r)Y_l^{m_l}(\theta, \phi) = Ef(r)Y_l^{m_l}(\theta, \phi) \tag{A.10}$$

$$or, -\frac{\hbar^2}{2m_e}\left[\frac{d^2f(r)}{dr^2} + \frac{2}{r}\frac{df(r)}{dr}\right] + \frac{l(l+1)\hbar^2}{2m_e r^2}f(r) - \left(\frac{e^2}{4\pi\epsilon_0 r} + E\right)f(r) = 0$$

If the radial part $f(r)$ of the wave function $\psi(r, \theta, \phi)$ is written as $f(r) = u(r)/r$, then the radial part of the Schrödinger equation (equation (A.10)) for the hydrogen atom becomes:

**Table A.1.** Some radial wave functions of a hydrogen atom with their energy eigenvalues

| $n$ | $l$ | $E$ (eV) | $R_{nl}(r)$ |
|---|---|---|---|
| 1 | 0 | $-13.6$ | $2(\frac{1}{a_0})^{3/2}e^{-r/a_0}$ |
| 2 | 0 | $-3.4$ | $(\frac{1}{2a_0})^{3/2}(2 - \frac{r}{a_0})e^{-r/2a_0}$ |
| 2 | 1 | $-3.4$ | $(\frac{1}{\sqrt{24}})(\frac{1}{a_0})^{3/2}(\frac{r}{a_0})e^{-r/2a_0}$ |

$$-\frac{\hbar^2}{2m_e}\frac{d^2u(r)}{dr^2} + \left[-\frac{e^2}{4\pi\epsilon_0 r} + \frac{l(l+1)\hbar^2}{2m_e r^2} - E\right]u(r) = 0 \qquad (A.11)$$

This differential equation (A.11) can be solved by the standard method with the condition that functions $u(r)$ should be finite everywhere [1]. This allows only some particular values of energy for hydrogen atoms given by [1]:

$$E = -\frac{m_e e^4}{2(4\pi\epsilon_o)^2\hbar^2 n^2} = -\frac{13.6\,eV}{n^2} \qquad (A.12)$$

Here $n = (l+1)$, $(l+2)$, $(l+3)$, and so on, with $l$ taking only positive integer values. Thus, $n$ takes vales, $n = 1, 2, 3, 4, ....$ The energy eigenvalues expressed in equation (A.12) are thus characterized by the parameter $n$, which takes only positive integer values and is known as a *principal quantum number*. The radial part of the wave function for hydrogen atom $f(r) = u(r)/r$ is therefore characterized by two quantum numbers $n$ and $l$, and is designated by $R_{nl}(r)$. Table A.1 lists some of these radial wave functions $R_{nl}(r)$ with their corresponding energy eigenvalues [1]. Here $a_0 = \frac{4\pi\epsilon_0\hbar^2}{m_e e^2} = 0.53$ Å is known as the Bohr radius.

The angular part of the Schrödinger equation (equation (A.10)) for the hydrogen atom can be written as:

$$\hat{L}^2 Y_l^{m_l}\left(\theta, \phi\right) = l\left(l+1\right)\hbar^2 Y_l^{m_l}\left(\theta, \phi\right) \qquad (A.13)$$

Here $\hat{L}^2$ is a differential operator involving $r$, $\theta$, and $\varphi$, and it represents the square of the angular momentum of the electron. The angular wave function $Y_l^{m_l}(\theta, \phi)$ will have acceptable solutions if the eigenvalue $l(l+1)\hbar^2$ on the right-hand side of equation (A.13) takes on positive integer values of $l$ less than the principal quantum number $n$ i.e $l = 0, 1, 2, 3, 4, ... , n-1$. The eigenvalue $l(l+1)\hbar^2$ must take on the possible values of the square of the angular momentum $L^2$ because the operator $\hat{L}^2$ represents the square of the angular momentum. The quantum number $l$ is termed as orbital angular momentum quantum number and

$$L^2 = l(l+1)\hbar^2 \qquad (A.14)$$

However, angular momentum $L$ is a vector quantity and has three Cartesian components $L_x$, $L_y$, and $L_z$, which in quantum mechanics are also represented by differential operators. The angular wave functions $Y_l^{m_l}(\theta, \phi)$ are an eigenfunction of $\hat{L}_Z$, which is the operator for the component of the angular momentum $L_z$ along the z-axis.

$$\hat{L}_z Y_l^{m_l}(\theta, \phi) = m_l \hbar Y_l^{m_l}(\theta, \phi) \qquad (A.15)$$

So that

$$L_z = m_l \hbar \qquad (A.16)$$

Here $m_l$ is an integer, and only takes values $-l, -l+1, -l+2, ... -1, 0, 1, ... l-1, l$. Thus, for any given value of $l$ there are $2l + 1$ possible solutions of equation (A.15). The integer $m_l$ is termed the orbital magnetic quantum number.

Summarizing the above discussion, we say that any wave function for the hydrogen atom is specified by three quantum numbers, the principal quantum number $n$, the orbital angular momentum quantum number $l$, and the magnetic quantum number $m_l$. The angular wave function $Y_l^{m_l}(\theta, \phi)$ is labeled by $l$ and $m_l$, and the radial wave function $R_{nl}(r)$ is labeled by $n$ and $l$. The complete energy eigenfunction corresponding to the quantum numbers $n$, $l$ and $m$ is expressed as:

$$\psi_l(r, \theta, \phi) = R_{nl}(r) Y_l^{m_l}(\theta, \phi) \qquad (A.17)$$

The three quantum numbers $n$, $l$, and $m$ correspond to energy, the square of the angular momentum, and z-component of angular momentum, respectively. In the state defined by equation (A.17), the energy is $E_n$, the square of the orbital angular momentum is $l(l+1)\hbar^2$, and the z-component of angular momentum is $m_l \hbar$ [1].

The lowest energy state corresponding to $n = 1$ is called the ground state of the hydrogen atom. The value of $l$ in the ground state is zero, and hence $m_l$ is also zero. Thus, the orbital angular momentum in the ground state is zero. This ground state wave function is expressed as [1]:

$$\psi_{000}(r, \theta, \phi) = R_{10}(r) Y_0^0(\theta, \phi) = \frac{2}{\sqrt{4\pi}} \left(\frac{1}{a_0}\right)^{3/2} e^{-r/a_0} \qquad (A.18)$$

The ground state is designated by atomic physicists and chemists as '1s' state. Here the number 1 denotes the principal quantum number $n = 1$ and the character 's' stands for the orbital angular momentum quantum number $l = 0$. In fact, $l = 0$ is termed as s-state, $l = 1$ as p-state, $l = 2$ as d-state, $l = 3$ as f-state, etc [1]. Thus, in the 2s state $n = 2$ and $l = 0$, and so on.

Apart from orbital angular momentum, the electron in the hydrogen atom has intrinsic spin angular momentum. The z-component of spin angular momentum is a measurable quantity and has an operator $\hat{S}_z$ associated with it. We also define an operator $\hat{S}^2$ defining the square of the intrinsic spin angular momentum of the electron. The eigenvalues of $\hat{S}^2$ are written as $s(s+1)\hbar^2$ and of $S_z$ as $m_s \hbar$. The number $s$ is called spin angular momentum quantum number, and for an electron $s$

has just one value, $s = 1/2$. The quantum number $m_s$ in general takes values $-s$ to $+s$ in steps of 1, and for electron $m_s$ takes values $+1/2$ and $-1/2$.

The Hamiltonian expressed in equation (A.7) as such does not contain any information on the spin in the hydrogen atom, hence the eigenfunction in equation (A.17) does not provide a complete description of the state of the hydrogen atom. One can, however, multiply $\psi_{nlm_l}(r, \theta, \phi)$ by any function $\chi$ from the spin-space to create a wave function $\psi_{nlm_l}(r, \theta, \phi)\chi$, which will still be an eigenfunction of the Hamiltonian. In the case of the electron, there is only one eigenvalue of $S^2$, which is $3\hbar^2/4$ [1]. However, the $z$-component of spin $S_Z$ can have two eigenvalues $+\hbar/2$ and $-\hbar/2$. Taking into account electron spin, the stationary state wave function is expressed as:

$$\psi_{nlm_lm_s} = R_{nl}(r)Y_l^{m_l}(\theta, \phi)\chi_{m_s} \tag{A.19}$$

The possible values of $m_s$ are $+1/2$ and $-1/2$. The eigenfunction is thus characterized by four quantum numbers $n$, $l$, $m_l$, and $m_s$. The stationary state is completely and uniquely described by these four quantum numbers. However, the energy of hydrogen atoms expressed in equation (A.12) depends only on the principal quantum number $n$, and not on $l$, $m_l$, and $m_s$. There are $n$ values of orbital angular momentum quantum number $l$ for each value of $n$, namely $0, 1, 2, \ldots, n-1$. In addition, for each $l$ there are $2l+1$ values of $m_l$, namely $-l, -l+1, -l+2, \ldots, l$. And then $m_s$ has value $\pm 1/2$. So, the total number of eigenstates corresponding to the energy $E_n$ is:

$$2 \times \sum_{l=0}^{n-1}(2l + 1) = 2n^2 \tag{A.20}$$

So, there is $2n^2$ fold degeneracy for the energy eigenvalue $E_n$.

## A.2 Radial probability density

We will now consider the probability $dP$ of finding the electron at a distance between $r$ and $r + dr$ from the nucleus of the hydrogen atom. Considering the range $dr$ to be infinitesimally small, this probability $dP$ is also infinitesimally small. However, the quantity $dP/dR$ is finite and it is a function of $r$. This quantity is termed *radial probability density* and is represented by $p(r)$. In a state defined by the quantum numbers $n$, $l$, $m_l$, and $m_s$ the probability of finding the electron in a volume $dr$ at $(r, \theta, \phi)$ is [1]

$$\left|\psi_{nlm_l}\right|^2 dr = \left|R_{nl}(r)\right|^2\left|Y_l^{m_l}(\theta, \phi)\right|^2 r^2 \sin\theta dr d\theta d\phi \tag{A.21}$$

It may be noted here that spin plays no part in the determination of the probability. In order to determine the probability of finding the electron at a distance between $r$ and $r + dr$ from the nucleus, we need to integrate the probability density function over all the values of $\theta$ and $\varphi$ [1]:

$$p(r)dr = \int_{\phi=0}^{2\pi} \int_{\theta=0}^{\pi} |R_{nl}(r)|^2 |Y_l^{m_l}(\theta, \phi)|^2 r^2 \sin\theta \, dr \, d\theta \, d\phi$$

$$= r^2 |R_{nl}(r)|^2 dr \int_{\phi=0}^{2\pi} \int_{\theta=0}^{\pi} |Y_l^{m_l}(\theta, \phi)|^2 \sin\theta \, d\theta \, d\phi \qquad \text{(A.22)}$$

$$= r^2 |R_{nl}(r)|^2 dr$$

## A.2.1 Ground state

The space part of the wave function of hydrogen atom expressed in equation (A.18) does not depend on $\theta$ and $\varphi$. The wave function is spherically symmetric, and hence the probability of finding the electron in a hydrogen atom is the same in all directions from the nucleus. Thus, the picture electron of in the ground state of hydrogen is that of a spherical cloud around the nucleus. This is in clear contrast to the flat circular orbit type picture within the framework of the Bohr atomic model.

The radial part of the ground state wave function as presented in table A.1 is:

$$R_{10}(r) = \frac{2}{a_0^{3/2}} e^{-r/a_0} \qquad \text{(A.23)}$$

Thus, using equations (A.22) and (A.23), we can determine the $p$, the Bohr radius. The probability of finding the electron at a distance between $r$ and $r + dr$ from the nucleus:

$$p(r)dr = r^2 |R_{10}(r)|^2 dr$$

$$= \frac{4}{a_0^3} r^2 e^{-2r/a_0} dr \qquad \text{(A.24)}$$

And the radial probability density is:

$$p(r) = \frac{4}{a_0^3} r^2 e^{-2r/a_0} \qquad \text{(A.25)}$$

Figure A.2 presents a plot of the radial probability density $p(r)$ as a function of $r$, which shows a maximum value at $r = a_0$.

## A.2.2 First excited state

The first excited state of the hydrogen atom corresponds to the principal quantum number $n = 2$, and with respect to that angular momentum quantum number $l$ can have values $l = 0$ and 1. When $l = 1$, there is only one value magnetic quantum number that can take $m_l = 0$. With $l = 1$ there are three values of $m_l = -1, 0$, and 1. The radial probability density $p(r)$, however, does not depend on $m_l$, and $p(r)$ in the $2s$ state with $n = 2$ and $l = 0$ is expressed as:

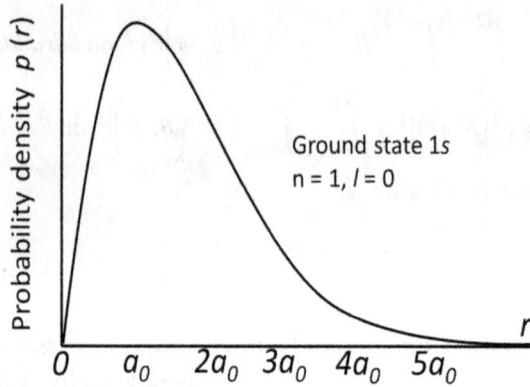

**Figure A.2.** The radial probability density $p(r)$ of finding an electron at a distance $r$ from the nucleus in the ground state 1s of hydrogen atom.

$$p(r) = r^2|R_{20}(r)|^2$$

$$= \frac{1}{8a_0^3}r^2\left(2 - \frac{r}{a_0}\right)^2 e^{-r/a_0} \tag{A.26}$$

Figure A.3 shows the probability density $p(r)$ for the 2s state as a function of $r$. There are two maxima in $p(r)$, one near $r = a_0$ and the other close to $5\, a_0$. In the case of 2p state that is $n = 2$ and $l = 1$, the probability density $p(r)$ is expressed as:

$$p(r) = r^2|R_{21}(r)|^2$$

$$= \frac{1}{24a_0^5}r^4 e^{-r/a_0} \tag{A.27}$$

### A.2.3 The angular probability density

From equation (A.8) we can see that the angular part of the wave function of hydrogen atom $Y_l^{m_l}(\theta,\ \phi) = P_l^{m_l}(\theta)e^{im_l\phi}$ is the product of two factors involving $\theta$ and $\varphi$ separately. Now $|e^{im_l\phi}|^2 = 1$, hence one can write:

$$|Y_l^{m_l}(\theta,\ \phi)|^2 = |P_l^{m_l}(\theta)|^2 \tag{A.28}$$

This is independent of $\varphi$. For $l = 0$, $|P_l^{m_l}(\theta)|^2 = |P_0^0(\theta)|^2$.

The angular dependence of the probability density is represented by a polar diagram, where the lengths of vectors from the origin are made proportional to the value of $|P_l^{m_l}(\theta)|^2$ in the direction of the vectors. Figure A.4 shows the polar diagram of angular probability density for $l = 0$, which indicates that the vector has the same length for any value of $\theta$, i.e., in any direction. The angular probability density of electron in hydrogen atom $|\psi(r,\ \theta,\ \phi)|^2$ is therefore spherically symmetrical. This has the same dependence on $r$ in every direction independent of $\theta$ and $\varphi$.

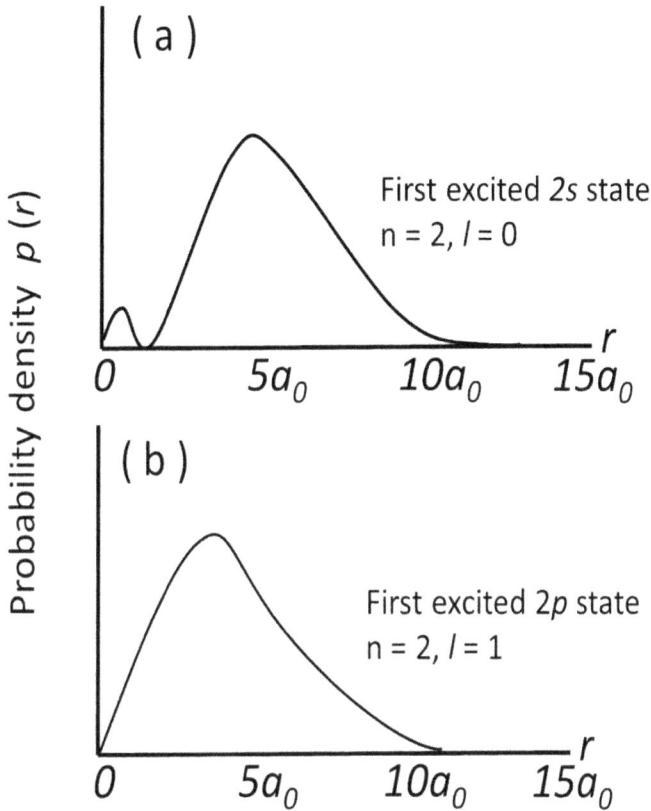

**Figure A.3.** The radial probability density $p(r)$ of finding an electron at a distance $r$ from the nucleus in the first excited state of hydrogen atom: (a) $2s$ state; (b) $2p$ state.

In the case of $l$ values other than zero, $P_l^{m_l}(\theta)$ depends on $\theta$ and also on $m_l$. Figure A.5(a) shows a polar diagram of angular probability density for $l = 1$, $m_l = 0$. This reveals that the angular probability density only has appreciable values in the region of positive and negative $z$-directions, i.e., near $\theta = 0$ and $\theta = \pi$. Figure A.5(b) shows a polar diagram of angular probability density for $l = 1$, $m_l = \pm 1$ in the $y$–$z$-plane where $\phi = 0$. In this case, the probability density is only appreciable in the region near $\theta = \pi/2$.

### A.2.4 Fine structures of hydrogen spectrum

The discussion so for on the energy eigenvalues and eigenfunctions of a hydrogen atom is based on the assumption that the proton in the nucleus of a hydrogen atom interacts with the electron only via Coulomb interaction. While it is true that Coulomb interaction is the major force in a hydrogen atom, there exists some more interaction, which needs to be taken into account to understand the fine structures observed in the hydrogen spectrum with modern high-resolution instruments.

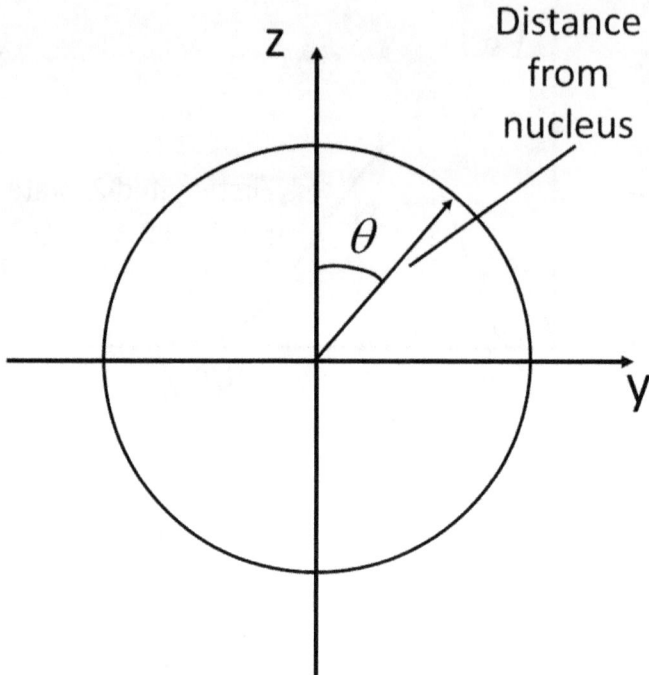

**Figure A.4.** Polar diagram of angular probability density for electron in $l = 0$ 1 $s$ state of hydrogen atom. The vector has the same length for any value of $\theta$.

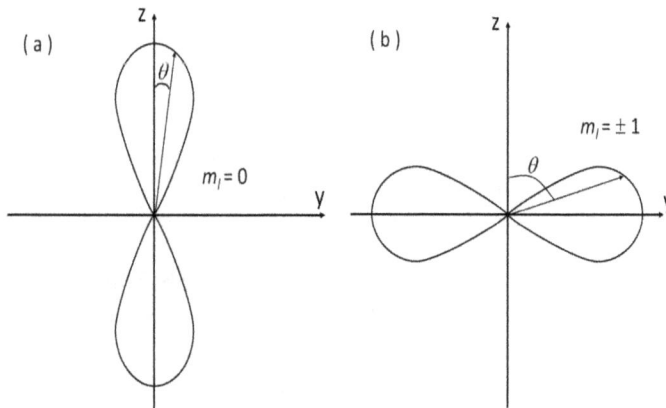

**Figure A.5.** Polar diagrams of the angular probability density for a $2p$ electron in hydrogen atom: (a) $m_l = 0$, (b) $m_l = \pm1$. Here $|P_l^{m_l}(\theta)|^2$ varies with $\theta$.

One such important interaction is magnetic interaction, which arises from the spin or intrinsic magnetic moment of the electron. This magnetic interaction is known as $L - S$ coupling and is described by the term $f(r)\mathbf{L}.\,\mathbf{S}$, which needs to be included in the Hamiltonian [1]. Here $\mathbf{L}.\,\mathbf{S}$ stands for $L_x S_x + L_y S_y + L_z S_z$.

In the presence of the $f(r)\mathbf{L} \cdot \mathbf{S}$, the Hamiltonian does not commute with $L_z$ and $S_z$. This is because $L_x$ and $L_y$ do not commute with $L_z$, and $S_x$ and $S_y$ do not commute with $S_z$ [1]. Thus, it is not possible to specify $m_l$ and $m_s$ in the energy eigenfunctions. However, the Hamiltonian still commutes with $L^2$ and $S^2$.

One can define an operator $\mathbf{J}$ as the total angular momentum, which is approximately given by the vector sum of the total orbital angular momentum $\mathbf{L}$ and the total spin angular momentum $\mathbf{S}$. The components of $\mathbf{J}$ are given by the relations [1]:

$$J_x = L_x + S_x$$
$$J_y = L_y + S_Y \qquad \qquad \text{(A.29)}$$
$$J_z + L_z + S_z$$

One can write the above relations in a single equation $\mathbf{J} = \mathbf{L} + \mathbf{S}$, and has the magnitude of $\sqrt{j(j+1)}\,\hbar$, in which $j$ can take any positive value in the range $|l - s|$ to $|l + s|$ in steps of 1[1]. The associated quantum number $m_j$, specifies the orientation of the total angular momentum of the hydrogen atom as a whole, and $m_j$ can take any value from $j$ to $-j$.

We can now write [1]:

$$J^2 = (\mathbf{L} + \mathbf{S})^2 = L^2 + S^2 + 2\mathbf{L} \cdot \mathbf{S}$$
$$or, \ \mathbf{L} \cdot \mathbf{S} = \frac{J^2 - L^2 - S^2}{2} \qquad \qquad \text{(A.30)}$$

This term $\mathbf{L} \cdot \mathbf{S}$ commutes with $J^2, L^2, S^2$, and $J_z$, and so does the Hamiltonian $H$. Thus, $J^2, L^2, S^2, J_z$, and $H$ commute with each other and it is possible to write the stationary state wave functions from the common eigenfunctions of these. Here $L_z$ and $S_z$ do not have definite values, but $J_z = L_z + S_z$ is definite and so are $l$, $s$, and $j$.

We now consider the case of $2p$ state of the hydrogen atom, where $l = 1$ and $s = 1/2$. In this case, $j$ can take values of $1/2$ and $3/2$. Associated with $j = 3/2$, there are four energy eigenstates with $m_j = -3/2, -1/2, 1/2, 3/2$. They all have the same energy eigenvalue. With $j = 1/2$, there are two energy eigenstates with $m_j = -1/2$ and $1/2$ having the same energy eigenvalues. However, the energy of four $j = 3/2$ states is slightly different from those of $j = 1/2$ states. If the $\mathbf{L} \cdot \mathbf{S}$ interaction is not considered, there are still six energy eigenstates corresponding to $l = 1$ and $s = 1/2$. The energy eigenvalues of all these states are the same. This degeneracy in energy is partly lifted when $\mathbf{L} \cdot \mathbf{S}$ interaction is taken into account, with four states associated with $j = 3/2$ having slightly different energy values from the two states with $j = 1/2$. The energy splitting of spectral lines due to $\mathbf{L} \cdot \mathbf{S}$ type interaction is termed fine structure splitting. The difference from the energy values of pure Coulombic origin is very small in the order of 1 part in $10^4$ [1], but can be measured easily with modern spectroscopic techniques.

The electron in a hydrogen atom can also interact with an externally applied magnetic field. In this case, the energy eigenvalue of the electron will depend on $m_l$, as well as on $l$. As a result, the transitions involving an initial and final state of given

$l$-values will result in several different energy eigenvalues depending on the initial and final $m_l$ values instead of a single energy eigenvalue. This splitting of the energy levels in an external magnetic field leads to the splitting of the observed spectral lines and is termed the Zeeman effect. Similarly, the energies of hydrogen atomic levels can also be affected by external electric fields. This effect of energy splitting caused by the application of a strong electric field is known as the Stark effect.

## Reference

[1]  Verma H C 2020 *Concepts of Physics* **vol 2** (New Delhi: Bharati Bhawan)

**IOP** Publishing

Hydrogen
Physics and technology
**Sindhunil Barman Roy**

# Appendix B

## Liquefaction of hydrogen

In this appendix we will briefly describe the various possible processes for the liquefaction of hydrogen. The texts and descriptions presented here have been borrowed significantly from an open-source review article entitled 'Liquid Hydrogen: A Review on Liquefaction, Storage, Transportation, and Safety' by Muhammad Aziz [1]. For a detailed information on the processes for liquefaction of hydrogen, the readers are referred to the book by Walter Peschka [2].

According to Faraday, six known gases—nitrous oxide, carbon monoxide, hydrogen, oxygen, and methane—are permanent gases because they cannot be liquefied by compression [3]. The boiling points of all permanent gases are lower than −110 °C. The critical pressure and temperature of hydrogen are 1.3 MPa and −240 °C, respectively. Hydrogen becomes liquid at a temperature of −253 °C and an ambient pressure of 1 atm. Cryogenic cooling reduces the volume of hydrogen by 1/848, which results in far more efficient hydrogen storage. Hydrogen liquefaction is now an established technology, and several developments have taken place over the years in this field to minimize energy consumption.

In 1898, Sir James Dewar accomplished the first-ever liquefaction of hydrogen [4]. Dewar employed a modest liquefaction apparatus with a capacity of 0.24 L h$^{-1}$, which was many years before a precooled Hampson–Linde cycle was tested in a laboratory-scale liquefaction system [1, 3, 5]. After being first pressurized to 18 MPa, the hydrogen gas was precooled to −250 °C using liquid air and carbolic acid. Around 1900, a number of further liquefaction techniques were developed, such as Claude, precooled Claude, and helium-refrigerated systems [1]. In 1957, a comparatively large hydrogen liquefaction facility based on a precooled Claude system was designed to meet the demands of the aerospace and chemical industries. In this system, hydrogen was first precooled with liquid nitrogen to a temperature of about −193 °C, which was followed by refrigeration using hydrogen until liquid hydrogen was formed [1].

doi:10.1088/978-0-7503-5172-0ch20
© IOP Publishing Ltd 2024

## B.1 Basic cycles of hydrogen liquefaction

The working principles of the basic cycles used for large hydrogen liquefaction and the simple configuration of such liquefaction schemes are described in detail in the literature (see reference [3] and references within). We will provide a brief description of some of these hydrogen liquefaction processes.

Two important concepts in the process of hydrogen liquefaction are throttling and Joule–Thomson (J–T) effects. Throttling is the common feature of the large gas liquefaction cycles, which is achieved by introducing nonideality into the cycle's working fluid. Usually, throttling is used after temperature reduction or pressure increase. With the J–T effect, liquefaction processes can be categorized according to the maximum inversion temperature of the gas to be liquefied relative to the ambient temperature [3]. The direction and magnitude of the temperature shift during the isenthalpic change of state are represented by the J–T coefficient. A reduction in temperature would follow a decrease in isenthalpic pressure if the J–T coefficient was positive. The concept of a positive J–T coefficient is used in the liquefaction of hydrogen, where a rapid pressure variation can be achieved using a nozzle [1]. A negative J–T coefficient causes a temperature increase during the fall in isenthalpic pressure [1]. This idea becomes useful while filling the high-pressure vessel with hydrogen.

### B.1.1 Linde process

The Hampson–Linde, or J–T, expansion process is the most fundamental and straightforward liquefaction method. Here, the ambient hydrogen gas is compressed before being cooled by heat exchange. Next, a throttling valve was used to conduct the isenthalpic J–T expansion for this compressed and cooled gas. A high pressure of hydrogen is often required because the system relies on the J–T effect for liquefaction [1]. The pressurized gas partially condenses into liquid. For the next cooling phase, the gas remaining in gaseous form is recirculated back. The performance of this method is better for gases such as nitrogen that can expand to cool at ambient temperature. At ambient temperature, however, hydrogen warms up during expansion. In order to cool down hydrogen after expansion, it must first be chilled to its inversion temperature, which is −73 °C at 1 bar, or below. The hydrogen can be precooled using liquid nitrogen, which has a boiling point of −195 °C at 1 bar. A pressure adjustment is necessary for producing appropriate precooling because the inversion temperature is a function of the pressure.

Figure B.1 presents a schematic of the Linde–Sankey process for hydrogen liquefaction. Hydrogen gas is compressed and cooled in subsequent heat exchangers for cooling with the use of compressed and liquid nitrogen. When compressed hydrogen reaches a temperature below its inversion temperature, J–T expansion is carried out. This causes some of the hydrogen gas to liquefy. Next, the remaining hydrogen gas is recirculated and mixed with the fresh supply of hydrogen gas [1].

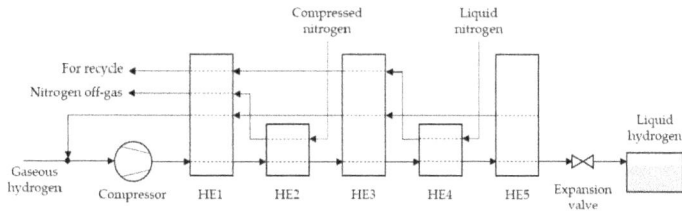

**Figure B.1.** Schematic presentation of Linde–Sankey process for liquid hydrogen production. Reproduced from [1]. CC BY 4.0.

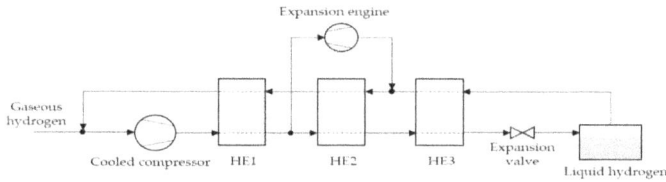

**Figure B.2.** Basic schematic diagram of Claude process for hydrogen liquefaction. Reproduced from [1]. CC BY 4.0.

### B.1.2 Claude process

Georges Claude developed a method in 1902 for liquefying air with a reciprocating expansion machine. The J–T expansion effect is combined with the expansion engine in this procedure. The low energy efficiency of the system is a disadvantage, but it has the simplicity of isenthalpic expansion. By including an expansion engine in the Claude process, a lower temperature can be reached prior to isenthalpic expansion [1]. If the expansion engine is the main source of refrigeration, then in principle no liquid nitrogen is needed for cooling. However, subsequent studies indicated that using liquid nitrogen for precooling could lead to 50–70% % higher exergy efficiency.

Figure B.2 shows a schematic representation of the Claude method for hydrogen liquefaction. This mechanism involves isenthalpic expansion and isothermal compression. The compressed hydrogen gas is cooled by multiple sets of heat exchangers, with an expansion engine placed between each row of heat exchangers. Part of the compressed hydrogen gas is fed into the expansion engine, which uses it to cool the remaining gas. Condensation is not realistically achievable with the expansion engine because there is a chance that the liquid material may harm the expansion engine. A part of the high-pressure hydrogen is expanded using this expansion engine, lowering the temperature of the remaining hydrogen gas. As shown in figure B.2, it is then mixed with cold hydrogen at a low temperature and heated in the heat exchanger using compressed hydrogen. Finally, the J–T expansion process takes place.

### B.1.3 Collins process

A schematic illustration of the Collins process is shown in figure B.3. Because of its resemblance to the Claude process, the Collins method (which was first created for

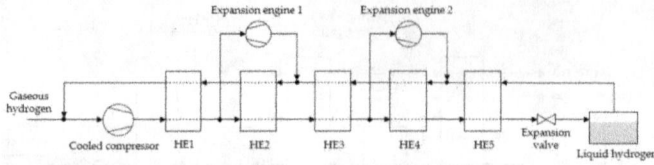

**Figure B.3.** Basic schematic diagram of Collins process for hydrogen liquefaction. Reproduced from [1]. CC BY 4.0.

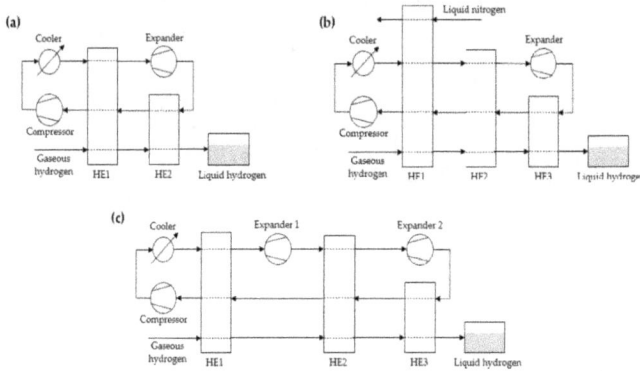

**Figure B.4.** Schematic diagram of Brayton cycle for hydrogen liquefaction: (a) simple helium Brayton cycle; (b) helium Brayton cycle with liquid nitrogen precooling, and (c) two-step helium Brayton cycle. Reproduced from [1]. CC BY 4.0.

helium liquefaction) is sometimes referred to as the modified Claude process [1]. After being compressed, the gaseous hydrogen is sent through a number of heat exchangers before being expanded by the J–T expansion valve. After the expansion, the pressure of the hydrogen drops and some of the hydrogen condenses. The residual gaseous hydrogen returns to the heat exchangers in the counter-current mode. Two adiabatic expansion engines with varying operating temperatures are employed for cooling. Cooling is continued until the temperature falls below the inversion temperature.

### B.1.4 Helium Brayton cycle

The helium Brayton cycle has several alternative forms: the basic helium Brayton cycle, the two-step helium Brayton cycle, and the helium Brayton cycle with liquid nitrogen precooling. In power production plants, gas turbines or jet engines work well with the Brayton cycle. Instead of being utilized as a liquefier, helium is mainly used as a refrigerant. Its purpose is to bring the temperature down below that of hydrogen liquefaction. The work of the compressor is minimized by precooling with liquid nitrogen. Figure B.4 shows schematic diagrams of the helium Brayton cycle hydrogen liquefaction cycles.

### B.1.5 Magnetic refrigeration/liquefaction system

The phenomenon known as the magnetocaloric effect describes how a variation in the magnetic field can cause a reversible change in the temperature of a working material. The magnetic refrigeration technique utilizes the magnetocaloric effect by employing a magnetic field to both magnetize and demagnetize a suitable magnetic material repeatedly [6]. Thus, this approach can be used to create a low temperature environment. Magnetic refrigeration was first documented in the 1930s. In 1976, Brown [7] developed a near-room-temperature magnetic refrigerator utilizing gadolinium as the working material. Research on magnetocaloric materials and magnetic refrigeration has continued at a rapid pace, with notable advances made both in technology and materials [8].

A schematic illustration of the hydrogen liquefaction process with magnetic refrigeration is presented in figure B.5. The four continuous steps in this process are adiabatic magnetization, isothermal magnetization, adiabatic demagnetization, and isothermal demagnetization [1]. It primarily employs the reversed Carnot cycle. The working material is put in a thermally insulated environment for the adiabatic magnetization stage. The application of an external magnetic field tends to align the magnetic moments of the magnetic medium. Thus, the entropy and heat capacity decrease as a result. The next step in the process is isothermal magnetic enthalpy transfer, which involves transferring heat by means of a different fluid while maintaining a steady magnetic field to avert heat absorption and magnetic moment misalignment. The working material is subjected to adiabatic conditions in the subsequent stage of adiabatic demagnetization, which maintains the total enthalpy constant. Entropy increases as a result of the aligned magnetic moments being misaligned with the decrease in the magnetic field. The thermal entropy is converted into the magnetic entropy because the system is insulated from the environment in the adiabatic condition. The working material cools as a result. Isomagnetic entropy transfer takes place in the final step, in which the magnetic field is kept constant to prevent material warming. In this stage, the hydrogen that needs to be cooled or liquefied comes into thermal contact with the working material. Heat is transmitted from hydrogen to the working material because the ambient temperature is higher than that of the working material.

The theoretical efficiency of the magnetic refrigeration cycle is better than that of the Carnot cycle-based systems for compressed-gas refrigeration. Compared to the 38% theoretical Carnot efficiency for the compressed-gas refrigeration system,

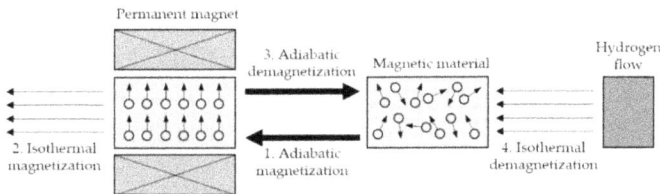

**Figure B.5.** Basic schematic diagram of magnetic refrigeration technique for hydrogen liquefaction. Reproduced from [1]. CC BY 4.0.

magnetic refrigeration can attain about 50% theoretical Carnot efficiency [9]. Because a solid magnetic substance has a larger entropy density than a gas, the liquefying plant will also be more compact.

### B.1.6 Catalyzed ortho-hydrogen to para-hydrogen conversion

The process of conversion from ortho-hydrogen to para-hydrogen is very slow. When this conversion takes place during hydrogen storage and transit, hydrogen loss from boil-off occurs. During liquefaction, a catalyst is frequently utilized to speed up the transition from ortho-hydrogen to para-hydrogen. In this process, the amount of heat released during storage is reduced, as well as the conversion of ortho-hydrogen to para-hydrogen. Consequently, a high concentration of para-hydrogen is achieved, and the boil-off associated with the ortho-hydrogen to para-hydrogen conversion is reduced [1]. The catalytic conversion of ortho-hydrogen and para-hydrogen can be expressed in the following reaction [1]:

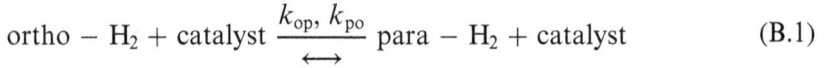

$$\text{ortho} - H_2 + \text{catalyst} \xrightleftharpoons{k_{op}, k_{po}} \text{para} - H_2 + \text{catalyst} \tag{B.1}$$

$$k^0 = k_{op} + k_{po} \tag{B.2}$$

Here $k_0$, $k_{op}$, and $k_{po}$ stand for the total conversion rate constant, ortho-hydrogen to para-hydrogen conversion rate constant, and para-hydrogen to ortho-hydrogen conversion rate constants, respectively.

Figure B.6 presents a schematic of an ortho-hydrogen to para-hydrogen conversion reactor in different liquefaction systems of the helium Brayton cycle with liquid nitrogen precooling and a two-step helium Brayton cycle [1]. A two-stage

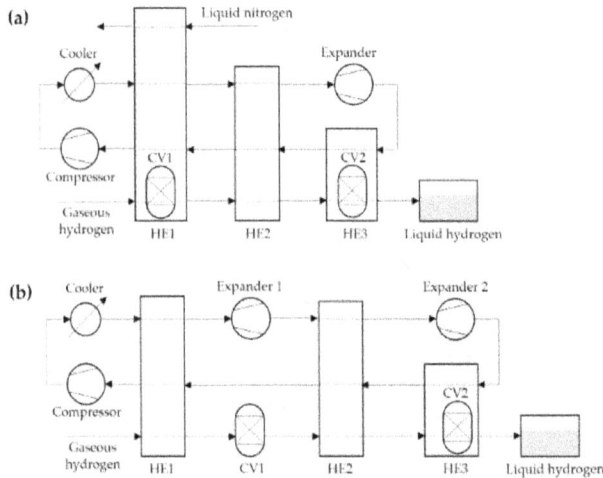

**Figure B.6.** Schematic presentation of a possible installation of ortho-hydrogen to para-hydrogen converter in: (a) helium Brayton cycle with liquid nitrogen precooling, and (b) two-step helium Brayton cycle. Reproduced from [1]. CC BY 4.0.

setup is employed with two ortho-hydrogen to para-hydrogen conversion reactors (CV1 and CV2), one at an intermediate temperature in the gas phase and the other at the point of liquefaction, where liquid hydrogen is generated. The first ortho-hydrogen to para-hydrogen conversion reactor in the helium Brayton cycle (figure B.6a) is situated in the first heat exchanger at liquid nitrogen temperature in an isothermal environment. Liquid nitrogen thus covers the conversion heat. The first conversion reactor in the two-step helium Brayton cycle is located in an adiabatic environment, where the temperature of the hydrogen rises after ortho-hydrogen to para-hydrogen conversion [1]. The second conversion reactor runs at $-253\,°C$ under isothermal conditions in both liquefaction systems.

For the efficient conversion of ortho-hydrogen to para-hydrogen, materials such as silver, copper, charcoal, graphite, chromium, and ferric oxides are used. Such targeted catalysts are developed and evaluated in terms of the efficiency of their ortho-hydrogen to para-hydrogen conversion. The following sequential steps are involved in the kinetics of the catalytic conversion of ortho-hydrogen to para-hydrogen [1]: (a) ortho-hydrogen diffuses from the liquid to the surface of the catalyst; (b) ortho-hydrogen diffuses in the catalyst pore until it reaches the active site; (c) ortho-hydrogen adsorption; (d) ortho-hydrogen and para-hydrogen react on the surface of the catalyst (ortho-hydrogen ↔ para-hydrogen); (e) para-hydrogen desorbs; (f) para-hydrogen diffuses in the catalyst pore to the catalyst surface; and (g) para-hydrogen diffuses to the liquid through the boundary film.

## References

[1] Aziz M 2021 *Energies* **14** 5917
[2] Peschka W 1992 *Liquid Hydrogen* (Wien: Springer)
[3] Aasadnia M and Mehrpooya M 2018 *Appl. Energy* **212** 57
[4] Dewar J 1898 *Science* **8** 3
[5] Krasae-in S, Stang J H and Neksa P 2010 *Int. J. Hydrogen Energy* **35** 4524
[6] Pecharsky V K and Gschneidner K A 2005 Magnetocaloric effect *Encyclopedia of Condensed Matter Physics* ed F Bassani, G L Liedl and P Wyder (Amsterdam: Elsevier)
[7] Brown G J 1976 *Appl. Phys.* **387** 3673
[8] Roy S B 2014 Magneto-caloric effect in intermetallic alloys and compounds *Handbook Magnetic Mater.* **vol 22** (Amsterdam: Elsevier)
[9] Institute of Slush Hydrogen the World's First Hydrogen Liquefaction by Magnetic Refrigeration. Available online: http://slush-ish-mag-english.com/technology2.html

www.ingramcontent.com/pod-product-compliance
Lightning Source LLC
Chambersburg PA
CBHW080513220326
41599CB00032B/6064